MW00341970

BACTERIOLOGY OF HUMANS

For the Wilson clan – Andrew, Anne, Caroline, Edie, Fionn, Joan, Liz, Pippa, and Sarah – a rare example of a complex community governed only by positive interactions.

BACTERIOLOGY OF HUMANS

AN ECOLOGICAL PERSPECTIVE

Michael Wilson

University College London

Blackwell
Publishing

© 2008 by Blackwell Publishing Ltd

BLACKWELL PUBLISHING
350 Main Street, Malden, MA 02148-5020, USA
9600 Garsington Road, Oxford OX4 2DQ, UK
550 Swanston Street, Carlton, Victoria 3053, Australia

The right of Michael Wilson to be identified as the Author of the Editorial Material in this Work has been asserted in accordance with the UK Copyright, Designs, and Patents Act 1988.

All rights reserved. No part of this publication may be reproduced, stored in a retrieval system, or transmitted, in any form or by any means, electronic, mechanical, photocopying, recording, or otherwise, except as permitted by the UK Copyright, Designs, and Patents Act 1988, without the prior permission of the publisher.

First published 2008 by Blackwell Publishing Ltd

1   2008

Library of Congress Cataloging-in-Publication Data

ISBN-13: 978-1-4051-6165-7 (hardback)

A catalogue record for this title is available from the British Library

Set in 9/11pt Photina
by Graphicraft Limited, Hong Kong
Printed and bound in Singapore
by Fabulous Printers Pte Ltd

The publisher's policy is to use permanent paper from mills that operate a sustainable forestry policy, and which has been manufactured from pulp processed using acid-free and elementary chlorine-free practices. Furthermore, the publisher ensures that the text paper and cover board used have met acceptable environmental accreditation standards.

For further information on
Blackwell Publishing, visit our website:
www.blackwellpublishing.com

CONTENTS

PREFACE

More than 50 years ago, Dylan's *The Times they Are A-Changing* reverberated around the planet and, with an unerring sense of timelessness, those words are as appropriate now as they were then. Unsurprisingly to me, and to many others, they are applicable to the science of bacteriology as well as to so many other aspects of our existence. Hence, for many years, the primary focus of the vast majority of bacteriologists studying the bacterial inhabitants of humans has been those species responsible for disease, e.g. *Streptococcus pyogenes*, *Haemophilus influenzae*, *Neisseria meningitidis*, *Staphylococcus aureus*, etc. What is now a-changing is that interest is shifting towards the vast majority of our microbial partners that do not cause disease and, indeed, are essential to our well being. The indigenous microbiota of healthy humans is now the subject of intense scrutiny, and its enormous diversity and the crucial roles that it plays in the development, protection, and maintenance of *Homo sapiens* are slowly being revealed. In this book, I have concentrated on the first of these, and have described the nature of the microbial communities that inhabit the various regions of the healthy human body. I have also attempted to explain their presence at a particular site in terms of the environmental factors that operate there. This book, I hope, will be useful to undergraduates and postgraduates on courses in microbiology, medical microbiology, microbial ecology, infectious diseases, immunology, human biology, medicine, dentistry, nursing, health sciences, biomedical sciences, and pharmacy and, indeed, to all those who have an interest in the complex microbial communities with which we have co-evolved.

I am not sure why books need a preface, and I cannot recall ever having read one. Nevertheless, it does give the author an opportunity for self-indulgence and I am, therefore, going to amuse myself by interweaving two important threads in my life – bacteriology and Dylan. Any reader who is not a Dylan fan should read no further and should turn immediately to the table of contents, which details what this book is about, or else flick through the book and take a look at some of the great figures that researchers have kindly supplied.

In mapping out the indigenous microbiota of humans, anyone whose life has been pervaded by Dylan's lyrics cannot fail to see parallels. Bob – how come you knew all this bacteriology? Which of the large variety of possible primary, secondary, and subsequent colonizers arriving at any site is certainly determined by *A Simple Twist of Fate*? But when a lonely bacterium's "gravity fails" (*Just Like Tom Thumb's Blues*) and it's deposited on the vast acidic, dry surfaces of the forearm, leg, or hand, does it cry out in desperation "oh my God am I here all alone?" (*Ballad of a Thin Man*). But it's unlikely to suffer for very long as the number of bacteria managing to survive there is minimized by "Shedding off one more layer of skin" (*Jokerman*).

To me, the colon, with its immense numbers of tightly packed microbes, is epitomized by that incredible line from *Leopard Skin Pill Box Hat* – "it balances on your head just like a mattress balances on a bottle of wine". How can so many microbes be fitted into such a small space, and how can such a sentence be fitted into the song without breaking its rhythm? *Idiot Wind*, with its acidic, bile-ridden lyrics just has to be the duodenum – one of the most acidic regions of the body and permeated by bile salts. The chances of bacteria surviving there are low, and the chances of love flourishing in that relationship are zero. Who could fail to *Pity the Poor Immigrant* bacteria in the urethra – repeatedly flushed away by a regular dowsing of urine. For them, there's nothing but *Trouble*. Is there nowhere they can get *Shelter from the Storm*? And spare a thought for those poor bacteria trying to colonize the conjunctiva who are washed away by "buckets of tears" (*Buckets of Rain*) – they must have the feeling that they're *Going, going, gone*. And lastly, of course, don't forget the microbial transients – those that are "the searching ones, on their speechless, seeking trail" (*Chimes of Freedom*) who have "no direction home, like a complete unknown, like a rolling stone" (*Like a Rolling Stone*). Someone needs to tell them that "Somewhere in this universe there's a place that you can call home" (*We Better Talk this Over*).

Finally, in gratitude to Bob Dylan for all his songs and the effect they've had on my life, I dedicate to him my own composition *Microbiota Row*. It's to be sung (if you're a great singer like me) to the tune of *Desolation Row* or, if you can't sing, then you can simply recite it to the rhythm of that truly amazing song.

MICROBIOTA ROW

The eyes have a microbiota
That's very sparse indeed
A few Gram-positive cocci
Scavenge from tears all that they need.
The skin has a denser population
P. acnes is plentifully found
While coryneforms and staphylococci
Are invariably around.
But molecular tools have shown us
There's much more still to know
About the microbes that live upon us
And even help us grow.

The respiratory tract is moist and inviting
With food aplenty there
But of the mucociliary escalator
All microbes must beware.
Yet haemophili and streptococci
And *Neisseria* can survive
While *Mollicutes* and *Moraxella*
Will there be found alive.
But there are pathogens among them
Most deadly, that's for sure
Armed with many deadly toxins
To bring us to death's door

Inside the terminal urethra
Staphylococci hold on tight
But most of the urinary tract is sterile
Thanks to innate immunity's might.
A male's reproductive system is arid
But with microbes a female's abounds
With lactobacilli and other genera
Their variety astounds.
But hormones have a great effect
On which microbes there can grow
And their relative proportions
Change as time's stream does flow.

From the mouth down to the rectum
The intestinal tract unwinds
Producing ecosystems so complex
And microbiotas of many kinds.
The oral cavity is aswarming
With 800 taxa there
While the hostile, acidic stomach
Apart from *H. pylori* is almost bare.
The small intestine is nearly sterile
But the colon is replete
With almost a thousand species
And without them we're incomplete.

Yes, I know you think they're nasty
Those minutest forms of life
Your mother said that they were dirty
And would only cause you strife.
But they're essential for your survival
Believe me you really must
They digest our food and protect us from
Pathogens that would make us dust.
Most of our indigenous microbes
Play a beneficial role, so please
Don't disturb or try to remove them
The result would be disease.

Michael Wilson

ABBREVIATIONS USED FOR MICROBIAL GENERA

A.	*Actinomyces*		*H.*	*Haemophilus*
Ab.	*Abiotrophia*		*Hel.*	*Helicobacter*
Ach.	*Acholeplasma*		*K.*	*Klebsiella*
Acin.	*Acinetobacter*		*Koc.*	*Kocuria*
Ag.	*Aggregatibacter*		*Kin.*	*Kingella*
All.	*Alloiococcus*		*L.*	*Lactobacillus*
An.	*Anaerococcus*		*Lep.*	*Leptotrichia*
At.	*Atopobium*		*Lis.*	*Listeria*
B.	*Bacteroides*		*M.*	*Micrococcus*
Bac.	*Bacillus*		*Mal.*	*Malassezia*
Bif.	*Bifidobacterium*		*Mob.*	*Mobiluncus*
Brev.	*Brevibacterium*		*Mor.*	*Moraxella*
C.	*Corynebacterium*		*Met.*	*Methylobacterium*
Camp.	*Campylobacter*		*Methyl.*	*Methylophilus*
Can.	*Candida*		*Myc.*	*Mycoplasma*
Cap.	*Capnocytophaga*		*N.*	*Neisseria*
Chlam.	*Chlamydia*		*P.*	*Propionibacterium*
Cl.	*Clostridium*		*Pep.*	*Peptostreptococcus*
Col.	*Collinsella*		*Por.*	*Porphyromonas*
D.	*Dermabacter*		*Prev.*	*Prevotella*
Des.	*Desulphovibrio*		*Ps.*	*Pseudomonas*
Dial.	*Dialister*		*R.*	*Rothia*
E.	*Escherichia*		*Ros.*	*Roseburia*
Eg.	*Eggerthella*		*Rum.*	*Ruminococcus*
Eik.	*Eikenella*		*Sal.*	*Salmonella*
Ent.	*Enterococcus*		*Sel.*	*Selenomonas*
Eub.	*Eubacterium*		*Staph.*	*Staphylococcus*
F.	*Fusobacterium*		*Strep.*	*Streptococcus*
Fil.	*Filifactor*		*T.*	*Treponema*
Fin.	*Finegoldia*		*Tan.*	*Tannerella*
G.	*Gardnerella*		*U.*	*Ureaplasma*
Gem.	*Gemella*		*V.*	*Veillonella*
Gran.	*Granulicatella*			

Chapter 1

THE HUMAN–
MICROBE SYMBIOSIS

While in the womb, the human fetus inhabits a sterile environment and, from a microbiological perspective, the most significant factor associated with the birth of a human being is that its environment is transformed from one that is free of microbes to one that is microbe-dominated. On delivery, the neonate encounters, for the first time, a wide range of microbes from a variety of sources. The main sources of microbes that colonize a neonate are:

• Vagina, gastrointestinal tract (GIT), skin, oral cavity, and respiratory tract of the mother;
• Skin, respiratory tract, and oral cavity of other individuals present at the delivery;
• Instruments and equipment used during delivery;
• The immediate environment.

Within a very short time of delivery, microbes are detectable on those surfaces of the baby that are exposed to the external environment, i.e. the eyes, skin, respiratory tract, genito-urinary system, GIT, and oral cavity. What is surprising, however, is that despite the neonate's exposure to such a variety of microbes, only a limited number of species are able to permanently colonize the various body sites available, and each site harbors a microbial community comprised of certain characteristic species, i.e. the microbes display "tissue tropism". The organisms found at a particular site constitute what is known as the indigenous (or "normal") microbiota of that site, and they are termed "autochthonous" (i.e. native to the place where they are found) species. The term "indigenous microbiota" encompasses all of the bacteria, archaea, viruses, fungi, and protoctists present on the body's surfaces. However, most studies have investigated only the bacterial component of the microbiota, and little is known about the identity of the other types of microbes present at any body site. This book, therefore, will concentrate on the bacterial

members of the indigenous microbiota, although it will also include those fungi (e.g. *Candida albicans* and *Malassezia* spp.) that are also indigenous to humans.

Colonization by microbes of the neonate at birth marks the beginning of a life-long human–microbe symbiosis. Symbiosis means "living together", and the term can be applied to any association between two or more organisms. When the species comprising a symbiosis differ in size, the larger member is known as the host while the smaller is termed a "symbiont". At least three types of symbiotic associations are recognized: (1) mutualism – when both members of the association benefit, (2) commensalism – when one member benefits while the other is unaffected, and (3) parasitism – when one member suffers at the expense of the other. During the course of his/her lifetime, a human being will experience all three types of symbiotic relationships with various members of their indigenous microbiota.

One of the many remarkable features of the microbiota of a particular body site is that its composition among human beings worldwide is, broadly speaking, similar despite the huge variations in the climate to which they are exposed, the diet that they consume, the clothes that they wear, the hygiene measures that they practice, and the lifestyle that they adopt. This implies that humankind has co-evolved with some of the microbial life forms that are present on Earth to form a symbiosis that is usually of mutual benefit to the participants.

1.1 OVERVIEW OF THE NATURE AND DISTRIBUTION OF THE MICROBIAL COMMUNITIES INHABITING HUMANS

The indigenous microbiota of humans consists of a large number of microbial communities each with a

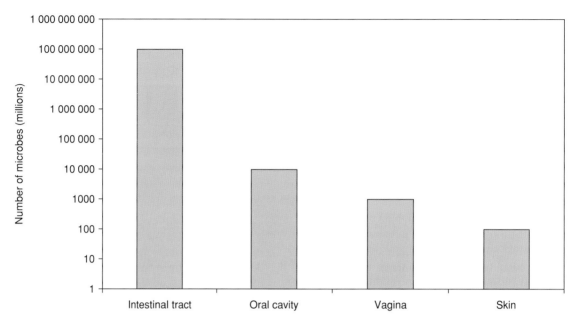

Fig. 1.1 Approximate number of microbes found in the most densely populated sites in an adult human. These numbers should be regarded as being only very rough approximations. Unfortunately, no estimates appear to have been made of the total number of microbes in the respiratory tract, which has several densely populated regions.

composition that is characteristic of a particular body site. In an adult human, microbes outnumber mammalian cells by a factor of ten – the average individual consists of 10^{13} mammalian cells and 10^{14} microbial cells. With few exceptions (the eyes, stomach, duodenum, and certain skin regions), the communities consist of large numbers of microbes and a large variety of species. Most of the microbes inhabiting humans are present in the GIT – the relative numbers in the GIT, in the mouth, in the vagina, and on the skin being approximately 1 000 000 : 100; 10; 1, respectively (Fig. 1.1).

Although most body surfaces that are exposed to the external environment are colonized by microbes, some are not (e.g. the lungs), and the population density of those sites that are colonized varies markedly from site to site (Fig. 1.2). Hence, the colon, oral cavity, and the vagina are densely populated by microbes, while much smaller numbers are present on the eyes, in the stomach, in the duodenum, and in the urethra. Even within an organ system, the density of colonization, as well as the community composition, can vary enormously from site to site. For example, in the respiratory tract the upper regions are more densely populated

than the lower regions – in fact, the bronchi and alveoli are usually sterile.

The complexity of the microbial community depends on the particular body site – only organisms able to grow under the conditions prevailing at the site being able to survive and grow there. The number of species present in the community can range from one or two at sites in which the environment is not conducive to microbial growth or survival (e.g. the conjunctiva) to more than 800 at sites such as the colon, which offers a microbe-hospitable environment containing a wide variety of nutrients.

1.1.1 Difficulties encountered in determining the composition of a microbial community

A major problem with trying to characterize the indigenous microbiota of a body site is the high species diversity of many of these communities. Until relatively recently, cultivation of the organisms present was the main approach used and, in general, this continues to be the case. However, this culture-based approach is

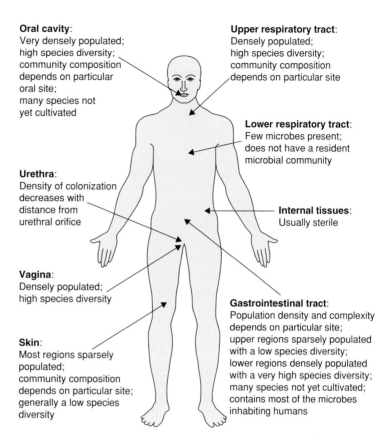

Oral cavity:
Very densely populated;
high species diversity;
community composition
depends on particular
oral site;
many species not
yet cultivated

Upper respiratory tract:
Densely populated;
high species diversity;
community composition
depends on particular site

Lower respiratory tract:
Few microbes present;
does not have a resident
microbial community

Urethra:
Density of colonization
decreases with
distance from
urethral orifice

Internal tissues:
Usually sterile

Vagina:
Densely populated;
high species diversity

Skin:
Most regions sparsely
populated;
community composition
depends on particular site;
generally a low species
diversity

Gastrointestinal tract:
Population density and complexity
depends on particular site;
upper regions sparsely populated
with a low species diversity;
lower regions densely populated
with a very high species diversity;
many species not yet cultivated;
contains most of the microbes
inhabiting humans

Fig. 1.2 General distribution of microbes at various sites within and on the human body.

fraught with problems (see section 1.4.2), the main one being that it fails to detect those organisms that are unable to grow under the culture conditions employed. These not-yet-cultivated species comprise a considerable proportion of the communities inhabiting some body sites e.g. up to 80% of the colonic microbiota. Fortunately, an increasing number of studies now employ culture-independent methods of identification, and these have added greatly to our knowledge (but not necessarily our understanding) of the composition of the microbial communities inhabiting humans (section 1.4.3). Disappointingly, most of these studies have been carried out only on samples from the GIT – mainly the oral cavity and colon.

In addition to the technical difficulties of analyzing complex communities, there are a number of other problems inherent in attempting to define the indigenous microbiota of a body site:
• Technical problems due to complexity of the microbial community;
• The very labor-intensive nature of the studies; generally only small numbers of samples can be processed – this limits the statistical reliability of the data obtained;
• Difficulties in obtaining appropriate, uncontaminated samples from many body sites;
• Variations between individuals related to factors such as genotype, age, sex, diet, hygiene practices, health status, type of clothing, occupation, and prevailing climate;
• Difficulties in comparing results obtained using different methodologies;
• Changes in microbial nomenclature – renders comparison with previous studies difficult.

Firstly, regardless of whether culture-based or culture-independent methodologies are used, the work involved in processing a single sample is considerable; this limits the number of samples that can be handled, thereby reducing the statistical reliability of the results obtained. Secondly, comparisons between studies are often difficult because of the different methodologies employed. Hence, culture-based studies carried out in different laboratories rarely use the same media and, therefore, will differ in their ability to detect and/or quantify particular organisms. In the case of culture-independent studies, the specificity and/or sensitivity of the primers or probes employed often differ between laboratories. Another problem is that changes in microbial nomenclature and taxonomy often make it difficult to compare the results of present-day studies with those obtained previously – even though these may be separated by only a few years. There are also a number of problems with obtaining samples. Hence, while it is relatively easy to obtain samples from sites such as the skin and oral cavity that are not contaminated with organisms from neighboring sites, this is not the case when sampling sites such as the stomach and duodenum. The attitude of the individuals being sampled can also be problematic. Hence, few individuals would object to having a swab taken from their forearm, but not many would happily volunteer to have their urethra swabbed or a nasogastric tube inserted to take samples of gastric juice.

There are also difficulties associated with selecting which individuals to sample. Studies have shown that the numbers and types of microbes present at a site may be affected by many attributes of the individual, including age, gender, sexual maturity, diet, hygiene practices, type of clothing worn, occupation, housing conditions, climate, etc. A properly designed study, therefore, should minimize such variations between the study participants, but this results in recruitment difficulties and invariably limits the number of individuals that match the inclusion criteria for the study. Finally, the investigators are faced with the problem of deciding whether or not a particular organism detected in an individual should be regarded as being a member of the indigenous microbiota of the site under investigation. This can be very difficult and, in the absence of any rigid rules, is sometimes a controversial issue. If a particular organism is detected at a body site in large numbers in every individual within a large group (matched for age, gender, etc.) and similar results are obtained on a number of different sampling occasions, then it is reasonable to regard it as being a member of the indigenous microbiota of that site. However, what if an organism is isolated from only 10% of these individuals – is it a member of the indigenous microbiota? Or, what is the status of an organism that is isolated from all individuals on one occasion but from only a few individuals on subsequent sampling occasions?

Attempts have been made to distinguish between microbes that are "residents" of a site and those that are "transients". Residents of a site should be able to grow and reproduce under the conditions operating at the site, whereas organisms that cannot do so, but are found at the site, are regarded as transients. However, the complexity of the microbial communities at many sites, the paucity of longitudinal studies of most sites, and the difficulties associated with trying to establish whether or not an organism is actively growing or reproducing at a site often make such distinctions difficult to establish.

Once an organism has been designated as being a member of the indigenous microbiota of a body site, it is important to try and understand why it is present at that site. In most cases (exceptions being the lumen of the lower regions of the GIT), the organism adheres to some substratum within the site – this may be a host cell, the extracellular matrix, some molecule secreted by the host, some structure produced by the host (e.g. a tooth or hair), or another microbe. The predilection of an organism for a particular host site is termed "tissue tropism". The presence of a receptor on a host tissue that is able to recognize the complementary adhesin on the bacterium is considered to be the mechanism underlying tissue tropism. However, this alone cannot explain the presence of an organism at a site because, as well as providing a substratum for adhesion, the site must also be able to satisfy all of the organism's growth requirements. Furthermore, the organism must also be able to withstand any of the host's antimicrobial defenses at that site. An understanding of such host–microbe interactions can be gained only by considering the anatomy and physiology of the site – and these are largely responsible for creating the unique environment that exists there. As Pasteur remarked more than 120 years ago, "The germ is nothing. It is the terrain in which it is found that is everything". The author has tried, therefore, to provide information on the environmental factors operating at each of the body sites colonized by microbes. Unfortunately, in many cases such data are not available – this being due to the difficulties in accessing the site or in analyzing the small quantities of fluid and/or tissue that can be obtained from the site.

Although the environment provided by the host is one of the most important factors dictating whether or not an organism can colonize a particular site, once colonization has occurred the environment is altered by microbial activity. This results in the phenomenon of microbial succession, in which organisms previously unable to colonize the original site are now provided with an environment suitable for their growth and reproduction. This process is fundamental to understanding the development of microbial communities at the various body sites and will be referred to repeatedly throughout this book.

1.1.2 Structural aspects of microbial communities

1.1.2.1 Microcolonies

Adhesion of an organism to a substratum is followed by its growth and reproduction – provided that the environment is suitable and that the organism can withstand the host defense systems operating there. This often results in the formation of an adherent microbial aggregate known as a microcolony, which is usually enclosed within a microbial polymer (Fig. 1.3). Microcolonies have been detected on the surface of the skin and on mucosal surfaces such as the respiratory, urogenital, and intestinal tracts (Fig. 1.4).

However, this does not happen in all cases as aggregate formation may be limited by mechanical and/or hydrodynamic forces operating at the site which will disrupt or dislodge such structures. Furthermore, if the organism is motile, one or more of the daughter cells may detach and move to another site within the habitat. Many epithelial cells, therefore, may only have small numbers of individual microbial cells on their surfaces (Fig. 1.5).

Another factor limiting the growth of microbial aggregates is that most of the surfaces exposed to the external environment (apart from the teeth) are comprised of epithelial cells which are continually being shed, taking the attached aggregates with them.

1.1.2.2 Intracellular colonization

There is increasing evidence that members of the indigenous microbiota of the oral cavity and respiratory tract can invade epithelial cells as well as simply adhering to them. Figure 1.6 shows oral epithelial cells containing large numbers of streptococci and other members of the indigenous oral microbiota – a single

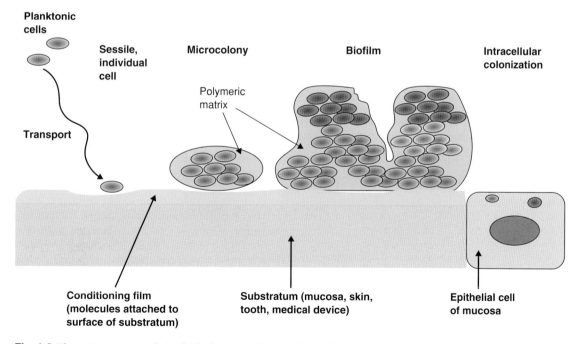

Fig. 1.3 The various patterns of microbial colonization that may be found on host tissues. Adhesion of single cells to the host tissue may lead to the production of microcolonies or biofilms. Bacteria may also reside within epithelial cells.

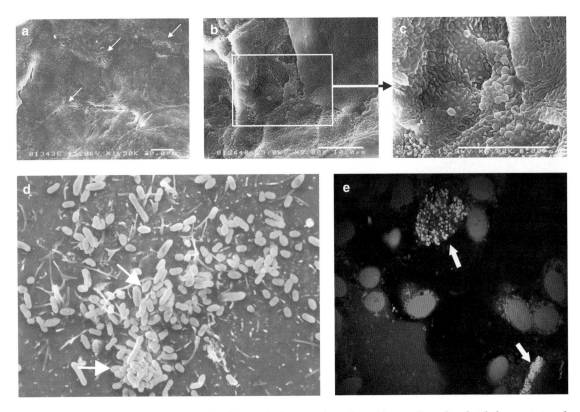

Fig. 1.4 (a–c) Scanning electron micrographs of bacterial colonies on the surface of the arm. Reproduced with the permission of The Japanese Society for Investigative Dermatology (© 2005) from: Katsuyama, M., Kobayashi, Y., Ichikawa, H., Mizuno, A., Miyachi, Y., Matsunaga, K. and Kawashima M. (2005) A novel method to control the balance of skin microflora, Part 2. A study to assess the effect of a cream containing farnesol and xylitol on atopic dry skin. *J Dermatol Sci* 38, 207–13. (d) Scanning electron micrograph of individual cells of *Haemophilus influenzae* and microcolonies (arrowed) growing on the surfaces of primary human bronchus cells, Magnification ×6000. Image kindly supplied by the Laboratory of Michael Apicella at the University of Iowa, Iowa City, USA. (e) Photomicrograph showing microcolonies (arrowed) on the rectal mucosa. The bacteria are stained with a live/dead stain – live cells are colored yellow while dead cells are red. Reproduced with the permission of Cambridge University Press, Cambridge, UK, from: Macfarlane, S. and Macfarlane, G.T. (2003) Bacterial growth on mucosal surfaces and biofilms in the large bowel. In: Wilson, M. and Devine, D. (eds) *Medical Implications of Biofilms*.

cell can contain as many as 100 bacteria. Such a habitat provides conditions that are very different from those existing on the epithelial cell surface and offers protection from many of the host's antimicrobial defense systems. However, shedding of the cell from the mucosa will, of course, remove any internalized or adherent organisms.

1.1.2.3 Biofilms

In certain circumstances, a microcolony can grow further (or several can merge) and develop into a larger structure known as a biofilm – this occurs on the teeth, which are nonshedding surfaces, and on mucosal surfaces with suitable anatomical features, e.g. within the crypts of the tongue and tonsils. They are also found on particulate matter in the colon and on medical devices and prostheses, e.g. catheters, artificial joints, limbs, and heart valves. A biofilm is defined as a matrix-enclosed microbial community attached to a surface. Because most surfaces in nature are coated with an adsorbed layer of macromolecules, the biofilm is usually attached to this layer (termed a "conditioning film") rather than directly to the surface itself. The

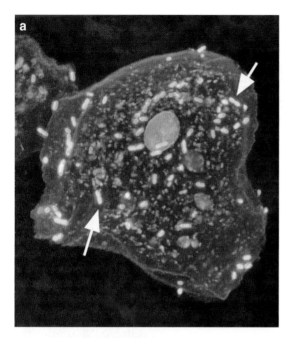

Fig. 1.5 Epithelial cells from the oral mucosa viewed by (a) confocal laser scanning microscopy and (b) scanning electron microscopy (scale bar, 20 μm). Pairs of bacteria (arrowed) and individual cells can be seen attached to the epithelial cells. Images kindly supplied by (a) Dr Chris Hope and (b) Mrs Nicola Mordan, UCL Eastman Dental Institute, University College London, London, UK.

matrix consists of polymers produced by the constituent microbes as well as molecules derived from the host. An organism growing within a biofilm has a phenotype different from that which it displays when it grows planktonically (i.e. in an aqueous suspension),

and the collective properties of a biofilm differ considerably from those of a simple aqueous suspension of the same organism(s). The general properties of biofilms are as follows:
• Biofilms contain a wide range of microhabitats as a result of gradients generated with respect to oxygen content, pH, Eh, nutrient concentration, concentration of metabolic end products, etc.
• Biofilms display reduced susceptibility to antimicrobial agents.
• Biofilms display reduced susceptibility to host defense systems.
• The constituent organisms of biofilms display novel phenotypes.
• The nutritional interactions between constituent organisms are facilitated by biofilms.
• Quorum sensing is facilitated by biofilms.
Furthermore, utilization of oxygen and nutrients in the environment by cells in the outermost layers of the biofilm, together with impeded diffusion of molecules by the biofilm matrix, results in chemical and physicochemical gradients within the biofilm (Figs 1.7 and 8.20). Other gradients will be generated with respect to metabolites produced by bacteria inside the biofilm. Within the biofilm, therefore, a wide variety of microhabitats exist, thereby providing conditions suitable for colonization by different physiological types of microbes.

Biofilms are highly hydrated structures, and the bacteria within them may occupy as little as 10% of the total volume. This means that the staining and dehydration techniques used to prepare biofilms for examination by light and/or electron microscopy grossly distort their structure. Confocal laser scanning microscopy (CLSM) enables the examination of biofilms in their native, hydrated state, and this has enabled a more accurate estimation of their structure and dimensions. CLSM (and other modern microscopic techniques such as differential interference contrast microscopy and two-photon excitation microscopy) have shown biofilms to have a more complex structure than that revealed by electron microscopy (Fig. 1.8).

As a number of factors can affect biofilm structure, there is no single, unifying structure that can be said to characterize all biofilms. The key variables involved include: the nature of the organism (or community), the concentration of nutrients present, the hydrodynamic properties of the environment, and the presence (and nature) of any mechanical forces operating at the site. Hence, the structure of a biofilm can range from a relatively featureless, flat type to one consisting

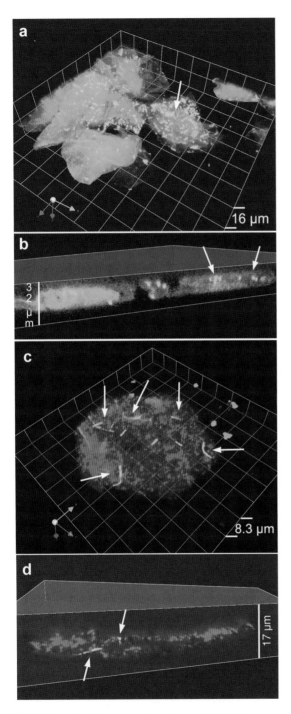

of a more complex organization involving tower-like "stacks" (consisting of microbes enclosed in an extracellular matrix) separated by water channels (Fig. 1.8). Depending on the microbial composition of the particular biofilm, the stacks may consist of a single species or of microcolonies of a number of different species. A microcolony forms at the particular location within a stack that has the appropriate combination of environmental factors suitable for the survival and growth of that organism. The water channels may function as a primitive circulatory system, bringing fresh supplies of nutrients and oxygen while removing metabolic waste products. Further details of the structure and composition of biofilms will be found in subsequent chapters.

1.1.3 Communication in microbial communities

Although it has been known for a long time that bacteria can sense, and respond to, their external environment, it has only recently been discovered that many species are also able to sense the presence of other bacterial cells. This phenomenon, known as "quorum sensing", involves the production of a low molecular mass auto-inducer which, on reaching a threshold concentration (due to the presence of a critical population density) in the external environment, can activate the transcription of certain genes. The result is a population-wide change in gene expression so that the community is, in effect, behaving as a multicellular organism.

Fig. 1.6 Confocal laser scanning micrographs of oral epithelial cells containing large numbers of intracellular and extracellular cocci. (a) A cluster of buccal cells containing intracellular bacteria. Some of the cocci are labeled by both a green *Streptococcus*-specific probe and a red universal probe and therefore appear yellow. Bacteria that are not streptococci appear red. The cell denoted by the arrow has been extensively invaded by streptococci. (b) A vertical section through the epithelial cells. Intracellular streptococci are denoted by arrows. (c) A buccal cell dominated by intracellular cocci (presumably streptococci) labeled using the red universal probe. The sample was also treated with a yellow *Fusobacterium nucleatum*-specific probe, and the buccal cell was found to contain several *F. nucleatum* cells (arrowed). (d) A vertical section of the cell shown in (c). Reproduced with the permission of the International Association for Dental Research, Alexandria, Virginia, USA, from: Rudney, J.D., Chen, R. and Zhang, G. (2005) Streptococci dominate the diverse flora within buccal cells. *J Dent Res* 84, 1165–71.

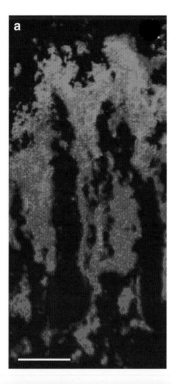

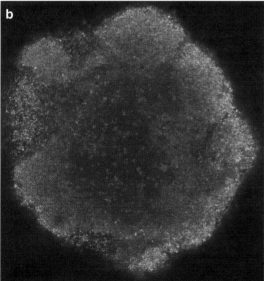

The nature of the auto-inducer depends on the particular species and there are, basically, three types of quorum-sensing systems (Fig. 1.9). Most Gram-negative bacteria use the LuxR/I system, in which the LuxI protein (or a homologue) produces the auto-inducer (an acyl-homoserine lactone – AHL), which diffuses freely out of, and into, the cell. The AHL binds to the intracellular LuxR protein (or a homologue), which regulates gene transcription. In the second system, which is used by Gram-positive organisms, the auto-inducer is an oligopeptide with between 6 and 17 amino acids that is usually post-translationally modified. The auto-inducer is actively transported out of the cell and activates a two-component signal transduction system (TCSTS). The auto-inducer is recognized by the sensor kinase of the TCSTS which initiates phospho-transfer to the response regulator that controls gene transcription. The third system, LuxS/AI-2, is found in a wide variety of Gram-positive and Gram-negative species and enables interspecies communication. LuxS produces the auto-inducer 4,5-dihydroxy-2,3-pentanedione (DPD), which spontaneously rearranges to produce a variety of molecules. Different species of bacteria recognize particular DPD rearrangements, and this enables bacteria to respond to AI-2 derived from their own DPD as well as to that produced by other species.

By using one or other of these systems, bacteria can regulate the expression of certain genes in a

Fig. 1.7 (a) confocal laser scanning micrograph of a biofilm of *Staph. epidermidis* stained with 4′,6-diamidino-2-phenylindole (which binds to DNA and stains all cells blue) and with 5-cyano-2,3-ditolyl tetrazolium chloride (which is converted to a red color by respiring cells). The substratum to which the biofilm is attached is at the bottom, while the biofilm/nutrient interface is at the top. The respiratory activity decreases from the top to the bottom of the biofilm as a result of the existence in the biofilm of various gradients with respect to environmental conditions (e.g. nutrient concentration, pH, etc.) that affect the metabolic activity and viability of the constituent cells. Reproduced with the permission of Cambridge University Press, Cambridge, UK, from: Zheng, Z. and Stewart, P.S. (2004) Growth limitation of *Staphylococcus epidermidis* in biofilms contributes to rifampin tolerance. *Biofilms* 1, 31–5. (b) Confocal laser scanning micrograph of a multi-species oral biofilm. The image shows a cross-section through a biofilm stack (viewed from above) and is stained to reveal live (green) and dead (blue) bacteria. A gradient exists within the biofilm stack with respect to cell viability – the proportion of viable cells decreases from the outer to the inner layers. Reproduced with the permission of Elsevier, Amsterdam, The Netherlands (© 2006) from: Hope, C.K. and Wilson, M. (2006) Biofilm structure and cell vitality in a laboratory model of subgingival plaque. *J Microbiol Methods* 66, 390–8.

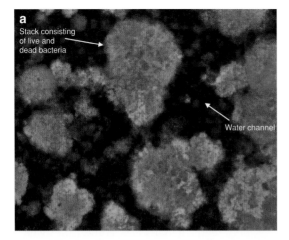

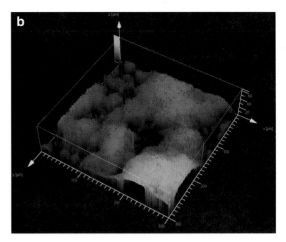

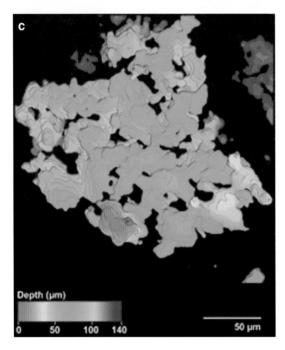

Fig. 1.8 Confocal laser scanning micrograph (low magnification) of multi-species oral biofilms grown in the laboratory under conditions similar to those that exist in vivo. (a) The biofilm has been stained to differentiate between live cells (green) and dead cells (blue) and is viewed from above. Bacterial stacks separated by water channels are clearly visible. (b) A fluorescent dye was used to stain the constituent organisms but not the extracellular biofilm matrix. Stacks of bacteria (up to 60 μm high) separated by water channels are visible. (c). Mapping of a confocal laser scanning micrograph of a multi-species oral biofilm with contours denoting the distance of the various regions of the biofilm from the microscope objective. Again, the biofilm can be seen to consist of numerous "stacks" of bacteria of different heights separated by fluid-filled regions. Images kindly supplied by Dr Chris Hope, School of Dental Sciences, University of Liverpool, Liverpool, UK.

population-dependent manner – a phenomenon of undoubted relevance to biofilms and microcolonies with their high bacterial densities (Fig. 1.10).

Genes controlled by quorum sensing include those involved in biofilm formation, competence, and conjugation as well as those encoding many virulence factors. Activities and processes of microbes indigenous to humans that have been shown to be regulated by quorum sensing include:

• Mixed-species biofilm formation by *Porphyromonas gingivalis* and *Streptococcus gordonii*;
• Biofilm formation and sugar metabolism in *Strep. gordonii*;
• Bacteriocin production by *Streptococcus pyogenes*, *Streptococcus salivarius*, and *Lactobacillus plantarum*;
• Iron acquisition and leukotoxin production by *Aggregatibacter actinomycetemcomitans*;
• Toxin production by *Clostridium perfringens*;

a

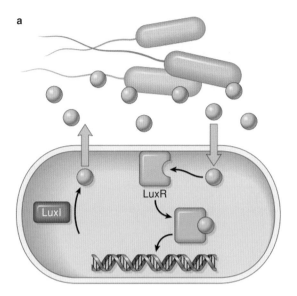

b

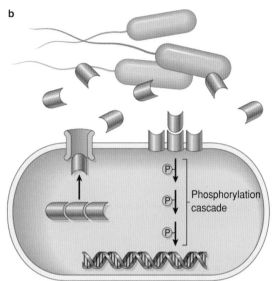

c

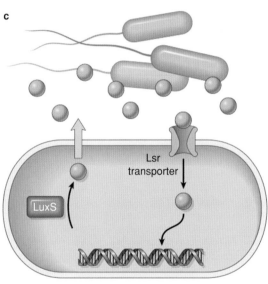

Fig. 1.9 Diagram showing the three main quorum-sensing systems. (a) LuxI/LuxR system in a Gram-negative organism. The auto-inducer (e.g. homoserine lactones, denoted by spheres) is synthesized through pathways involving LuxI, diffuses out of the cell and, on re-entering, binds to LuxR, thereby altering gene transcription. (b) In Gram-positive bacteria, an oligopeptide (denoted by linked shapes) is exported and then binds to a sensor kinase of a TCSTS. This phosphorylates a response regulator which regulates gene transcription. (c) The LuxS/AI-2 system is found in Gram-negative and Gram-positive bacteria. DPD is synthesized by a pathway involving LuxS and then undergoes re-arrangement to produce an AI-2 (denoted by spheres), the exact structure of which depends on the particular bacterial species that is released from the cell. AI-2 gains entry to the cell via a specific transporter (Lsr in the case of *Salmonella typhimurium*) and then regulates the transcription of certain genes. Reproduced with the permission of the American Society for Pharmacology and Experimental Therapeutics, Bethesda, Maryland, USA, from: Raffa, R.B., Iannuzzo, J.R., Levine, D.R., et al. (2005) Bacterial communication ("quorum sensing") via ligands and receptors: A novel pharmacologic target for the design of antibiotic drugs. *J Pharmacol Exp Ther* 312, 417–23.

• In *Escherichia coli*, the transcription of 242 genes (nearly 6% of its genome) is affected by the concentration of AI-2;
• Cell division and morphology in *E. coli*;
• Competence in *Streptococcus pneumoniae*;

• Protease and cytolysin production by *Enterococcus faecalis*;
• Transfer of conjugative plasmids by *Enterococcus* spp.;
• Biofilm formation and competence in *Streptococcus mutans*;

a

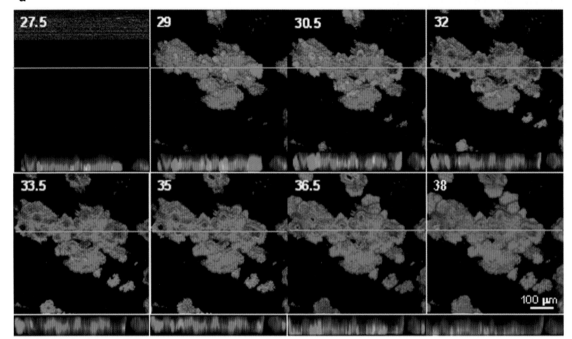

b

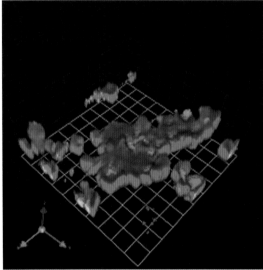

Fig. 1.10 Confocal scanning laser micrographs showing expression of the accessory gene regulator (*agr*) quorum-sensing system in biofilms of *Staph. aureus*. Expression of *agr* was monitored using a reporter plasmid containing a green fluorescent protein (gfp) reporter gene linked to the P3 promoter region of *agr*. (a) Time-lapse expression of the agr P3-gfp reporter in a biofilm of *Staph. aureus*. The images were acquired at 90-min intervals with the number of hours post-inoculation indicated. Cells expressing the reporter are labeled green, while the rest of the cells are stained red with propidium iodide. The larger images represent a compressed *z* series, where multiple *x–y* planes from top to bottom of the biofilm are combined. The smaller images represent *x–z* (side) views of the biofilm at the location indicated by the line. Expression occurred in different regions of the biofilm and oscillated with time. (b) Three-dimensional reconstruction of a *z* series taken 36.5 h after inoculation. Each side of a grid square represents 60 μm. Reproduced with permission from: Yarwood, J.M., Bartels, D.J., Volper, E.M. and Greenberg, E.P. (2004) Quorum sensing in *Staphylococcus aureus* biofilms. *J Bacteriol* 186, 1838–50.

• Biofilm formation and bacteriocin production by *Staphylococcus epidermidis*;
• Biofilm formation and exotoxin production by *Staphylococcus aureus*;

• Biofilm formation in *Strep. pyogenes*.

The ability to limit gene expression until a large population has been reached is advantageous to the organism in a number of ways. Bacteria often derive their nutrients from complex polymers, and the degradation of such polymers requires the concerted secretion of enzymes from large numbers of cells. An individual cell, or a population in which only some of the members are secreting the appropriate enzymes, would not constitute an effective means of utilizing the available nutrient resources. The advantage of competence and conjugation being regulated by a population-dependent process is obvious – DNA transfer is not possible in the absence of other cells. The ability to limit virulence factor secretion until a large number of bacteria are present could be a protective measure against host defense systems. Hence, if only a few bacteria were to secrete a particular virulence factor (small concentrations of which would be unlikely to cause serious damage), this could alert the host, which may then be able to dispose of this threat effectively – something that it is less likely to be able to do if a large population is present.

1.2 ENVIRONMENTAL DETERMINANTS THAT AFFECT THE DISTRIBUTION AND COMPOSITION OF MICROBIAL COMMUNITIES

In order for an organism to become a member of the indigenous microbiota of a body site, the environment of that site must be able to satisfy its nutritional and physicochemical requirements. Furthermore, the organism must be able to withstand the host defense systems operating at the site as well as the various mechanical and hydrodynamic microbe-removing systems that may be present, e.g. urination, coughing, and the mucociliary escalator.

The nutritional and physicochemical constraints on microbial growth within an ecosystem are codified in two laws – Liebig's law of the minimum and Shelford's law of tolerance. Liebig's law of the minimum states that the total yield of an organism is determined by the nutrient present in the lowest concentration in relation to the requirements of the organism. In any ecosystem, some nutrient will be present at a concentration that will limit the growth of a particular species. If the concentration of this nutrient is increased, the population of that species will increase until another nutrient becomes limiting, and so on. The nutrients available to an organism at a site on the human body are derived, in most cases, from two sources – the host and other microbes inhabiting the site. Organisms inhabiting the GIT have an additional source of nutrients – food ingested by the host.

Shelford's law of tolerance relates to the non-nutritional factors that govern growth of an organism (e.g. pH, temperature, and Eh), and states that every species requires a complex set of conditions for survival and growth and that there are bounds for these factors outside of which an organism cannot survive or grow. A species will only survive within an ecosystem, therefore, if each of the physicochemical conditions operating there remains within its tolerance range. While the host "sets the scene" in terms of many of the physicochemical factors operating at a body site, the microbes present at that site often alter these factors substantially.

As well as being governed by nutritional and physicochemical factors, the ability of an organism to become established at a particular site is also influenced by biological factors such as the production of antimicrobial compounds by the host and by other microbes present at the site. The organism must also be able to withstand any mechanical removal forces operating at the site such as the flow of saliva, urine, and other fluids produced by the host.

All of the factors governing the ability of an organism to maintain itself at a body site, and which determine the composition of the microbiota colonizing that site, are collectively termed "environmental determinants" or "environmental selecting factors" and are described in the following sections.

Although the environmental conditions prevailing at a particular site dictate to a large extent the types of microbes that can colonize that site, subsequent changes to the environment brought about by microbial activities, as well as beneficial and antagonistic interactions between the species present, then influence the composition of the community which evolves until it reaches a stable climax state (see Fig. 1.14). However, this is not the whole story and community composition is affected by a number of other factors. Colonization history, for example, also plays a role in shaping community composition, i.e. which of the many possible primary colonizers of a body site happens, by chance, to arrive first in a particular individual. Subsequent community development could also be influenced by the arrival of any of a number of possible secondary colonizers able to cope with the altered environment – and so on. Which particular organism arrives at a site, and when, are largely chance events. It may well be that, while environmental factors dictate which of the higher-level taxa are present at a body site, the particular species or strains found in an individual could be the result of colonization history and other chance events. This would help to explain the person-to-person variation that exists in the microbiota of a particular body site.

1.2.1 Nutritional determinants

The main elements required for the growth of all microbes include carbon, oxygen, nitrogen, hydrogen, phosphorus, sulfur, potassium, sodium, calcium, magnesium, chlorine, and iron. In addition, small quantities of a number of trace elements (e.g. cobalt, copper, manganese, zinc, and molybdenum) are also needed as co-factors for various enzymes and as constituents of proteins and other cell components. Many organisms also require small quantities of organic growth factors

such as amino acids, vitamins, fatty acids, or lipids. There is enormous diversity with regard to the nature of the compounds (i.e. nutrients) that an organism can utilize as a source of each particular element.

Microbes also differ with respect to the types of compounds that can serve as an energy source. A microbe will be able to colonize only those sites that can provide a suitable energy source and that can supply all of the elements needed for its growth in a form that it can utilize. The host provides a range of potential nutrients in the form of the excretory and secretory products of live cells, transudates from the fluid which permeates tissues (interstitial fluid), and the constituents of dead cells (Fig. 1.11). In addition, food ingested by the host can act as a source of nutrients for microbes inhabiting

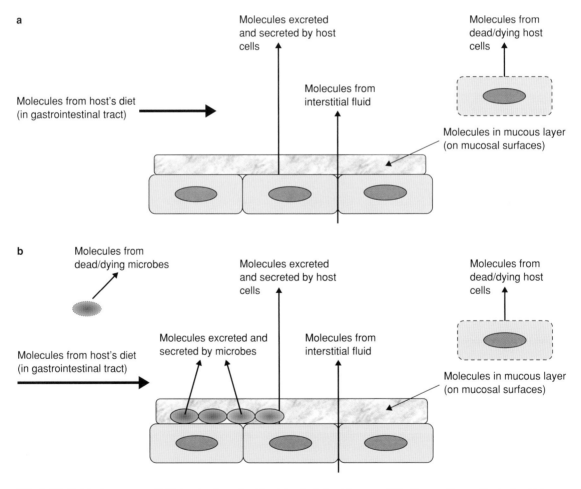

Fig. 1.11 Nutrient sources available to microbes colonizing (a) a sterile host tissue and (b) a tissue with a resident microbiota.

Table 1.1 Important host-derived nutrients present at various body sites. This is not a detailed catalogue of the nutrients present at each body site – in each case, many other molecules capable of acting as nutrients are also present and these are described in subsequent chapters.

Body site	Important host-derived nutrients
Skin surface	Lipids, proteins
Respiratory mucosa	Mucins, proteins
Caecum and ascending colon	Carbohydrates derived from the host's diet
Descending colon	Dietary proteins
Urethra	Mucins, proteins, urea
Vagina	Mucins, proteins, polysaccharides, amino acids, sugars
Oral cavity	Mucins, proteins, dietary constituents

the GIT. Although the primary colonizers of any body site are entirely dependent on host-derived nutrients, once a site has been colonized, molecules produced by the organisms present can serve as additional nutrient sources. Such nutrients will differ from those supplied by the host, and this enables species with different nutritional requirements to establish themselves at the site; in other words, microbial succession takes place.

The nutrients provided by the host vary with the body site (Table 1.1), and these will be described in detail in subsequent chapters. This has a profound influence on the types of organisms that can colonize a particular site and become established as members of the site's microbiota.

Often the host-derived carbon sources available at a body site are complex macromolecules such as polysaccharides, proteins, glycoproteins, fats, glycolipids, lipoproteins, etc. In order to utilize these, the microbe must first hydrolyze the molecule to smaller units that can be transported into the cell. In some cases, the complexity of the carbon source is such that the cooperation of several species is needed to achieve its breakdown. Nitrogen sources available at many sites include proteins, glycoproteins, lipoproteins, amino acids, urea, ammonium ions, and nitrate ions. Oxygen can be obtained from the atmosphere and/or host tissues at some body sites and is present in water and most organic compounds. Phosphorus is available in nucleic acids, phosphoproteins, phospholipids, phosphates, etc. Many of the other major elements needed for microbial growth are present as inorganic salts or low molecular mass organic compounds.

The dependence of microbes on small quantities of a variety of metal ions provides an opportunity for the host to control microbial growth by depriving them of these micronutrients. This is achieved by converting them to forms that many microbes are unable to use. An important example of this is iron deprivation. This element is an essential constituent of cytochromes and other enzymes and, in order to grow and reproduce, most microbes need iron to be available in their environment at a concentration of approximately 10^{-6} M. Most of the iron in the human body is bound to hemoglobin, myoglobin, and cytochromes, and any excess iron is complexed with iron-binding proteins (transferrin, lactoferrin, ovalbumin, and ferritin), leaving only minute quantities of free iron (approximately 10^{-15} M) available to microbes. In order to scavenge the limited amounts of iron available, microbes produce high-affinity iron-binding compounds known as siderophores. Some species produce enzymes able to degrade host iron-binding compounds thereby liberating iron; others express receptors for these compounds and then remove iron from the bound complex. Furthermore, many organisms produce hemolysins which lyse red blood cells, thereby liberating iron-containing compounds such as hemoglobin from which iron can be obtained.

The host is the sole source of nutrients only for the pioneer organisms of a body site. Once that site has been colonized, substances excreted and secreted by microbes, as well as dead microbes themselves, will serve as additional sources of nutrients, and so the range of nutrients available is extended, thereby creating opportunities for colonization by other organisms. Hence, species which would otherwise not have been

Table 1.2 The range of possible interactions that can occur between two microbial populations. The interactions involved may be nutritional, physicochemical, or biological.

Type of interaction	Examples
Positive for one or both organisms	Commensalism, synergism (or protocooperation), mutualism
Negative for one or both organisms	Competition, amensalism (or antagonism), predation, parasitism
No interaction between organisms	Neutralism

able to inhabit the site become established there (such organisms are termed "secondary colonizers"), and this results in "autogenic succession", i.e. a change in the composition of a microbial community arising from microbial activities. A change in the composition of the community due to external, nonmicrobial factors is termed "allogenic succession". As the community increases in complexity, the variety of interactions between its members increases, and these may have negative as well as positive outcomes (Table 1.2). Such nutritional interactions (along with other physicochemical interactions described later) contribute to the development of a microbial community at that site and ensure its stability, although gross changes imposed on the site by external forces can lead to its disruption.

Commensalism is a relationship between two organisms in which one benefits while the other is unaffected. Examples of this include: (1) the production of amino acids or vitamins by one organism which are then used by another organism; (2) the degradation of polymers to release monomers which can be utilized by another organism; (3) the production, by one organism, of substances that solubilize compounds; these compounds can then be utilized by another organism; (4) the utilization, or neutralization, by an organism of a molecule that is toxic to another species.

Synergism is a relationship in which both members benefit – when this involves a nutritional interaction, the term syntrophism is often used. An important example of syntrophism is cross-feeding. Cross-feeding occurs when one organism (A) utilizes a compound which the other organism (B) cannot, but in doing so produces a new compound which organism B can utilize. The removal of the new compound by B is usually beneficial to A, because it removes the negative feedback control exerted by the compound. Examples of syntrophism are given in Fig. 1.12.

Important sources of nutrients for microbes at most body sites are mucins. These are complex glycoproteins whose complete breakdown requires a variety of enzymes – sulfatases, sialidases, glycosidases, proteases, and peptidases. Very few organisms produce this full complement of enzymes, therefore the cooperation of a number of species is usually necessary to achieve complete degradation of this valuable source of nutrients (see section 1.5.3). The liberation of sulfate from mucins in the GIT (e.g. by *Bacteroides* spp.) enables the growth of *Desulfovibrio* spp. which use sulfate as an energy source. These are simple examples of what are termed "food chains". However, these often involve more than two organisms and can result in complex "food webs" (see Figs 2.41 and 8.17).

Mutualism is a form of synergism in which the relationship between the two organisms is obligatory. However, the terms mutualism and synergism are often used interchangeably. Further confusion arises because the term "symbiosis" which, strictly speaking, refers to any long-term relationship between two populations, is also often used in the sense of a mutually beneficial relationship, i.e. synergism or mutualism.

Competition is an interaction in which both microbes are adversely affected and occurs when each organism utilizes the same nutrient. Ultimately, one of the organisms will be excluded from the habitat, and this effect is an example of "competitive exclusion". For example, hydrogen is produced in the colon as a byproduct of microbial metabolism, and the two main users of the gas are the methanogens and the sulfate-reducing bacteria. These two groups of organisms compete for hydrogen, and there is always an inverse relationship between their proportions in the colon. In continuous-culture models of the colon, competition for a carbohydrate has been shown to result in suppression of the growth of both *E. coli* and *Clostridium difficile*.

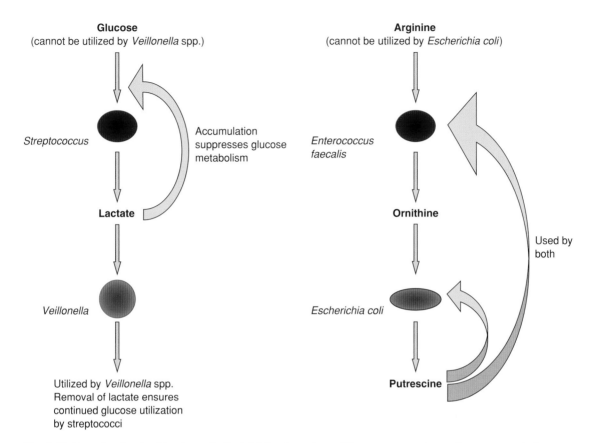

Fig. 1.12 Examples of syntrophism (cross-feeding). (a) Streptococci can utilize sugars as a carbon and energy source but *Veillonella* spp. cannot. One of the end products of sugar metabolism by streptococci is lactate, the accumulation of which inhibits glycolysis. However, lactate can be used by *Veillonella* spp. as a carbon and energy source, and this utilization of lactate removes its inhibitory effect on streptococcal glycolysis. (b) *Ent. faecalis* can convert arginine to ornithine (something that *E. coli* is unable to do), which is then converted by *E. coli* to putrescine, a compound that both organisms can utilize.

Amensalism is not a nutritional interaction but will be described briefly as an example of a negative interaction – it is a topic that will be dealt with in more detail in subsequent chapters. This occurs when one organism produces a substance that is toxic to another or creates an environment that is unsuitable for others. Examples include the production of: (1) antimicrobial proteins (bacteriocins) by many organisms at all sites; (2) hydrogen peroxide by lactobacilli in the vagina; and (3) acidic end products of metabolism – these may exert a direct toxic effect and/or lower the pH of the environment making it unsuitable for the growth of many species. The low pH on the skin surface is due, in part, to acid production by members of the skin microbiota,

while the low pH of the vagina is partly attributable to the lactobacilli that reside there.

Parasitism involves one organism (the parasite) deriving its nutritional requirements from another (the host), which ultimately is harmed in some way, e.g. the bacteriophages (or phages) that parasitize bacteria. Although phages that are able to attack most members of the human microbiota have been detected, there is little information regarding their ability to affect the composition of the microbial community at a body site. Predation involves the engulfing of one organism (the prey) by another (the predator), e.g. bacteria are the major food source for protoctists, i.e. eucaryotic microbes without a cell wall. Protoctists are found in the oral cavity (e.g. *Entamoeba gingivalis*) and the intestinal tract (e.g. *Dientamoeba fragilis*), but little is known of their effect on the bacterial communities of these sites.

Positive interactions tend to create new niches (the term "niche" means the functional role of an organism in a community, i.e. the sum of the processes it performs, and does not refer to its location in an ecosystem) to be filled by other potential colonizers of the site and thereby enable the utilization of all of the nutritional resources available at the site. Negative interactions act as feedback mechanisms that ultimately limit the population of an organism in the community. The combination of these negative and positive nutritional interactions, together with interactions affecting the physicochemical features of the site (Fig. 1.11), ultimately gives rise to a "climax community" in which the constituent organisms are in a state of dynamic equilibrium. Such a community is characterized by homeostatic mechanisms that enable it to resist change and render it stable in terms of its composition – unless it is subjected to some catastrophic externally generated change. An important aspect of such stability is "colonization resistance", which prevents the establishment of a microbe that is not normally a member of the microbiota of that body site, i.e. an "allochthonous" species. Each organism within a climax community occupies a niche (i.e. has a functional role) that contributes towards the stability of that community.

1.2.2 Physicochemical determinants

A number of physicochemical factors affect the growth of microbes, but not all of these (e.g. atmospheric pressure, magnetic fields) have much relevance to the composition or growth of microbial communities inhabiting humans. This is because we ourselves are limited with respect to the type and range of environments that can be tolerated, and homeostatic mechanisms ensure a relatively constant environment at most of the sites colonized by microbes. These homeostatic mechanisms are augmented by factors such as clothing, housing, and air-conditioning, which help to protect against the more extreme environments to which humans may be exposed. However, despite the effectiveness of our homeostatic mechanisms, the conjunctiva and exposed regions of the skin will experience more dramatic fluctuations in temperature and humidity, for example, than sites such as the mucosal surfaces of the respiratory system and GIT. The most important

physicochemical determinants affecting growth of the indigenous human microbiota are temperature, pH, Eh, atmospheric composition, water activity, salinity, and light.

The temperature of most body sites in human beings is approximately 37°C. This limits the type of microbe that can colonize humans to those species that are mesophiles, i.e. they grow over the temperature range 25–40°C and have an optimum growth temperature of approximately 37°C. The temperature of exposed areas of the skin (e.g. face and hands) tends to be a few degrees lower (approximately 33°C) than most other body sites and, although temperatures far lower than 37°C have been recorded at these sites in those living in polar and temperate climates, this does not appear to affect the types of organism colonizing such sites. The temperature of conjunctival surfaces may be even lower than those found on the skin, and there is some evidence that climatic temperature change affects the composition of the ocular microbiota, although concomitant humidity changes may also be partially responsible (section 3.4.2).

In contrast to temperature, the pH of different body sites varies enormously and ranges from as low as pH 1–2 to alkaline values (Fig. 1.13). This has a considerable effect on the types of organism that inhabit these sites. While very acidic pHs are found at a number of body sites (e.g. stomach, duodenum, caecum, and skin), alkaline environments are infrequently encountered and are generally present only in the tear film, the ileum, and in subgingival regions of the oral cavity.

Depending on the range of pHs over which they can grow and their optimum pH for growth, microbes are broadly classified as being acidophiles, neutrophiles, or alkaliphiles. Acidophiles generally have an optimum pH for growth of less than 5.5, neutrophiles grow best over the pH range 5.4 to 8.0, and alkaliphiles have an optimum pH for growth of greater than 8.0. Most members of the indigenous microbiota of humans are neutrophiles. Some organisms can tolerate extremely acidic and alkaline pHs even though they may not be able to grow optimally at such pHs – such organisms are termed aciduric and alkaliduric, respectively. The term "acidogenic" refers to those organisms that produce acidic end products of metabolism, e.g. streptococci, lactobacilli, and propionibacteria. Although aciduric and acidophilic organisms, not surprisingly, are isolated from body sites with a low pH, not all organisms colonizing acidic sites belong to either of these categories. Hence, *Hel. pylori*, which colonizes the most acidic

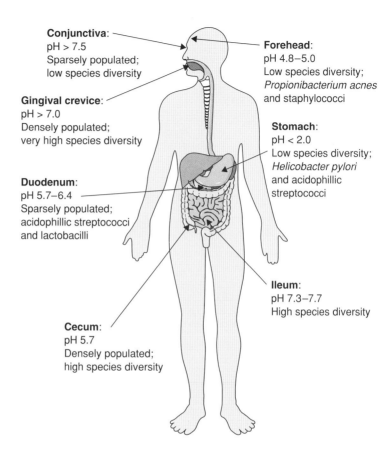

Conjunctiva:
pH > 7.5
Sparsely populated;
low species diversity

Forehead:
pH 4.8–5.0
Low species diversity;
Propionibacterium acnes
and staphylococci

Gingival crevice:
pH > 7.0
Densely populated;
very high species diversity

Stomach:
pH < 2.0
Low species diversity;
Helicobacter pylori
and acidophillic
streptococci

Duodenum:
pH 5.7–6.4
Sparsely populated;
acidophillic streptococci
and lactobacilli

Ileum:
pH 7.3–7.7
High species diversity

Cecum:
pH 5.7
Densely populated;
high species diversity

Fig. 1.13 Regions of the body that do not have a neutral pH and their associated microbiotas.

region of the body – the stomach – is a neutrophile, not an acidophile. This organism creates a microenvironment with a pH near neutral due to the production of ammonia from urea present in gastric juices.

While the pH of a body site is initially dictated by the host, the activities of the microbes colonizing that site can also have a profound effect on its pH. Hence, the metabolic activities of microbes present on the skin and in the vagina play a major role in lowering the pH of these sites. The generation of these low pHs affects the composition of the resident microbial communities by preventing the colonization of species unable to survive at such pHs and is an example of amensalism. The pH of the colon also gradually increases along its length to a more neutral value because of the production of ammonia due to amino acid fermentation by resident bacteria. The presence of sucrose in the diet enables acid formation by oral streptococci and lactobacilli, resulting in low pHs in the biofilms that form on the tooth surface. Although the growth of a particular organism is limited to a certain pH range, studies have shown that when the organism is part of a mixed species community, the range of pHs over which it can grow is often extended considerably.

The oxygen content of the atmosphere present at a site is an important determinant of which species will be able to reside there and, in particular, of the type of microbes that can function as pioneer organisms. It is possible to distinguish a number of microbial groups on the basis of their relationship to molecular oxygen, and these are listed in Table 1.3.

Although human beings inhabit an aerobic environment, most members of their indigenous microbiota are either obligate or facultative anaerobes rather than obligate aerobes. Obligate aerobes are not frequently encountered on any body surface, even at sites such as the skin and oral cavity, which are in constant contact with the atmosphere. At birth, all surfaces of an individual are aerobic, so that only obligate aerobes or facultative anaerobes can colonize any body site.

Table 1.3 Classification of microbes on the basis of their oxygen requirements.

Microbial group	Description	Examples
Obligate aerobes	Require oxygen to grow	*Acinetobacter, Moraxella, Micrococcus, Brevibacterium*
Capnophiles	Aerobes that grow best at CO_2 concentrations of between 5 and 10% – this is greater than concentrations encountered in air	*Neisseria, Haemophilus, Aggregatibacter*
Obligate anaerobes	Do not grow in the presence of oxygen	*Bacteroides, Clostridium, Porphyromonas, Fusobacterium, Veillonella, Bifidobacterium, Propionibacterium* (some species), *Eubacterium, Ruminococcus*
Facultative anaerobes	Can grow in the presence or absence of oxygen	*Staphylococcus, Streptococcus, Enterococcus, Escherichia, Proteus*
Microaerophiles	Grow best at low concentrations of oxygen (2–10%)	*Helicobacter, Campylobacter, Lactobacillus* (some species), *Propionibacterium* (some species)

However, once a pioneer community has become established, microbial consumption of oxygen and the production of carbon dioxide alters the composition of the atmosphere. This creates a range of microhabitats at the site, enabling colonization by capnophiles, microaerophiles, and obligate anaerobes, the growth of which will, in turn, bring about further alterations in the composition of the atmosphere. The ability of aerobes and facultative anaerobes to create environments suitable for the growth of obligate anaerobes and microaerophiles serves to illustrate the operation of both commensalism and autogenic succession within a habitat (Fig. 1.14). Specific examples of this include the colonization of the tooth surface by facultatively anaerobic streptococci followed by the establishment of obligate anaerobes such as *Veillonella* spp. and *Fusobacterium* spp. Many other examples of autogenic succession induced by changes in the gaseous composition of a site will be found in subsequent chapters.

Related to the oxygen content of a site is the redox potential (Eh). This is a measure of the reducing power of a system and has an important influence on the functioning of those enzymatic reactions that involve the simultaneous oxidation and reduction of compounds. Some organisms can accomplish such reactions only in an oxidizing environment while, for others, a reducing environment is essential. Obligate aerobes are metabolically active only in environments with a positive Eh, while obligate anaerobes require a negative Eh. Facultative anaerobes can function over a wide range of Eh values. Because of its powerful oxidizing properties, the presence of oxygen in an environment exerts a dramatic effect on its Eh. Nevertheless, because of the rapid consumption of oxygen by respiring microbes, a site may be in contact with atmospheric oxygen but have a low oxygen content and an Eh that is sufficiently negative to permit the survival of obligate anaerobes. Examples of such sites include biofilms on tooth surfaces (section 8.3.3), many regions of the respiratory mucosa (section 4.3.1), and hair follicles (section 2.3). The production of a negative Eh by aerobic and facultative organisms, thereby creating an environment suitable for the growth of obligate anaerobes, is another example of commensalism resulting in autogenic succession.

Carbon dioxide stimulates the growth of capnophilic organisms such as *Haemophilus* spp. and *Neisseria* spp. Such organisms grow poorly in atmospheres with a normal concentration of the gas and are dependent on CO_2 generated by host or microbial respiration.

The water activity (a_w) of a site is an indication of the proportion of water available for microbial activity and is invariably less than the total amount of water present as it is affected by the concentration of solutes and also by the presence of surfaces. Pure water has an a_w of 1.0,

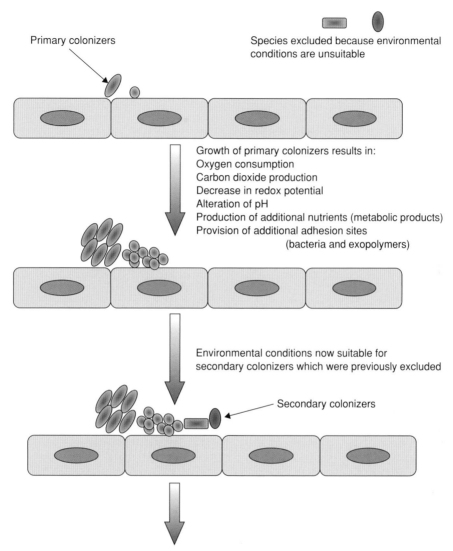

Fig. 1.14 Autogenic succession within a habitat. The primary colonizers of a habitat alter the environmental conditions in a number of ways, e.g. by consuming nutrients and oxygen and by producing metabolic end products and secreting macromolecules. Additional adhesion sites and nutrients together with different environmental conditions (e.g. pH, Eh, or oxygen concentration) are therefore created, enabling other organisms (secondary colonizers) to establish themselves at the site. The latter can themselves then alter the environment, allowing different organisms to colonize the habitat. Eventually a stable "climax community" is established at the site.

and human cells require an a_w of 0.997 for growth. Many of the microbes colonizing humans require an a_w of at least 0.96 for active metabolism (Table 1.4), and most sites can satisfy this requirement, with the exception of many regions of the skin. Staphylococci, unlike many microbes, can grow when the a_w is as low as 0.85 and so are able to colonize dry regions of the skin such as the arm, leg, and palm of the hand (section 2.3).

Table 1.4 Water activity required to support the growth of various groups of microbes.

Organism category	Examples	Water activity (a$_w$)
Gram-negative bacilli	Range for the group	0.94–0.97
	E. coli	0.95
	Klebsiella aerogenes	0.94
Gram-positive bacilli	Clostridium spp.	0.95–0.97
	Lactobacillus spp.	0.95–0.97
Gram-positive cocci	Range for the group	0.83–0.95
	Micrococcus luteus	0.93
	Staph. aureus	0.86
	Most coagulase-negative staphylococci	0.85
	Streptococcus spp.	0.9–0.95
	Corynebacterium spp.	0.9–0.95
	Micrococcus spp.	0.9–0.95
Halophiles	Range for the group	0.75–0.83

High salt concentrations are detrimental to many microbes as they cause dehydration and denaturation of proteins. Owing to the evaporation of sweat, they are found on many regions of the skin and this selects for colonization by halotolerant organisms such as staphylococci (section 2.3). The salt content of human body sites other than the skin does not exert a selective effect as it is within the limits tolerated by most microbes.

Sunlight contains potentially damaging ultraviolet radiation and also can induce the generation of toxic free radicals and reactive oxygen species from compounds known as photosensitizers. Both the skin and the eyes are exposed to light, but there is little evidence that light exerts any selective pressure on the microbes inhabiting these sites.

1.2.3 Mechanical determinants

In some regions of the body, mainly the GIT and urinary tract, mechanical forces are generated that affect the ability of microbes to colonize such sites (Fig. 1.15). Hence, in the oral cavity, the stomach, and the upper regions of the small intestine, the flow of saliva or intestinal secretions create hydrodynamic shear forces that can remove microbes not attached to mucosal surfaces. In the lower regions of the GIT (stomach and small and large intestines), peristalsis and other gut movements will also tend to remove unattached microbes. These mechanical forces exert a selection pressure that favors those organisms that are able to adhere to host surfaces. Similarly, in the urinary tract, the periodic flushing action of urine removes microbes that are not attached to the uroepithelium. In the oral cavity, tongue movements and the chewing action of teeth generate considerable mechanical forces that may be sufficient to dislodge bacteria attached to the mucosa and to teeth. This encourages colonization of sites protected from these forces such as those between the teeth and in the gingival crevice (section 8.3.1).

While no strong hydrodynamic shear forces operate in the respiratory tract, the production of a mucous "blanket" that is continually propelled towards the oral cavity traps and expels microbes arriving in this region – this system is known as the "mucociliary escalator" (section 4.2). Only organisms able to adhere to the underlying epithelium or to the more static periciliary layer beneath the mucous layer are able to colonize the respiratory tract.

The microbial communities colonizing any body site (other than those on the tooth surface) are attached to epithelial surfaces which are continually being shed. This means that the microbiota of any site (other than that on the teeth) is subjected to continual mechanical erosion. Microbes on the skin and urogenital tract are

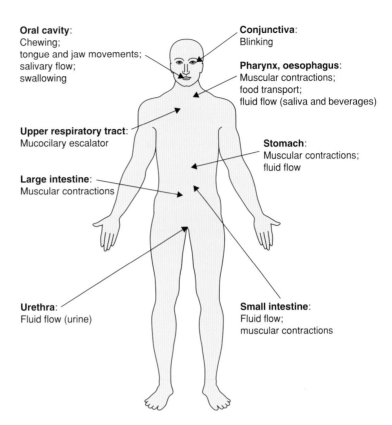

Oral cavity:
Chewing;
tongue and jaw movements;
salivary flow;
swallowing

Conjunctiva:
Blinking

Pharynx, oesophagus:
Muscular contractions;
food transport;
fluid flow (saliva and beverages)

Upper respiratory tract:
Mucocilary escalator

Stomach:
Muscular contractions;
fluid flow

Large intestine:
Muscular contractions

Urethra:
Fluid flow (urine)

Small intestine:
Fluid flow;
muscular contractions

Fig. 1.15 Body sites at which mechanical determinants have a major influence on the indigenous microbiota.

shed directly into the environment – although they may be retained temporarily by clothing – while those shed from the intestinal mucosa become part of the lumenal microbiota for many hours before being delivered to the environment. In the respiratory tract, the mucociliary escalator transports microbes to the oral cavity from where they are swallowed or expelled into the external environment.

1.2.4 Biological determinants

The innate and acquired immune systems of humans produce a variety of molecules and activated cells that kill microbes, inhibit their growth, prevent their adhesion to epithelial surfaces, and neutralize the toxins they produce. The effector molecules and effector cells of these systems, which are present at surfaces colonized by indigenous microbes, will be mentioned briefly here and will be described in more detail in appropriate sections of Chapters 2 to 9.

The innate immune system involves many different types of cells including epithelial cells, monocytes, macrophages, polymorphonuclear leukocytes (PMNs), natural killer (NK) cells, dendritic cells, and various lymphocyte subpopulations which link the innate and acquired immune systems. Those responses that involve the cooperation of different cell types are co-ordinated by cytokines and other signaling molecules. This system results in the release of a variety of effector molecules onto the external surfaces of epithelia including antimicrobial peptides, lysozyme, lactoferrin, lactoperoxidase, secretory phospholipase A_2, and collectins. The activities and functions of these molecules will be described in greater detail below (section 1.5.4). However, in order to mount a defensive response against microbial pathogens, the innate immune system must first of all recognize the presence of such organisms and be able to discriminate between them and members of the indigenous microbiota. Only recently have we begun to understand the mechanisms underlying these recognition and discriminatory processes.

Table 1.5 TLRs and the microbial components recognized by them.

Toll-like receptor(s)	MAMPs recognized
TLR1 (with TLR2)	Bacterial tri-acyl lipopeptides
TLR2 (often with TLR6)	Bacterial lipopeptides; lipoteichoic acids; peptidoglycan; lipoarabinomannan (mycobacteria); a phenol-soluble modulin from *Staph. epidermidis*; glycoinositolphospholipids (*Trypanosoma cruzi*); glycolipids (*Treponema maltophilum*); porins (*Neisseria* spp.); zymosan (fungi); hemagglutinin (measles virus)
TLR3	Double-stranded viral DNA
TLR4	Lipopolysaccharide; fusion protein (respiratory syncytial virus); heat-shock protein 60 (*Chlamydia pneumoniae*)
TLR5	Flagellin from bacterial flagella
TLR6 (with TLR2)	Bacterial di-acyl lipopeptides
TLR7	Single-stranded viral RNA
TLR8	Single-stranded viral RNA
TLR9	Bacterial and viral DNA
TLR10	Unknown
TLR11	Unknown

Recognition is based on the ability of human cells to detect conserved microbial structural components known as "microbe-associated molecular patterns" (MAMPs). MAMPS were originally termed "pathogen-associated molecular patterns" (PAMPs). However, as the molecules included in this term (see below) are also present in the vast majority of non-pathogenic species, the term MAMPs is more appropriate. MAMPs include molecules such as lipopolysaccharide (LPS), lipoteichoic acid (LTA), peptidoglycan, lipoproteins, and proteins (i.e. modulins). These are recognized by "pattern-recognition receptors" (PRRs), of which a group of trans-membrane proteins known as Toll-like receptors (TLRs) are particularly important. These are found on neutrophils, monocytes, macrophages, dendritic cells, and epithelial cells (see Fig. 1.16 and also other Figures in Chapters 2 to 9). Currently, 11 different TLRs have been identified in human cells, and each recognizes one or more MAMPs, although sometimes two TLRs are necessary to recognize some MAMPs (Table 1.5).

Interaction of the TLR with its ligand activates a signaling pathway in the host cell, resulting in the activation of several transcription factors, including NF-κB and interferon regulatory factors. The net result is the induction of inflammation and the establishment of adaptive immune responses. Inflammatory cytokines, including IL-1, IL-6, IL-12, and TNFα are released, and these activate surrounding cells to produce chemokines or adhesion molecules, resulting in the recruitment of inflammatory cells to the site. Recruited macrophages and neutrophils are activated and can kill the microbes they encounter. Antigen-presenting cells are stimulated, and this results in the activation and priming of antigen-specific, naive T cells, thereby triggering the adaptive immune response. Activation of TLRs can also result in increased production of antimicrobial peptides.

Another set of PRRs is present inside mammalian cells – these are the two soluble cytoplasmic proteins NOD1 and NOD2 (see Fig. 8.8). NOD1 recognizes peptidoglycan from Gram-negative bacteria as well as from *Listeria* spp. and *Bacillus* spp., whereas NOD2 recognizes muramyl dipeptide and so can detect the presence of both Gram-negative and Gram-positive species. Recognition of its complementary ligand by a NOD protein results in activation of the transcription factor NF-κB and the release of the pro-inflammatory cytokines IL-6, IL-8, IL-1β, and TNF. Such an inflammatory response will be generated by this system only when host

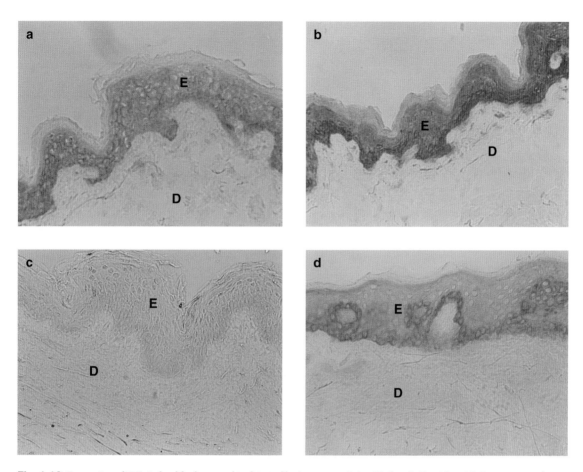

Fig. 1.16 Expression of TLRs in healthy human skin detected by immunostaining (D, dermis; E, epidermis). Orange-stained areas indicate the expression of (a) TLR 1, (b) TLR 2, (c) TLR 3, and (d) TLR 5. Magnification ×250. Reproduced with the permission of Blackwell Publishing Ltd, Oxford, UK, from: Baker, B.S., Ovigne, J.M., Powles, A.V., Corcoran, S. and Fry, L. (2003) *Br J Dermatol* 148, 670–9.

cells are invaded by microbes or when the MAMP itself is taken up by the cell.

So much for "recognition", but what about "discrimination"? Much less is known about how the innate immune system discriminates between members of the indigenous microbiota, which it allows to remain, and other, potentially harmful species, which it excludes. Combinations of TLRs may be used by host cells to recognize a specific microbe, or type of microbe, and so may be able to distinguish between autochthonous and allochthonous species.

Depending on the nature of the microbe, a specific set of TLR signaling pathways would be activated and the net signal generated may be recognized by the host as being characteristic of a "friendly" or "aggressive" microbe and an appropriate response generated. There is also evidence that some members of the indigenous microbiota, unlike exogenous species, are able to inhibit the inflammatory responses that would otherwise be induced following recognition of their MAMPs by TLRs. Furthermore, not all host cells express all TLRs, and the expression of these PRRs is different on the apical and basolateral surfaces. In intestinal epithelial cells, expression of many TLRs is strongly down-regulated on their apical surfaces, and therefore these cells do not respond to some of the MAMPs present on bacteria in the lumen so that no inflammatory response is elicited. However, invasion of such cells by pathogens will induce a protective inflammatory response following the recognition of peptidoglycan by Nod proteins. TLR5 is expressed only on the basolateral surface of intestinal epithelial cells and, consequently, no signaling pathways will be activated by the presence of flagellin on bacteria in the lumen. However, breaching of the epithelial barrier results in organisms gaining access to the basolateral surface of the epithelial cells, which will result in the induction of a protective inflammatory response via activation of TLR5.

More recently, it has been reported that activation of TLR9 on the apical and basolateral surfaces of intestinal epithelial cells results in different responses – the former induces a down-regulation of the signaling cascades responsible for promoting an inflammatory response, while the latter does the opposite. It has also been suggested that activation of TLR signaling pathways by exogenous organisms may require the presence of specific microbial virulence factors as co-factors.

The acquired immune system also involves a large number of different cell types and includes some of those involved in innate immunity, but the prime movers in this system are the B and T lymphocytes. Again, the response is coordinated mainly by cytokines. The most important effector molecule of the acquired immune response with respect to indigenous microbes is secretory immunoglobulin A (sIgA), which is able to reach the external surfaces of epithelia and so accumulate at sites of microbial colonization (Fig. 1.17). IgG is also present on mucosal surfaces but, except for the female reproductive tract, is present at a lower concentration than IgA. Of all the antibodies produced by humans, sIgA is produced in the greatest quantity – between 5 and 15 g per day in adults.

IgA is only a weak activator of complement and is a poor opsonin. However, one of its main functions is to prevent bacteria adhering to host structures, and this is achieved by the sIgA binding to bacterial adhesins, thereby preventing their interaction with receptors on host tissues. The binding of IgA to microbes not only prevents them from adhering to the mucosal surface, but also can result in the formation of aggregates, which are easier to remove than individual microbes being removed by fluids such as urine, saliva, or tears. Effector cells of the innate and acquired immune systems that may be present on epithelial surfaces in the absence of an infection include PMNs and macrophages, but little is known of their ability to affect the composition of the indigenous microbiota. Macrophages are thought to be important in preventing colonization

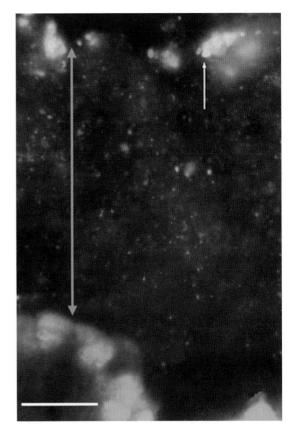

Fig. 1.17 Photomicrograph showing the colon stained with FITC-anti IgA (green). Large amounts of IgA (green) are present in the mucus. Bacteria (orange-yellow, arrowed) are present within the mucous layer (grey, double-headed arrow), and many of them are coated with IgA. Scale bar, 10 μm. Reproduced with the permission of Lippincott, Williams and Wilkins, Philadelphia, Pennsylvania, USA, from: van der Waaij, L.A., Harmsen, H.J., Madjipour, M. et al. (2005) Bacterial population analysis of human colon and terminal ileum biopsies with 16S rRNA-based fluorescent probes: Commensal bacteria live in suspension and have no direct contact with epithelial cells. *Inflamm Bowel Dis* 11, 865–71.

of the lower regions of the respiratory tract by inhaled microbes.

While the innate and acquired immune systems are important in defense against pathogenic microbes, their exact roles in the regulation of the indigenous microbiota is uncertain. This is because most work on the immune responses to microbes has focused on their role in combating pathogenic microbes rather than on their interactions with indigenous species. Little is known, therefore, of the mechanisms which enable the survival and long-term tolerance of indigenous microbial communities by their host or why these organisms do not elicit a damaging chronic inflammatory response. It is certainly surprising that no inflammatory response is generated by such enormous numbers of microbes which possess an array of modulins that are able to stimulate the release of pro-inflammatory cytokines from host cells.

It has been proposed that the immune systems of the host do not respond to members of the indigenous microbiota, i.e. the host exhibits "tolerance". This is supported by the results of a number of in vivo studies, which have demonstrated that lymphocytes do not proliferate in response to their indigenous gut microbiota, whereas they do proliferate when challenged with microbes from the gut of other individuals of the same species. Furthermore, the immune response provoked in mice injected with murine strains of *E. coli* or *Bacteroides* spp. is considerably reduced compared to that generated by injecting the corresponding human strains of these organisms.

It has been suggested that this tolerance is a consequence of the close similarities that exist between the surface antigens of many indigenous microbes and those of host tissues. This results in such organisms being recognized as "self" by the immune system, which is beneficial both for the microbes (as they are provided with a habitat suitable for their growth) and for the human host as it is provided with an indigenous microbiota that confers a number of benefits. However, other studies have shown that some indigenous microbes can provoke humoral and other immunological responses and that some (but not all) organisms in the gut and on the skin surface are covered in sIgA. Furthermore, antibodies against a variety of indigenous microbes (e.g. *Veillonella* spp., *Bacteroides fragilis*, and oral streptococci) have been detected in the sera of healthy individuals. While indigenous microbes may be able to induce the production of antibodies, these effector molecules do not appear to have a major role in regulating the composition of the indigenous microbiota. Hence, although deficiencies in sIgA production are relatively common, the composition of the communities colonizing such individuals do not differ significantly from those found in individuals without such deficiencies. Similarly, loss of T cell function appears to have little effect on the composition of the intestinal microbiota.

In contrast, the innate immune system may play a dominant role in controlling colonization and in regulating the composition of the microbiota at a body site. Mucosal surfaces secrete a number of antimicrobial compounds, and the mucosa of a particular site produces a characteristic range of such compounds, each of which has a distinct antimicrobial spectrum. Some of these compounds are constitutively expressed, while others are produced in response only to certain cytokines or to the presence of a particular organism (section 1.5.4). It is likely, therefore, that the characteristic mixture of antimicrobial compounds produced at a particular body site would influence the type of microbes able to colonize that site. Other innate immune responses of epithelial cells to the indigenous microbiota are described in section 1.5.4.

The production of hormones, and fluctuations in their concentrations, can exert a profound effect on the environment of certain body sites. For example, the increased production of sebum at puberty leads to dramatic changes in the skin environment. The production of estrogen and progesterone also alters the vaginal environment at different stages in the life of females (at the menarche and menopause) as well as during the menstrual cycle of post-menarcheal/pre-menopausal women (see section 6.3).

Microbes arriving at a body site must cope not only with host defense systems, but also with antimicrobial compounds produced by those organisms already present. As mentioned in section 1.2.1, an extensive range of antimicrobial substances (bacteriocins, fatty acids, hydrogen peroxide, etc.) is produced by microbes indigenous to humans, and these will be described in greater detail in the appropriate sections of Chapters 2 to 9. There is great interest in employing microbes that produce such compounds to prevent or treat infectious diseases, i.e. as probiotics or in replacement therapy.

1.3 HOST CHARACTERISTICS THAT AFFECT THE INDIGENOUS MICROBIOTA

The effects that host factors have on the composition of the microbial community occupying a particular body site will be emphasized throughout the rest of the book. Because of the variations that exist between individual human beings, the environmental conditions provided by the host at a particular body site will vary from person to person. Some of the many factors affecting the environmental conditions at a body site include age, gender, genotype, nutritional status, diet, health status, disability, hospitalization, emotional state, stress, climate, geography, personal hygiene, living conditions, occupation, and lifestyle. Many of these factors are, of course, inter-related, and some will affect all body sites while others are more likely to influence only particular sites. Such person-to-person variations result in differences in the composition of the microbial community resident at a site and make it difficult to define the "indigenous microbiota" of a body site. Unfortunately, the effects on the indigenous microbiota of most of the factors listed above have not been extensively investigated. Nevertheless, some information is available, and some general points will now be made here, while specific details will be included in the appropriate sections of Chapters 2 to 9.

1.3.1 Age

Most of the data on the composition of the indigenous microbiota of a body site described in Chapters 2 to 9 will relate to "healthy adults". However, the microbiota of many body sites appears to be different in very young and very old individuals (Table 1.6). In the very young, such differences are a consequence of many factors including an immature immune system, a milk-based diet, the absence of teeth, and behavioral factors.

In the elderly, differences arise as a result of a decrease in the effectiveness of the immune system, dysfunctioning of many organ systems, poor nutrition, poor hygiene, and increasing use of medical devices and prostheses such as catheters, dentures, etc. There are several markers of the existence of a dysfunction of the immune response in the elderly (i.e. immunosenescence)

Table 1.6 Examples of the effect of age on the composition of the indigenous microbiota of a body site.

Body site	Age-related differences in microbiota
Colon	In breast-fed infants prior to weaning, the fecal microbiota is dominated by *Bifidobacterium* spp., whereas in adults *Bifidobacterium* spp. are only minor constituents The fecal microbiota of elderly individuals has decreased proportions of *Veillonella* spp. and bifidobacteria but increased proportions of clostridia, lactobacilli, and enterobacteria
Oral cavity	Prior to tooth eruption, *Streptococcus sanguinis* is absent in children, whereas in adults this is one of the dominant organisms in the oral cavity In elderly individuals, there is an increase in the frequency of isolation of staphylococci and enterobacteria After puberty, the prevalence of *Prevotella intermedia*, *Prevotella denticola*, *Prevotella loescheii*, and *Prevotella melaninogenica* in the gingival crevice increases
Nasopharynx	The prevalence of *Neisseria meningitidis* is lower in infants than in young adults The nasopharynx of infants is dominated by *Moraxella catarrhalis*, *Strep. pneumoniae*, and *Haemophilus influenzae*, whereas these organisms comprise much lower proportions of the nasopharyngeal microbiota of adults
Oropharynx	In elderly individuals, there is increased colonization by Gram-negative bacteria (e.g. *Klebsiella* spp., *E. coli*, and *Enterobacter* spp.) as well as by *C. albicans*
Urinary tract	Increased microbial colonization in elderly individuals
Skin	Increased prevalence of streptococci and enterobacteria in elderly individuals Increased prevalence of *Propionibacterium avidum* in axillae of young adults compared with children Low prevalence of *Propionibacterium acnes* in young children
Eye	In elderly individuals, there is an increase in the frequency of isolation of coryneforms and Gram-negative bacilli

Table 1.7 Factors that may contribute to alterations in the indigenous microbiota of elderly individuals.

Factor	Consequences
Immunosenescence	Possibly influences microbiota at all sites
Malnutrition	Impaired immune response; affects composition of host secretions at most sites thereby influencing microbiota
Decreased mucociliary clearance	Increased microbial colonization of respiratory tract, including by species that are normally expelled
Decreased gastric acid production	Increased colonization of stomach, including by species that are normally killed by the low pH
Decreased urinary flow rate, post-void residual urine, increased bacterial adhesion to uroepithelium, prostatic hypertrophy in men	Increased colonization of urinary tract including by species that are normally expelled
Lack of estrogen in females	Decreased acid production in vagina, thereby enabling colonization by species that are usually excluded
Decreased intestinal motility; alterations in mucus composition	Enables colonization of small intestine

including decreased antibody production, decreased numbers of circulating lymphocytes, impaired T cell proliferation, and impaired phagocytosis and microbial killing by PMNs. However, whether these contribute to alterations in the composition of the microbiotas of body sites is not known. Malnutrition is a common problem in the elderly, and it has been estimated that between 10 and 25% of the elderly in industrialized countries have some nutritional defect. This will affect not only immune function but also the composition of host secretions which would certainly have an impact on indigenous microbes. Other factors that are likely to affect the microbiotas of specific body sites are listed in Table 1.7.

1.3.2 Host genotype

While a number of studies have investigated the effect of genotype on the susceptibility of individuals to infectious diseases, little is known about its effect on the indigenous microbiota. The environment of a particular body site would be expected to be more alike in individuals who have a high degree of genetic relatedness, because of anatomical and physiological similarities,

than in those who are more distantly related. The fecal microbiotas of monozygotic twins (analyzed by means of denaturing gradient gel electrophoresis [DGGE] profiles of the amplicons from 16S rRNA genes – see section 1.4.3) have been shown to have a significantly higher degree of similarity than those of unrelated individuals. Furthermore, the degree of similarity of the microbiotas shows a significant correlation with the genetic relatedness of individuals. The nasal microbiota of identical twins has a much higher degree of similarity than that of nonidentical twins. There are also a few examples of correlations between host genotype and carriage of members of the indigenous microbiota (Fig. 1.18).

1.3.3 Gender

Several differences between males and females have been observed with regard to the composition of the microbiota of a number of body sites. Unfortunately, the reasons underlying these have often not been determined but are likely to be multifactorial and involve anatomical, behavioral, hormonal, and other physiological factors. Some examples are shown in Fig. 1.19.

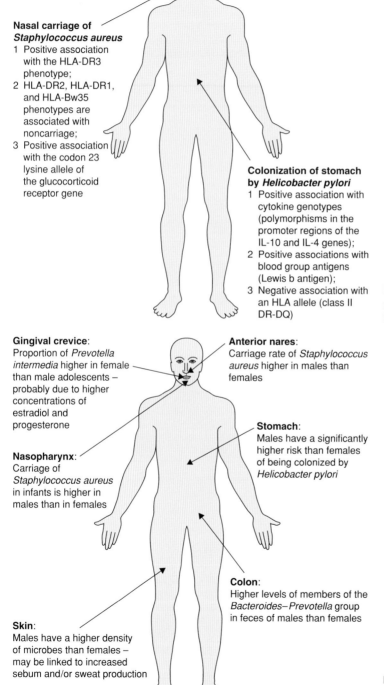

Nasal carriage of
Staphylococcus aureus
1 Positive association
  with the HLA-DR3
  phenotype;
2 HLA-DR2, HLA-DR1,
  and HLA-Bw35
  phenotypes are
  associated with
  noncarriage;
3 Positive association
  with the codon 23
  lysine allele of
  the glucocorticoid
  receptor gene

Colonization of stomach
by *Helicobacter pylori*
1 Positive association with
  cytokine genotypes
  (polymorphisms in the
  promoter regions of the
  IL-10 and IL-4 genes);
2 Positive associations with
  blood group antigens
  (Lewis b antigen);
3 Negative association with
  an HLA allele (class II
  DR-DQ)

Fig. 1.18 Association between host genotype and carriage of members of the indigenous microbiota.

Gingival crevice:
Proportion of *Prevotella
intermedia* higher in female
than male adolescents –
probably due to higher
concentrations of
estradiol and
progesterone

Anterior nares:
Carriage rate of *Staphylococcus
aureus* higher in males than
females

Nasopharynx:
Carriage of
Staphylococcus aureus
in infants is higher in
males than in females

Stomach:
Males have a significantly
higher risk than females
of being colonized by
Helicobacter pylori

Skin:
Males have a higher density
of microbes than females –
may be linked to increased
sebum and/or sweat production

Colon:
Higher levels of members of the
Bacteroides–Prevotella group
in feces of males than females

Fig. 1.19 Examples of the effect of gender on the indigenous microbiota of various body sites.

1.4 TECHNIQUES USED TO CHARACTERIZE THE MICROBIAL COMMUNITIES INHABITING HUMANS

A variety of techniques have been used to study the indigenous microbiota, and each has its advantages and disadvantages.

1.4.1 Microscopy

Figure 1.20 shows microscopy techniques that are used in the analysis of microbial communities indigenous to humans.

Light microscopy is one of the simplest and most direct approaches used to study microbial communities. One of its advantages is that it can reveal details of the physical structure of a community and the spatial arrangement of the constituent organisms. It also serves as a "gold standard" with respect to the total number of microbes that are present within a sample – this is often used as a yardstick for assessing the ability of other, less direct, techniques to detect all of the organisms present in a community. In this way, it has been revealed that analysis of samples of feces by culture-based approaches may detect as few as 20% of the organisms that are present. Differential counts of the various morphotypes in a sample give an indication

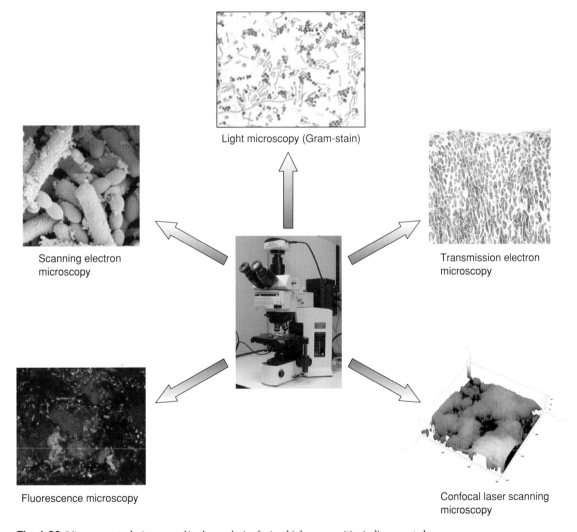

Light microscopy (Gram-stain)

Scanning electron microscopy

Transmission electron microscopy

Fluorescence microscopy

Confocal laser scanning microscopy

Fig. 1.20 Microscopy techniques used in the analysis of microbial communities indigenous to humans.

Fig. 1.21 Light micrograph of the mucosal surface of the colon stained with 16S rRNA oligonucleotide probes targeted against *Bacteroides* (red, Cy 3), *Bifidobacterium* (blue, Cy 5) and *Escherichia* (green, fluorescein isothiocyanate). Reproduced with the permission of Cambridge University Press, Cambridge, UK, from: Macfarlane, S. and Macfarlane, G.T. (2003) Bacterial growth on mucosal surfaces and biofilms in the large bowel. In: Wilson, M. and Devine, D. (eds), *Medical Implications of Biofilms.*

of the diversity of the microbiota, and this has proved useful for many years to ascertain whether the composition of the vaginal and subgingival microbiotas in an individual are indicative of health or disease. The analytical power of light microscopy can be enhanced in a number of ways. For example, the use of vital stains can reveal the relative proportions of live and dead cells that are present. Furthermore, information regarding the identity of the organisms that are present (and their spatial relationships) can be obtained by using labeled antibodies or oligonucleotide probes – the label often used is a fluorescent molecule (Fig. 1.21).

CLSM is a technique that enables the examination of communities in their living, hydrated state and which provides valuable information concerning the true spatial organization of the constituent cells as well as the overall shape and dimensions of the community (see Figs 1.7 and 1.8). It is a technique that has revolutionized our understanding of the structures of biofilm communities. Additional information can be obtained by using vital stains, fluorescent-labeled antibodies, and labeled oligonucleotide probes. Furthermore, information pertaining to the nature of the environment within the biofilm (e.g. pH, Eh, etc.) can be obtained using appropriate probes. It is also possible to monitor gene expression within biofilms using reporter genes such as green fluorescent protein (see Fig. 1.10).

The use of molecular techniques in conjunction with light microscopy is proving to be very rewarding in studies of the indigenous microbiota. One such technique is fluorescence in situ hybridization (FISH), which involves the use of fluorescent-labeled oligonucleotide probes to target specific regions of bacterial DNA (Fig. 1.21). Most of the probes currently used are those that recognize genes encoding 16S ribosomal RNA (16S rRNA). The gene encoding 16S rRNA in a bacterium consists of both constant and variable regions. Within the molecule, there are regions that are highly specific for a particular bacterial species as well as regions that are found in all bacteria, in only one bacterial genus, or in closely related groups of bacteria. Therefore probes can be designed to identify individual species, individual genera, certain related microbial groups, or even all bacteria. More recently, probes recognizing mRNA have been used to identify the genes that are being expressed in communities. An important advantage of this approach is that it can be automated, and the resulting data can be processed using computerized image analysis software.

Transmission and scanning electron microscopy can provide information that is not obtainable by ordinary light microscopy, and the high magnifications that are possible can be used to reveal details of microbial adhesins and adhesive structures (Fig. 1.22). The organisms that are present can be identified using antibodies conjugated to electron-dense markers (e.g. gold or ferritin). However, a major disadvantage of electron microscopy is that specimen processing and the accompanying

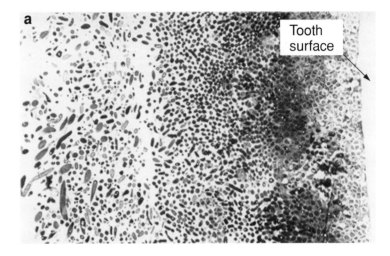

Tooth surface

Fig. 1.22 Transmission (a) and scanning (b) electron micrographs of dental plaques. Images kindly supplied by Mrs Nicola Mordan, UCL Eastman Dental Institute, University College London, London, UK.

dehydration alters the structure of the sample, thereby preventing elucidation of the exact spatial arrangement of cells within communities.

1.4.2 Culture-dependent approaches

Most of our knowledge of the composition of microbial communities that are indigenous to humans has come from using qualitative and quantitative culture techniques. Given the complexity of the communities at most body sites, such procedures are very labor-intensive. There are also a number of problems with this approach. First of all, if a nonselective medium is to be used, then one must be chosen that is capable of supporting the growth of all of the species likely to be present – this is virtually impossible given the disparate, and often very exacting, nutritional requirements of the members of such communities. Furthermore, it is difficult to provide the optimum environmental conditions (e.g. pH, oxygen content, CO_2 content, etc.) necessary to enable the growth of all of the different types of microbes present.

Problems arise as a result of some organisms growing faster than others, resulting in overgrowth of plates and failure to isolate slow-growing organisms. In samples taken from sites with a very dense microbiota (e.g. the colon and vagina), or dental plaque, it is essential to use dilutions of the sample to obtain isolated colonies for subsequent identification. This means that organisms present in low proportions are "diluted out" and so are rarely isolated. Many studies have used selective media instead of, or in addition to, nonselective media. These can be useful, but analysis of a complex microbiota requires the use of a number of media that are selective for the various groups of organisms present. However, no medium can be relied upon to be truly selective, and the inhibitory constituents may also have some adverse effect on the organisms for which the medium is supposedly selective. These problems all contribute to a greater workload, which means that it is expensive and inevitably results in a decrease in the number of samples that can be processed – and hence a decrease in the statistical reliability of the data obtained.

Comparison of samples analyzed by culture and by microscopy have revealed that even the best cultivation methods seriously underestimate the number of organisms present in the microbiotas of certain body sites – particularly those from the GIT and oral cavity. The reasons for this are many, and the following are the most important causes of problems associated with culture-dependent analysis of microbial communities.

• Failure to satisfy the nutritional requirements of some of the organisms present;
• Failure to satisfy the environmental requirements of some of the organisms present;
• Failure to detect organisms in a "viable-but-not-cultivable" state;
• Failure to disrupt chains or clusters of organisms prior to plating out – this results in the production of only one "colony-forming unit" from a cluster or chain consisting of many viable bacteria, thereby underestimating their proportions;
• Death of viable cells during transportation and processing of the sample;
• Overgrowth of culture plates by fast-growing organisms;
• "Diluting-out" of organisms present in small proportions;
• Possible inhibitory effect of selective media;
• The labor-intensive nature of the whole process.
Collectively, these difficulties have resulted in a serious underestimation of the number, and variety, of organisms in a sample taken from any environment, and it has been estimated that we are able to culture in the laboratory no more than 1–2% of the microbial species (which are thought to number between 10^5 and 10^7) present on planet Earth. Once individual isolates have been obtained, the next task is to identify each. Traditionally, this has involved the use of a battery of morphological, physiological, and metabolic tests which is very labor-intensive and often not very discriminatory. The use of commercially available kits for this purpose has made the process less technically demanding.

Other phenotypic tests that have been used for identification purposes include cell wall protein analysis, serology, and fatty acid methyl ester analysis. During the past few years, there has been a trend towards increasing the use of molecular techniques for identifying the organisms isolated, and one of these is based on the sequencing of genes encoding 16S rRNA. The gene is amplified by polymerase chain reaction (PCR), and the sequence of the resulting DNA is determined and then compared with the sequences of the 16S rRNA genes of organisms that have been deposited in databases. If the sequence demonstrates at least a 98% similarity to the sequence of a gene that is already in the database, then it is assumed that the gene is from the same species, and hence the identity of an unknown organism can be established. The procedure is much simpler to perform than a battery of phenotypic identification tests and has the great advantage of enabling phylogenetic comparisons of the isolated organisms. However, some taxa are recalcitrant to PCR, and some taxa (e.g. many viridans streptococci) are so closely related that they cannot be differentiated using this approach.

Another approach to identifying the isolated organisms involves colony hybridization with nucleic acid probes. Basically, the technique involves lysing an isolated colony and exposing it to a labeled oligonucleotide probe. Hybridization is recognized by detection of the probe after washing – the label may be radioactive, enzymatic, or fluorescent. The probe can be designed to recognize a species or genus or a group of organisms. In practice, probing is carried out simultaneously on many colonies that have been transferred to a nitrocellulose membrane. Other molecular techniques used for the identification of isolated colonies include pulsed-field gel electrophoresis (PFGE), ribotyping, multiplex PCR, and arbitrary-primed PCR.

1.4.3 Culture-independent, molecular approaches

Many of the problems inherent in culture-dependent approaches to analyzing microbial communities can be circumvented by the use of molecular techniques, although these have their own problems. The first stage in the analysis of a microbial community by a molecular technique is to isolate either DNA or RNA from the sample – and herein lies the first problem. Extraction of nucleic acids from microbes requires that the cells are lysed, and the ease of lysis varies significantly among

different organisms. Numerous protocols for the lysis of microbes present in samples have been devised and include the use of enzymes, chemicals, and mechanical methods. Once lysed, care must be taken to avoid shearing and degradation of the nucleic acids. The extracted nucleic acids can then be used in a variety of ways to reveal the identity of the microbes originally present in the sample and/or to produce a "profile" or "finger-print" of the microbial community (Fig. 1.23).

Universal primers can be used to amplify all of the 16S rRNA genes that are present in the DNA extracted from the sample, and the amplified sequences are then

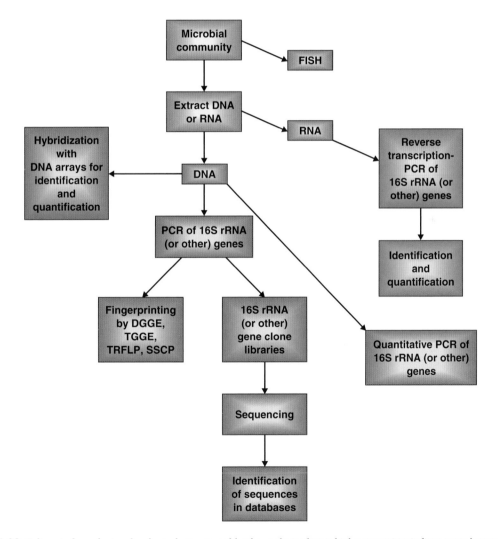

Fig. 1.23 Culture-independent molecular techniques used for the analysis of microbial communities indigenous to humans.

cloned. The sequences of the cloned amplicons are then determined and compared to sequences deposited in databases such as the Ribosomal Database Project (http://rdp.cme.msu.edu/index.jsp) or in BLAST (http://www.ncbi.nlm.nih.gov/BLAST/). In this way, the sequences of the 16S rRNA genes of all of the organisms present in the community (including those that cannot be grown in the laboratory) can be determined and, if these sequences match those of known organisms in databases, then the identities of all the organisms present will be revealed. However, studies of the fecal microbiota, for example, have revealed that not all of the sequences of the 16S rRNA genes obtained correspond to sequences in databases – in fact as many as 75% of the sequences do not match those of known organisms. Some of the problems with this approach are that both the PCR and cloning steps in the procedure have their biases, and the technique is expensive and labor-intensive – which limits the number of samples that can be processed.

Very few studies have used quantitative real-time PCR to analyze microbial communities that are indigenous to humans, but the use of this technique is likely to increase.

Another useful approach is to separate the amplicons on a denaturing gradient gel. Although all of the amplicons have the same length, their different base compositions results in them having different melting points, and so each will melt at a different point when run on a gel along which there is either a temperature gradient (temperature gradient gel electrophoresis – TGGE) or a gradient in the concentration of a denaturing agent such as urea or formamide (DGGE). The altered conformation of the DNA due to denaturation slows their migration, and this results in separation of the various amplicons. Staining of the DNA in the resulting gel reveals a banding pattern or "fingerprint" that is characteristic of that particular community (Fig. 1.24). The individual bands can be cut out and each amplicon eluted, re-amplified, sequenced, and identified using databases as described above. Alternatively, the fingerprints produced from samples from the same individual obtained on different occasions can be compared and analyzed for differences. Hence, bands appearing or disappearing with time can be sequenced to determine the gain or loss of an organism from the community. The method is also useful for comparing the microbiotas present at the same body site in different individuals and has been used extensively for comparing the microbiotas in different

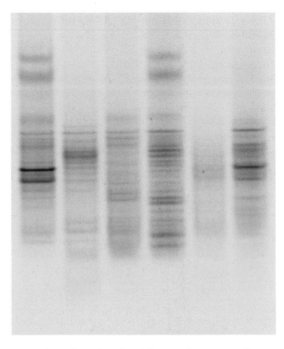

Fig. 1.24 Polyacrylamide gel showing the DGGE profiles of supragingival dental plaque samples from six individuals. The figure shows bands corresponding to the 16S rRNA gene amplicons of the same molecular mass that have been separated by DGGE on a 10% polyacrylamide gel with a denaturant gradient ranging from 40 to 80% (100% denaturant corresponds to 7 M urea and 40% deionized formamide). Image kindly supplied by Dr Gavin Gaffan, UCL Eastman Dental Institute, University College London, London, UK.

regions of the GIT. Comparisons are facilitated by computer and statistical analysis of the banding patterns obtained. Two additional community fingerprinting techniques available are single-strand conformation polymorphism (SSCP) and terminal-restriction fragment length polymorphism (TRFLP).

The DNA extracted from a microbial community can also be used in dot-blot hybridization assays. In such assays, aliquots of the extracted DNA are spotted onto a nitrocellulose membrane to form a gridded array. This array can then be probed with labeled oligonucleotide probes that are designed to recognize a single species, a genus, or a group of related organisms. Alternatively, gridded arrays of DNA from a range of organisms can be prepared, and these can be probed with the DNA extracted from the sample – once this has been labeled in some way.

DNA-based methods of analyzing the composition of microbial communities, however, are not without their drawbacks. The DNA extracted from a body site is likely to consist not only of DNA from the resident community, but also that derived from allochthonous species and from dead organisms that had gained access to the site but which were unable to survive there. Consequently, such analyses could produce large numbers of "false-positives".

1.4.4 Functional analysis of microbial communities

Determining the species composition of a community is only the first step in studying any ecosystem; this needs to be followed by establishing the functions of the community and of each of its members. However, given the complexity of most communities and the fact that many species have not yet been cultured, little progress has been made in this respect.

One method of determining the metabolic capability of the community as a whole, as well as its potential functional diversity, involves incubating samples of the community with a range of substrates (e.g. carbon sources) and determining which of these can be utilized. This results in a metabolic "fingerprint" indicating the range of substrates that can be utilized by the community. The technique is known as community-level physiological profiling (CLPP). In addition to providing information on the substrates actually used by a community, CLPP may be useful in monitoring the community's response to altered environmental conditions. There are, however, a number of problems with the technique. Firstly, as the conditions of incubation provided in the laboratory are different from those of the community's natural environment, it is very unlikely that the observed response would represent the metabolic capabilities of the community in its natural environment. Secondly, the observed substrate-utilization pattern will reflect the activities of only those organisms in the community that are metabolically active under the particular set of incubation conditions employed. Furthermore, the relative proportions of the constituent members of the community are likely to change during the period of incubation. Despite these drawbacks, this simple technique can provide some insight into the functional abilities of a community.

This type of approach has been used in a more limited way to establish some of the "microflora-associated characteristics" (MACs) of the host. Hence, a number of metabolic capabilities of the colonic microbiota have been assessed, including its protease, urease, glycosidase, bile salt hydrolase, and azoreductase activities, as well as its ability to produce short-chain fatty acids (SCFAs), to convert cholesterol to coprostanol, to break down mucin, to inactivate trypsin, and to produce gases such as methane, carbon dioxide, and hydrogen. The determination of one, or a limited number, of such activities has been employed by investigators interested in a particular function of the colonic microbiota. However, a more extensive range of these MACs (i.e. "profiling") has been determined in some studies, particularly those studies involving the effects of antibiotics and diet on the colonic microbiota or those investigating the development of such communities.

A number of techniques are available for monitoring gene expression in situ, although there are few reports of their application to the study of communities that are indigenous to humans. Some of the most promising techniques include reverse transcription (RT)-PCR, in vivo expression technology (IVET), DNA microarrays, in situ isotope tracking, and subtractive hybridization. There is also considerable interest in using metagenomic and metaproteomic approaches to ascertain the functional capabilities of microbial communities. For example, in a recent analysis of the fecal microbiota of two adults, the collective genome (microbiome) was found to be significantly enriched (compared with an average derived from all sequenced microbial genomes) with respect to the metabolism of a variety of sugars and polysaccharides, the generation of SCFAs, and the synthesis of essential amino acids and vitamins. Such findings are in keeping with the known role of the gut microbiota in degrading complex plant polysaccharides, providing SCFAs as an energy source for colonocytes and supplying the host with vitamins (see section 9.4.2.4.2).

1.5 THE EPITHELIUM – SITE OF HOST–MICROBE INTERACTIONS

The epithelium forms a continuous covering on all surfaces that are exposed to the external environment (i.e. the skin and eyes and the respiratory, gastrointestinal, urinary, and genital tracts) and consists of a layer of specialized cells, sometimes only one cell thick. Although one of its functions is to exclude microbes from the underlying tissues, it is itself colonized by those microbes which are the subject of this book.

Because the epithelium is the primary site of the interactions that occur between microbes and their human host (as well as the interactions that occur between the microbial colonizers themselves), it is important to consider this tissue in greater detail. The only other surfaces on which host–microbe interactions occur are the teeth, and their structure is described in Chapter 8.

1.5.1 Structure of epithelia

Basically two main types of epithelial surfaces are recognized – the dry epithelium (known as the epidermis) comprising the skin, and the moist epithelia which cover the eyes and those internal body surfaces that are in contact with the external environment. Moist epithelia are also referred to as mucosa (or mucous membranes) because they are invariably coated in a layer of mucus which consists primarily of glycoproteins known as mucins. The total surface area of the mucosa is more than 100 times greater than that of the skin (Fig. 1.25).

The main cellular element of epithelia is the epithelial cell. In an epithelium, the constituent cells are joined together by a variety of junctions (Table 1.8).

The epithelium is attached to the underlying tissue via a layer of extracellular matrix (mainly collagen and laminin) known as the basal lamina. Attachment to this layer is mediated by hemidesmosomes. Epithelial cells vary in shape and may be flattened (squamous), cuboidal, or columnar and their apical surface is sometimes covered in fine, hair-like processes known as cilia. They may be present as a single layer of cells, or they may form several layers and some may undergo a process known as keratinization, in which the protein keratin is deposited in the cell which eventually dehydrates and dies. Epithelial cells that undergo keratinization are known as keratinocytes and are particularly important constituents of skin on the surface of which they form a protective barrier; they are also found on certain mucosal surfaces, e.g. the tongue. The structure of skin differs in several respects from that of the mucosae and is described in greater detail in Chapter 2. The rest of this chapter will be concerned mainly with the mucosae found lining all of the other surfaces that are exposed to the external environment.

The main types of epithelia comprising mucosal surfaces are shown in Fig. 1.26. The epithelium has an apical surface facing the lumen of the particular body cavity, and so is in contact with the external

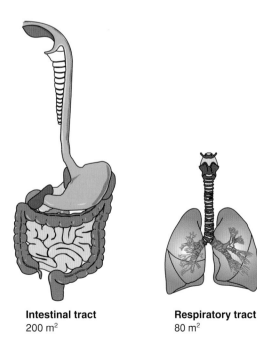

Intestinal tract
200 m²

Respiratory tract
80 m²

Skin
1.8 m²

Fig. 1.25 Relative surface areas of the main epithelial surfaces of the human body. Other mucosal surfaces include the urinary tract, reproductive system, and the eyes, but their surface areas are considerably smaller.

Table 1.8 Types of junctions between epithelial cells.

Junction type	Function
Tight junctions	To seal adjacent cells just below their apical surfaces (i.e. surfaces that are exposed to the external environment), thereby preventing the passage of molecules between cells – they are impermeable to all but the smallest of molecules
Gap junctions	Consisting of channels between adjacent cells, gap junctions allow the passage of molecules of less than 1000 Da
Adherens junctions	Adherens junctions provide strong mechanical attachment between cells and are formed by linkage of adjacent trans-membrane proteins known as cadherins

environment, and a basal (or basolateral) surface which is in contact with the basal lamina (or basement membrane) and underlying connective tissue. Similarly, each cell has an apical and a basolateral surface. Simple squamous and columnar epithelia consist of a single layer of epithelial cells and are found lining many body cavities. In some cases, the cells have numerous microvilli to increase the surface area, and such epithelia are found in regions designed for absorption of molecules, e.g. the small intestine. Epithelia consisting of several layers of cells with different shapes are known as transitional epithelia and are found lining cavities that expand and contract, e.g. the bladder. When the layers of cells progressively flatten, the epithelium is known as a "stratified squamous epithelium" and is characteristic of mucosae that are subjected to mechanical abrasion, e.g. the oral cavity and vagina. Keratinization of the outer layers of cells may occur, and the epithelium is then said to be a "keratinized, stratified, squamous epithelium", examples of which include the skin and the hard palate of the oral cavity.

Simple squamous epithelia consist of a single layer of flattened cells and line surfaces that are designed for absorption but do not experience wear and tear, e.g. the alveoli of the lungs. Simple columnar epithelia consist of a single layer of columnar cells and are found lining much of the digestive tract. They also line some regions of the respiratory tract where they may be covered in short, hair-like projections known as cilia (see Fig. 4.7). Ciliated epithelia constitute an important defense system because they propel a layer of mucus containing trapped bacteria along the epithelial surface and eject this from the body cavity (section 4.2). When more

than one layer of cells is present, the epithelium is described as being "stratified" and is found in regions that are subjected to wear and tear. Sometimes an epithelium consists of a single layer of different types of cells, some of which are small and do not reach the epithelial surface, while others are large and do reach the surface. The overall appearance often resembles a stratified epithelium but, as all of the constituent cells are in contact with the basement membrane, such an epithelium is termed a "pseudo-stratified epithelium".

As well as providing a barrier against microbes, epithelial cells have a number of other functions including the secretion and absorption of molecules and ions and the production of antimicrobial peptides and other effector molecules of the host defense system. In regions where absorption of molecules is an important function (e.g. the small intestine), the epithelium is folded into many finger-like projections known as villi in order to increase the surface area and so enhance absorption (Fig. 1.27).

In addition to epithelial cells, the epithelium of most mucosal surfaces contains one or more additional cell types. Cells capable of secreting the constituents of mucus are invariably present, and these are known as goblet cells (see Figs 3.3 and 9.7). A number of other cells involved in host defense may also be found including intra-epithelial lymphocytes and dendritic cells, both of which are involved in the acquired immune response. Furthermore, the epithelial surface is often punctuated by the openings of a variety of glands. Other cells may also be present depending on the particular mucosal surface, and these will be mentioned in subsequent chapters.

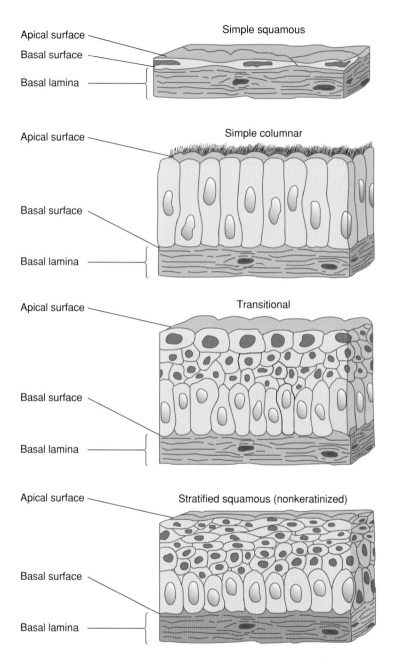

Fig. 1.26 The different types of mucosal surfaces present in the human body. Each of these is described in the text. Reproduced with the permission of Cambridge University Press, Cambridge, UK, from: Wilson, M., McNab, R., Henderson, B. (2002) *Bacterial Disease Mechanisms; An Introduction to Cellular Microbiology.*

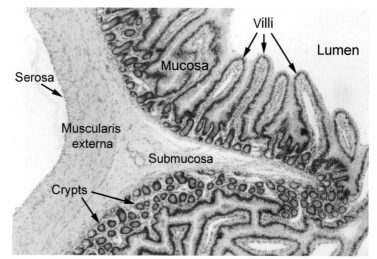

Fig. 1.27 Cross-section through the human jejunum. The epithelium is highly folded to form numerous villi, which results in an increased surface area for absorption and secretion of molecules. Image © 2006, David King, Southern Illinois University School of Medicine, Springfield and Carbondale, Illinois, USA, used with permission.

1.5.2 The epithelium as an excluder of microbes

There are a number of ways in which the epithelium excludes microbes from the rest of the body (Fig. 1.28). Firstly, it acts as a physical barrier preventing the penetration of microbes to the underlying tissues. Secondly, on mucosal surfaces the mucous layer secreted by the epithelium hinders access to the underlying epithelial cells, thereby helping to prevent attachment of microbes to these cells (Fig. 1.29). The mucous layer is continually expelled from the body along with any entrapped microbes.

Thirdly, the outermost cells of the skin and mucosa are continually being shed and replaced from below so that any microbes that do become attached are physically ejected from the body along with the shed epithelial cell (see Fig. 1.5). On the surface of the skin, the epithelial cells undergo keratinization which results in a layer of dry, dead cells (corneocytes) which microbes find difficult to colonize (Fig. 1.30).

Finally, the epithelium secretes a range of antimicrobial peptides and proteins which are able to either kill microbes or to inhibit their growth. Further details of these antimicrobial mechanisms will be provided in the appropriate sections of succeeding chapters.

1.5.3 Mucus and mucins

Although the mucous layer is an important component of the host defense system, it is also a major source of nutrients for microbes at sites such as the respiratory, genital, and urinary tracts. This illustrates an important aspect of the interaction between humans and microbes; no matter what defense system has been generated by the host, some microbes will have evolved a means of neutralizing it or utilizing it for their own benefit. Mucus, the constituents of which are produced by goblet and epithelial cells, has a number of important protective functions, and these are summarized in Table 1.9.

The main constituents of mucus are mucins which are usually present at a concentration of between 2 and 10% (w/v). These are glycoproteins containing high proportions of carbohydrates – usually between 70 and 85% (w/w). The mucins are unusual glycoproteins in that most of the carbohydrate side chains are linked to the protein at serine and threonine residues via an oxygen atom (i.e. they are "O-glycosylated"), although N-glycosylation also occurs (Fig. 1.31). The structure of a typical mucin molecule consists of a protein to which carbohydrate side chains are linked by O-glycosylation and/or N-glycosylation. The protein backbone consists of several thousand amino acid residues and contains regions with many oligosaccharide side chains and other regions without such side chains. The oligosaccharide-rich regions are resistant to proteases whereas the other regions are protease-sensitive. The oligosaccharide-containing regions of the protein are rich in serine, threonine, and proline.

The side chains usually consist of between two and

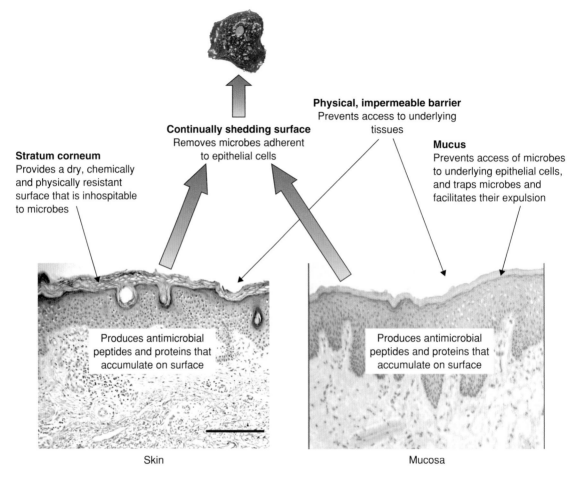

Stratum corneum
Provides a dry, chemically
and physically resistant
surface that is inhospitable
to microbes

Continually shedding surface
Removes microbes adherent
to epithelial cells

Physical, impermeable barrier
Prevents access to underlying
tissues

Mucus
Prevents access of microbes
to underlying epithelial cells,
and traps microbes and
facilitates their expulsion

Produces antimicrobial
peptides and proteins that
accumulate on surface

Produces antimicrobial
peptides and proteins that
accumulate on surface

Skin

Mucosa

Fig. 1.28 The epithelium as an excluder of microbes. Image on the left: Section of human skin from the lower leg, showing various layers, including the stratum corneum. Scale bar, 200 μm. Reprinted with the permission of Macmillan Publishers Ltd, London, UK, from: Glaser, R., Harder, J., Lange, H., Bartels, J., Christophers, E. and Schroder, J.M. (2005) Antimicrobial psoriasin (S100A7) protects human skin from *Escherichia coli* infection. *Nat Immunol* 6, 57–64, © 2005. Image on the right: Cross-section through the vaginal mucosa. Reprinted from: Pivarcsi, A., Nagy, I., Koreck, A., Kis, K., Kenderessy-Szabo, A., Szell, M., Dobozy, A. and Kemeny, L. (2005) Microbial compounds induce the expression of pro-inflammatory cytokines, chemokines and human β-defensin-2 in vaginal epithelial cells. *Microbes and Infection* 7, 1117–27, © 2005, with permission from Elsevier, Amsterdam, The Netherlands.

12 residues from a limited range of sugars – usually galactose, fucose, N-acetylglucosamine, N-acetylgalactosamine, mannose, and sialic acids. The region of the side chain involved in linkage to the protein is known as the "core", which is itself linked to the "backbone" which consists of a number of repeating disaccharide units containing galactose and N-acetylglucosamine (Fig. 1.31). The terminal sugars are known as "peripheral" residues and often consist of sialic acid or sulfated sugars. Sulfation occurs to the greatest extent in those mucins that are present in body sites colonized by microbes. The peripheral regions of the molecule are often antigenic and contain the ABH or Lewis blood group determinants.

The individual glycoprotein molecules are large (usually approximately 10^5–10^6 Da) but they are

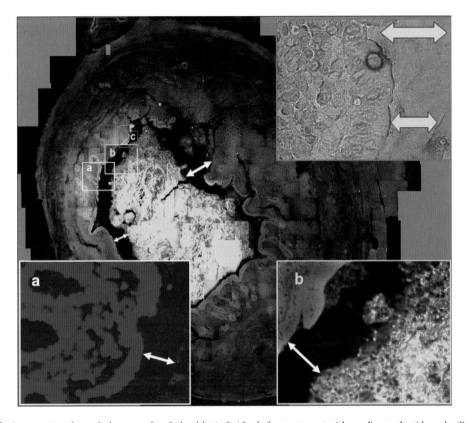

Fig. 1.29 Cross-section through the appendix of a healthy individual after treatment with an oligonucleotide probe (Eub 338) which hybridizes with all bacteria. In the main figure and insert (b), the mucosal tissues are visible as a result of the normal background fluorescence of human tissues. The mucous layer can be distinguished as a continuous gap (double-headed arrows) between the mucosal surface and bacteria (stained yellow/green) located in the lumen of the appendix. Bacteria are separated from the epithelial surface along the whole circumference. In insert (a) the tissues and bacteria are stained blue with 4,6-diamidino-2-phenylindole (DAPI) – a DNA stain. No bacteria can be seen within the mucous layer (dark region, double-headed arrows). Magnification ×400. In insert (c), the tissue and mucus are stained with alcian blue/periodic acid Schiff. No bacteria can be seen in the mucous layer (double-headed arrows) between the tissues and the bacteria in the lumen. Reproduced with the permission of the BMJ Publishing Group, London, UK, from: Swidsinski, A., Loening-Baucke, V., Theissig, F., Engelhardt, H., Bengmark, S., Koch, S., Lochs, H. and Doerffel, Y. (2007) Comparative study of the intestinal mucous barrier in normal and inflamed colon. *Gut* 56, 343–50.

generally present as even larger molecular mass polymers due to the formation of disulfide bonds linking the protein constituents of neighboring molecules (Fig. 1.32). These polymers may be several micrometers in length and form a viscoelastic gel in an aqueous medium.

The whole polymer has a "bottle-brush" structure with the carbohydrate side chains projecting as "bristles" from the central protein backbone. This structure is maintained by repulsive forces operating between the negatively charged carbohydrate side chains and between the side chains and the protein backbone. The side chains also protect the protein from degradation by proteases and so help to preserve the integrity of the mucous gel. Some of the mucins produced by the epithelium do not form polymers and remain covalently attached to the epithelial cell membrane where they form what is known as the "glycocalyx", which interacts noncovalently with the mucous layer, thereby helping it to remain associated with the mucosal surface (Fig. 1.33).

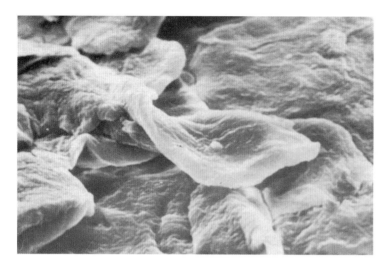

Fig. 1.30 Scanning electron micrograph showing squames on the skin surface. These are continually being shed into the environment taking any attached microbes with them. Reproduced with the permission of Elsevier, Amsterdam, The Netherlands, from: Noble, W.C. (ed.). (1974) *Microbiology of Human Skin*, Volume 2 in the series of *Major Problems in Dermatology*.

Table 1.9 The properties of mucus that contribute to its role in protecting epithelia.

Property	Effect
Contains receptors for the adhesins of a variety of microbes	Traps microbes, thereby preventing them from reaching the underlying epithelium and colonizing this site
Prevents diffusion of antimicrobial compounds (antibodies, enzymes, peptides, etc.) produced by mucosa, thereby increasing local concentrations	Kills many of the entrapped organisms
Continually produced and expelled from the body	Results in the ejection of microbes, their components, and their products
Protects the epithelium against chemical and physical damage by materials	Prevents damage by gastric acid, enzymes, and particulate matter
Lubricates and moistens the mucosa	Facilitates movement of materials such as food, fluids, chyme, etc.

The mucous layer often has a monomolecular layer of lipids on its external surface which renders it hydrophobic. Apart from mucins and lipids, mucus also contains exfoliated cells, the contents of dead, lysed host cells, antibodies, and a range of antimicrobial compounds produced by the mucosa. In some regions (e.g. the cervix and respiratory tract), the mucous gel is continually propelled along the mucosal surface by mean of cilia – this is described in greater detail in sections 4.2 and 6.2.1.

The protein backbones of the various mucins produced are encoded by a family of nine *MUC* genes, and the expression of the genes depends on the particular body site (Table 1.10). Hence, MUC2 is the main type of mucin protein produced in the intestinal tract, whereas expression of six *MUC* genes has been detected in the cervix – with different patterns of expression at different phases of the menstrual cycle. Differences also exist with regard to the glycosylation patterns of a particular type of mucin protein – hence the composition of

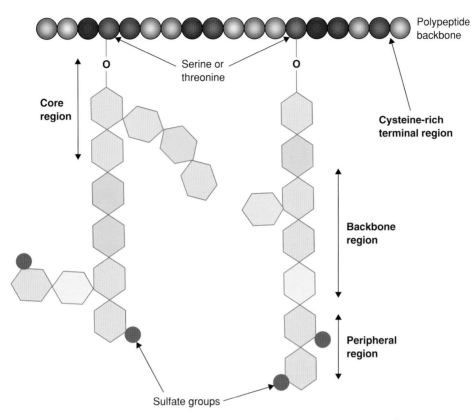

Fig. 1.31 Generalized structure of a mucin molecule.

the carbohydrate side chains of a mucin with a MUC1-protein backbone will be different in, for example, the respiratory and intestinal tracts. Certain mucin protein types are found only in mucus (e.g. MUC5AC – Fig. 1.34) while others are usually present only in mucins attached to epithelial cells (e.g. MUC1).

Although mucus plays an important role in host defense, a number of microbes have evolved the ability to use mucins as a source of carbon, nitrogen, and energy. This is particularly important at those body sites where there may be few alternative nutrient sources, e.g. the urinary and respiratory tracts. However, the structural complexity of these polymers means that their complete breakdown by a single microbial species is unlikely. Such degradation requires the production of a range of enzymes in a certain order because regions of the molecule become accessible only once others have been removed – this is more readily accomplished by microbial consortia than by individual species. Nevertheless, a limited number of microbes can achieve complete degradation of a mucin, e.g. the intestinal organisms *Ruminococcus torques, Ruminococcus gnavus,* a *Bifidobacterium* sp. and *Akkermansia muciniphila.* Many other species possess a more limited repertoire of enzymes and accomplish partial degradation. In doing so, not only obtain sufficient nutrients for their own needs but also produce a mucin fragment that can be utilized by another organism and so on until, ultimately, the consortium has degraded the whole molecule. The range of enzymes needed to achieve complete degradation of a mucin is shown in Table 1.11.

The ability to degrade mucin, entirely or partially, has been detected in microbes or microbial consortia inhabiting all mucosal sites of the body, and these will be described further in the appropriate sections of Chapters 2 to 9. The complete removal of mucus from a

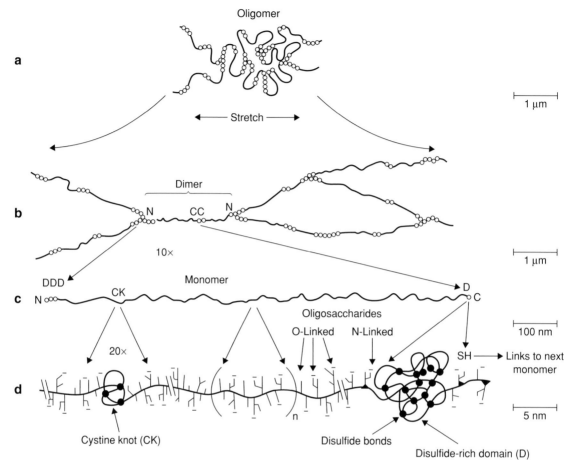

Fig. 1.32 Monomeric and oligomeric structures of mucin. (a) A number of mucin monomers (denoted by lines) are linked together (linkages denoted by circles) in an oligomeric gel. (b) The gel has been stretched to show more clearly the linkages between the individual monomers. N and C denote the N- and C-termini of the individual mucin monomers. (c) An individual monomer is schematically denoted showing the presence of the D domains which are involved in forming disulfide bonds between monomers. (d) The structure of the monomer is shown in more detail. It should be noted that the monomer contains many O- and N-linked oligosaccharides. Reproduced with the permission of Cambridge University Press, Cambridge, UK, from: Wilson, M., McNab, R. and Henderson, B. (2002) *Bacterial Disease Mechanisms; An Introduction to Cellular Microbiology*.

mucosal surface would leave the host vulnerable to microbial colonization and would have other harmful consequences. However, this does not appear to be a very common event, which means that mucus utilization by the indigenous microbiota must occur at the same rate as mucus production by the host – another example of the balanced relationship that exists between the host and its indigenous microbes.

1.5.4 Innate and acquired immune responses at the mucosal surface

Epithelial cells are in continuous contact with complex microbial communities which, for most of the time, consist of members of the indigenous microbiota of the particular site. All such microbes contain modulins and therefore have the potential to induce the release of

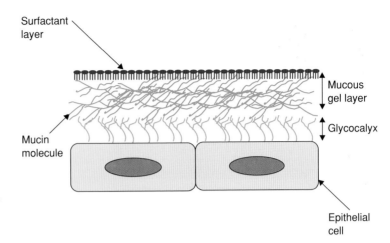

Fig. 1.33 A typical mucosal surface is covered with a layer of mucus which often has a thin lipid layer on its outer surface. The mucins of the mucous layer interact with the membrane-bound mucins (often referred to as the "glycocalyx") of the epithelial cells.

Table 1.10 Identity of mucins present at various body sites.

Gene	Type	Expression
MUC1	Membranous	Intestinal tract, genital tract, eye, respiratory tract
MUC2	Secreted	Intestinal tract, genital tract, respiratory tract
MUC3A	Membranous	Intestinal tract, genital tract, respiratory tract
MUC3B	Membranous	Intestinal tract, genital tract, respiratory tract
MUC4	Membranous	Intestinal tract, genital tract, eye, respiratory tract
MUC5AC	Secreted	Intestinal tract, genital tract, respiratory tract
MUC5B	Secreted	Intestinal tract, genital tract, respiratory tract
MUC6	Secreted	Intestinal tract, genital tract, respiratory tract
MUC12	Membranous	Intestinal tract, genital tract
MUC13	Membranous	Intestinal tract, genital tract
MUC17	Membranous	intestinal tract

pro-inflammatory cytokines from epithelial cells and to instigate an inflammatory response. In order to avoid the detrimental consequences of a constant inflammatory state in the mucosae, the epithelium at a particular site must be able to recognize the indigenous microbiota of that site and suppress any inflammatory response to it, i.e. it must display "tolerance" to its indigenous microbiota. However, the mucosa also needs to respond rapidly to the arrival of potentially harmful organisms and so must be able to recognize such organisms and to distinguish them from members of its indigenous microbiota. The mechanisms responsible for such tolerance and discrimination are only now beginning to be elucidated, and what little is known

has generally come from studies involving the intestinal mucosa. The role played by TLRs in discriminating between endogenous and exogenous organisms, and hence helping to prevent an inflammatory response by the former, has been described in section 1.2.4. Other ways in which this occurs will now be outlined briefly.

Intestinal pathogens can induce an inflammatory response by activating the transcription factor NF-κB, one of the consequences of this being the release of the neutrophil chemokine IL-8 which attracts PMNs to the site. In contrast, some residents of the GIT do not activate NF-κB and they inhibit its activation by inflammatory stimuli such as TNFα; consequently they do not

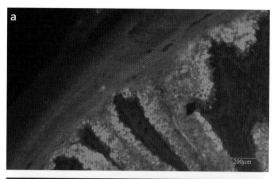

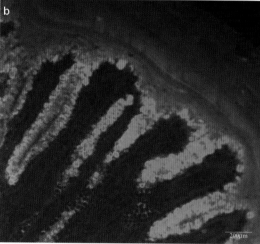

Fig. 1.34 Immunohistochemical identification of the mucins present in the mucus covering the gastric epithelium of an adult male. MUC6 is labeled with an antibody displaying green fluorescence, while MUC5AC is labeled with an antibody which fluoresces red–orange. The mucous layer on the gastric surface consists primarily of MUC5AC extending in layered sheets with MUC6 protein layered in between. MUC5AC is particularly abundant in the gastric pits. Reproduced with kind permission of Springer Science and Business Media, Berlin, Germany, from: Ho, S.B., Takamura, K., Anway, R. et al. (2004) The adherent gastric mucous layer is composed of alternating layers of MUC5AC and MUC6 mucin proteins. *Dig Dis Sci* 49, 1598–1606.

Table 1.11 Enzymes required for the complete degradation of mucins.

Type of enzyme	Role in mucin degradation
Sulfatases	Removal of terminal sulfate residues, thereby exposing underlying sugars and thus rendering them more susceptible to the action of glycosidases
Sialidases (neuraminidases)	Removal of terminal sialic acid residues exposes underlying sugars to the action of glycosidases; the sialic acid itself can be further degraded by acetylneuraminate pyruvate lyase to *N*-acetylmannosamine which can be used as a carbon and energy source by some bacteria
Exoglycosidases	To cleave sugars from side chains, e.g. β-D-galactosidase, *N*-acetyl-β-D-galactosaminidase, α-fucosidase, *N*-acetyl-β-D-glucosaminidase
Endoglycosidases	To cleave entire side chain from the peptide backbone or attack the side chain at sites other than the terminal residue – this may occur before or after the side chain has been cleaved from the protein
peptidases/ proteases	To cleave at nonglycosylated regions; degrade protein backbone after side chains have been removed

release IL-8. Whether this is due to the production of an immunosuppressive factor by nonpathogenic microbes or whether it is a host cell product induced in response to stimulation by such organisms remains to be established. The net effect is that the epithelium is maintained in a hypo-responsive state by the presence of members of the indigenous microbiota, i.e. it exhibits tolerance to these microbes.

Microbes that are indigenous to the GIT have been found to induce only a transient release of pro-inflammatory cytokines from intestinal epithelial cells. This is quickly suppressed by macrophages from the lamina propria. In contrast, macrophages are unable to suppress the release of pro-inflammatory cytokines induced by exogenous pathogens.

Certain indigenous intestinal species, such as

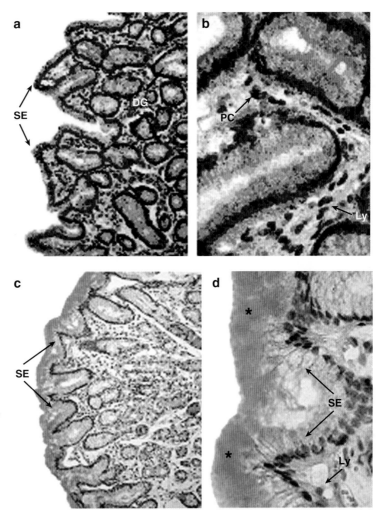

Fig. 1.35 Immunohistochemical detection of secretory leukocyte proteinase inhibitor (SLPI) in the gastric mucosa of the antrum (a, b) and the corpus (c, d). Red staining denotes the presence of SLPI, whereas cell nuclei are counterstained with hematoxylin. SLPI can be seen to be present in the surface epithelium (SE), plasma cells (PC), and lymphocytes (Ly). Asterisks denote the mucous layer which is considerably thicker in the corpus. Magnification: (a) and (c), ×100; (b) and (d), ×400. Reproduced with the permission of Lippincott, Williams and Wilkins, Philadelphia, Pennsylvania, USA, from: Hritz, I., Kuester, D., Vieth, M., Herszenyi, L., Stolte, M., Roessner, A., Tulassay, Z., Wex, T. and Malfertheiner, P. (2006) Secretory leukocyte protease inhibitor expression in various types of gastritis: A specific role of *Helicobacter pylori* infection. *Eur J Gastroenterol Hepatol* 18, 277–82.

Lactobacillus spp., are able to induce the release of TGFβ from intestinal epithelial cells. TGFβ is a key regulator of the immune response and has an overall anti-inflammatory effect.

Finally, antimicrobial peptides and proteins are now recognized as being an important means by which mucosal surfaces regulate their indigenous microbiotas (Fig. 1.35). A wide range of antimicrobial peptides are produced by epithelial cells and these are listed in Table 1.12. Although these molecules are best known for their ability to inhibit or kill microbes, some are also able to exert other effects such as neutralizing the biological activities of LPS. Each antimicrobial peptide

has a particular antimicrobial spectrum, and each mucosal surface produces only some of these peptides, Furthermore, some peptides are produced constitutively, while others are induced (or up-regulated) in response to certain microbial species, and some peptides display synergy with others. It is likely, therefore, that the particular mixture of peptides secreted by a particular mucosal surface plays some role in dictating the composition of the microbial community at that site.

It is likely that a combination of the mechanisms described above confers on mucosal surfaces the ability to control the composition of their resident microbial communities.

Table 1.12 Antimicrobial peptides/proteins produced by epithelia. For those peptides/proteins that do not currently show any activity other than as an antimicrobial, it is likely that some of these molecules will eventually be shown to have other activities, as many are newly discovered, and their activities have not been fully investigated.

Antimicrobial peptide/protein	Sites of production	Antimicrobial spectrum	Activities other than as an antimicrobial
Human β-defensin-1 (HBD-1)	E, S, OC, RT, St, I, UT, FGT	Gram-positive and Gram-negative species including *Staph. aureus*, *E. coli*, *Listeria monocytogenes*, *Klebsiella pneumoniae*, *Pseudomonas aeruginosa*, and *Can. albicans*.	—
HBD-2	E, S, OC, RT, I, FGT	*Strep. pyogenes*, *E. coli*, *Ent. faecalis*, *Can. albicans*, and *Ps. aeruginosa*; synergic activity with lysozyme and lactoferrin against a range of bacteria	Neutralizes some biological activities of LPS
HBD-3	S, OC, RT, FGT	*Staph. aureus*, *Strep. pyogenes*, *Ps. aeruginosa*, *Enterococcus faecium*, *E. coli*, and *Can. albicans*	Chemoattractant for monocytes
HBD-4	St, RT, FGT, MGT	*Staph. aureus*, *E. coli*, *Strep. pneumoniae*, *Burkholderia cenocepacia*, *Staphylococcus carnosus*, and *Ps. aeruginosa*; synergic activity with lysozyme and HBD-3	Chemoattractant for monocytes
Human epididymis 2 protein (HE2)	OC, RT, MGT	*E. coli*	—
Human α-defensin-5	I, FGT	*Sal. typhimurium*, *Lis. monocytogenes*, *E. coli*, *Can. albicans*, *Staph. aureus*, *Enterobacter aerogenes*, *Bacillus cereus*	—
Human α-defensin-6	I	*Bac. cereus*, *E. coli*	—
Histatins	OC	*Can. albicans* and other yeasts; some activity against Gram-positive species; *Por. gingivalis*	Suppression of bacteria-induced cytokine induction; inhibition of co-aggregation between *Por. gingivalis* and *Streptococcus mitis*; chemoattractant for monocytes
Peptidoglycan-recognition proteins (PGLYRP-1, -2, -3, -4)	S, OC, RT, I	*Lis. monocytogenes*, *Staph. aureus*, *Bacillus* spp., *Lactobacillus* spp.	—
Human cathelicidin (LL-37/hCAP-18)	S, OC, RT, I, FGT, MGT	*Strep. pyogenes*, *E. coli*, *Staph. aureus*, *Strep. pneumoniae*, *Staph. epidermidis*, *Ent. faecalis*, *Ps. aeruginosa*, *Lis. monocytogenes*, *Sal. typhimurium*, *Bacillus megaterium*, *Ag. actinomycetemcomitans*, *F. nucleatum*, and *Capnocytophaga* spp.; synergic activity with lysozyme against a range of bacteria	Neutralizes some biological activities of LPS and LTA; chemotactic for human peripheral monocytes, neutrophils, and CD4 T lymphocytes

Table 1.12 (*Cont'd*)

Antimicrobial peptide/protein	Sites of production	Antimicrobial spectrum	Activities other than as an antimicrobial
Hepcidin (LEAP-1; liver-expressed antimicrobial peptide-1)	UT, E	*E. coli, Staph. epidermidis, Staph. aureus, Can. albicans, Streptococcus agalactiae*	—
β-Lysin	E	Micrococci; synergic activity with lysozyme	—
Lysozyme	S, E, OC, RT, I, FGT, MGT	Effective primarily against Gram-positive bacteria	Agglutinates bacteria
Lactoferrin	E, OC, RT, I, FGT, MGT	*Strep. mutans, Streptococcus bovis, Vibrio cholerae, E. coli, Bacillus subtilis, Lis. monocytogenes*; synergic activity with lysozyme, IgA, LL-37, and human β-defensins	Iron-binding; prevents bacterial adhesion; enhances activity of NK cells
Lactoferricin	E, OC, RT, I, FGT, MGT	*E. coli, Can. albicans, Salmonella enteritidis, K. pneumoniae, Proteus vulgaris, Yersinia enterocolitica, Ps. aeruginosa, Campylobacter jejuni, Staph. aureus, Strep. mutans, Strep. bovis, Bac. subtilis, C. diphtheriae, Lis. monocytogenes, Cl. perfringens*	—
Adrenomedullin	S, OC, St, I	*P. acnes, Staph. aureus, M. luteus, B. fragilis, E. coli, Por. gingivalis, Ag. actinomycetemcomitans, Eikenella corrodens, Actinomyces naeslundii,* and *Strep. mutans*	—
Histones	I, FGT	*Salmonella* spp., *Ent. faecalis, Ent. faecium, Staph. aureus,* and *E. coli*	—
Secretory leukocyte proteinase inhibitor (SLPI)	E, OC, RT, FGT, MGT, St, I	*E. coli, Ps. aeruginosa,* and *Staph. aureus*	Neutralizes some biological activities of LPS
Secretory phospholipase A$_2$ (SPLA$_2$)	E, RT, I, MGT	Many Gram-positive organisms including staphylococci, streptococci, micrococci, enterococci, and *Lis. monocytogenes*	—
Dermcidin	S	*E. coli, Ent. faecalis, Staph. aureus,* and *Can. albicans*	—
Elafin (skin-derived anti-leukoprotease)	RT, S, FGT, MGT	*Staph. aureus* and *Ps. aeruginosa*	Neutralizes some biological activities of LPS; inhibits human neutrophil elastase
Lactoperoxidase (catalyzes the reaction between hydrogen peroxide and thiocyanate, resulting in the production of hypothiocyanite)	E, RT, OC, I	*Staph. aureus, Staph. epidermidis,* various streptococci, *H. influenzae, E. coli, Prevotella* spp., *Por. gingivalis,* and *Pseudomonas* spp.	—

Table 1.12 (*Cont'd*)

Antimicrobial peptide/protein	Sites of production	Antimicrobial spectrum	Activities other than as an antimicrobial
Anionic peptide	RT	*E. coli, Ps. aeruginosa, K. pneumoniae, Serratia marcescens, Staph. aureus,* and *Ent. faecalis*	—
RegIIIγ	I	*Lis. monocytogenes, Listeria innocua, Ent. faecalis*	—
Statherin	RT, OC	*Ps. aeruginosa*	—
Glandulin	RT	Some Gram-negative species	—
Calprotectin	FGT, OC,	*E. coli, Klebsiella* spp., *Staph. aureus, Staph. epidermidis, Can. albicans*	—
Psoriasin	S	*E. coli*	Chemokine
RNase7	S, RT, OC	*E. coli, Ps. aeruginosa, Staph. aureus, P. acnes, Can. albicans*	—
Bactericidal/permeability-inducing protein (BPI)	Present in neutrophils; recently shown to be secreted by some mucosa including E, RT, OC, I	Active against a wide variety of Gram-negative species including *E. coli, Salmonella* spp., *Shigella* spp., *K. pneumoniae, Ps. aeruginosa*	Neutralizes biological activities of LPS
Hemocidins (hemoglobin fragments)	FGT	*E. coli, K. pneumoniae, Salmonella* spp., *Ent. faecalis*	—
SPLUNC1 protein (short palate, lung, and nasal epithelium clone 1)	RT, E, I, OC	*Ps. aeruginosa* and possibly other Gram-negative species	Neutralizes the activities of LPS
Chemokine ligand 20 (CCL20)	RT	Active mainly against Gram-negative species	Stimulates the migration of B-cells, immature dendritic cells, and a subset of memory T cells

E, eye; FGT, female genital tract; I, intestines; MGT, male genital tract; OC, oral cavity; RT, respiratory tract; S, skin; St, stomach, UT, urinary tract.

1.6 FURTHER READING

1.6.1 Books

Atlas, R.M. and Bartha, R. (1997) *Microbial Ecology: Fundamentals and Applications.* Addison-Wesley, Boston, Massachusetts, USA.

Grubb, R., Midtvedt, T. and Norin, E. (eds). (1988) *The Regulatory and Protective Role of the Normal Microflora.* Stockton Press, New York, New York, USA.

Hill, M.J. and Marsh P.D. (1990) *Human Microbial Ecology.* CRC Press, Boca Raton, Florida, USA.

Nataro, J.P., Cohen, P.S., Mobley, H.L.T. and Weiser, J.N. (2005) *Colonization of Mucosal Surfaces.* ASM Press, Washington DC, USA.

Osborn, A.M. and Smith, C.J. (eds). (2005) *Molecular Microbial Ecology.* Taylor & Francis, New York, New York, USA.

Rosebury, T. (1962) *Microorganisms Indigenous to Man.* McGraw-Hill Book Company, New York, New York, USA.

Shafer, W.M. (ed.). (2006) *Antimicrobial Peptides and Human Disease.* Springer Science and Business Media, Berlin, Germany.

Skinner, F.A. and Carr, J.G. (eds). (1974) *The Normal Microbial Flora of Man.* Academic Press, London, UK.

Tannock, G.W. (1995) *Normal Microflora.* Chapman and Hall, London, UK.

Tannock, G.W. (ed.). (1999) *Medical Importance of the Normal Microflora.* Kluwer Academic Publishers, Dordrecht, The Netherlands.

Wilson, M. (2005) *Microbial Inhabitants of Humans: Their Ecology and Role in Health and Disease.* Cambridge University Press, Cambridge, UK.

1.6.2 Reviews and papers

Agerberth, B. and Gudmundsson, G.H. (2006) Host antimicrobial defence peptides in human disease. *Curr Top Microbiol Immunol* 306, 67–90.

Albiger, B., Dahlberg, S., Henriques-Normark, B. and Normark, S. (2007) Role of the innate immune system in host defence against bacterial infections: Focus on the Toll-like receptors. *J Intern Med* 261, 511–28.

Backhed, F., Ley, R.E., Sonnenburg, J.L., Peterson, D.A. and Gordon, J.I. (2005) Host-bacterial mutualism in the human intestine. *Science* 307, 1915–20.

Bae, J.W. and Park, Y.H. (2006) Homogeneous versus heterogeneous probes for microbial ecological microarrays. *Trends Biotechnol* 24, 318–23.

Battin, T.J., Sloan, W.T., Kjelleberg, S., Daims, H., Head, I.M., Curtis, T.P. and Eberl, L. (2007) Microbial landscapes: New paths to biofilm research. *Nat Rev Microbiol* 5, 76–81

Bayles, K.W. (2007) The biological role of death and lysis in biofilm development. *Nat Rev Microbiol* 5, 721–6

Beisswenger, C. and Bals, R. (2005) Functions of antimicrobial peptides in host defense and immunity. *Curr Protein Pept Sci* 6, 255–64.

Blaut, M., Collins, M.D., Welling, G.W., Dore, J., van Loo, J. and de Vos, W. (2002) Molecular biological methods for studying the gut microbiota: The EU human gut flora project. *Brit J Nutr* 87 (Suppl 2), S203–11.

Bohannan, B.J., Kerr, B., Jessup, C.M., Hughes, J.B. and Sandvik, G. (2002) Trade-offs and coexistence in microbial microcosms. *Antonie Van Leeuwenhoek* 81, 107–15.

Boix, E. and Nogues, M.V. (2007). Mammalian antimicrobial proteins and peptides: Overview on the RNase A superfamily members involved in innate host defence. *Mol Biosyst* 3, 317–35.

Boman, H.G. (2000) Innate immunity and the normal microflora. *Immunol Rev* 173, 5–16.

Bottari, B., Ercolini, D., Gatti, M. and Neviani, E. (2006) Application of FISH technology for microbiological analysis: Current state and prospects. *Appl Microbiol Biotechnol* 73, 485–94.

Branda, S.S., Vik, S., Friedman, L. and Kolter, R. (2005) Biofilms: The matrix revisited. *Trends Microbiol* 13, 20–6.

Brown, K.L. and Hancock, R.E. (2006) Cationic host defense (antimicrobial) peptides. *Curr Opin Immunol* 18, 24–30.

Cario, E. (2005) Bacterial interactions with cells of the intestinal mucosa: Toll-like receptors and NOD2. *Gut* 54, 1182–93.

Cheesman, S.E. and Guillemin, K. (2007) We know you are in there: Conversing with the indigenous gut microbiota. *Res Microbiol* 158, 2–9.

Clavel, T. and Haller, D. (2007) Molecular interactions between bacteria, the epithelium, and the mucosal immune system in the intestinal tract: Implications for chronic inflammation. *Curr Issues Intest Microbiol* 8, 25–43

Corthesy, B. (2007) Roundtrip ticket for secretory IgA: Role in mucosal homeostasis? *J Immunol* 178, 27–32.

Dethlefsen, L., Eckburg, P.B., Bik, E.M. and Relman, D.A. (2006) Assembly of the human intestinal microbiota. *Trends Ecol Evol* 21, 517–23.

Dethlefsen, L., McFall-Ngai, M. and Relman, D.A. (2007) An ecological and evolutionary perspective on human-microbe mutualism and disease. *Nature* 449, 811–18.

Domka, J., Lee, J., Bansal, T. and Wood, T.K. (2007) Temporal gene-expression in *Escherichia coli* K-12 biofilms. *Environ Microbiol* 9, 332–46.

Donskey, C.J., Hujer, A.M., Das, S.M., Pultz, N.J., Bonomo, R.A. and Rice, L.B. (2003) Use of denaturing gradient gel electrophoresis for analysis of the stool microbiota of hospitalized patients. *J Microbiol Methods* 54, 249–56.

Dunn, A.K. and Stabb, E.V. (2007) Beyond quorum sensing: The complexities of prokaryotic parliamentary procedures. *Anal Bioanal Chem* 387, 391–8.

Durr, U.H., Sudheendra, U.S. and Ramamoorthy, A. (2006) LL-37, the only human member of the cathelicidin family of antimicrobial peptides. *Biochim Biophys Acta* 1758, 1408–25.

Dziarski, R. and Gupta, D. (2006) Mammalian PGRPs: Novel antibacterial proteins. *Cell Microbiol* 8, 1059–69.

Eckburg, P.B., Bik, E.M., Bernstein, C.N., Purdom, E., Dethlefsen, L., Sargent, M., Gill, S.R., Nelson, K.E. and Relman, D.A. (2005) Diversity of the human intestinal microbial flora. *Science* 308, 1635–8.

Elson, C.O. and Cong, Y. (2002) Understanding immune-microbial homeostasis in the intestine. *Immunol Res* 26, 87–94.

Emonts, M., Hazelzet, J.A., de Groot, R. and Hermans, P.W. (2003) Host genetic determinants of *Neisseria meningitidis* infections. *Lancet Infect Dis* 3, 565–77.

Finegold, S.M. (2004) Changes in taxonomy, anaerobes associated with humans, 2001–2004. *Anaerobe* 10, 309–12.

Fujimoto, C., Maeda, H., Kokeguchi, S., Takashiba, S., Nishimura, F., Arai, H., Fukui, K. and Murayama, Y. (2003) Application of denaturing gradient gel electrophoresis (DGGE) to the analysis of microbial communities of subgingival plaque. *J Periodontal Res* 38, 440–45.

Gavazzi, G. and Krause, K.H. (2002) Ageing and infection. *Lancet Infect Dis* 2, 659–66.

Gill, S.R., Pop, M., Deboy, R.T., Eckburg, P.B., Turnbaugh, P.J., Samuel, B.S., Gordon, J.I., Relman, D.A., Fraser-Liggett, C.M. and Nelson, K.E. (2006) Metagenomic analysis of the human distal gut microbiome. *Science* 312, 1355–9.

Goodacre, R. (2007) Metabolomics of a superorganism. *J Nutr* 137, S259–66.

Greene, E.A. and Voordouw, G. (2003) Analysis of environmental microbial communities by reverse sample genome probing. *J Microbiol Methods* 53, 211–19.

Guan, R. and Mariuzza, R.A. (2007) Peptidoglycan recognition proteins of the innate immune system. *Trends Microbiol* 15, 127–34.

Gupta, G. and Surolia, A. (2007) Collectins: Sentinels of innate immunity. *Bioessays* 29, 452–64.

Hall-Stoodley, L., Costerton, J.W. and Stoodley, P. (2004) Bacterial biofilms: From the natural environment to infectious diseases. *Nat Rev Microbiol* 2, 95–108.

Hebuterne, X. (2003) Gut changes attributed to ageing: Effects on intestinal microflora. *Curr Opin Clin Nutr Metabolic Care* 6, 49–54.

Iweala, O.I and Nagler, C.R. (2006) Immune privilege in the gut: The establishment and maintenance of non-responsiveness to dietary antigens and commensal flora. *Immunol Rev* 213, 82–100.

Kapetanovic, R. and Cavaillon, J.M. (2007) Early events in innate immunity in the recognition of microbial pathogens. *Expert Opin Biol Ther* 7, 907–18.

Keller, L. and Surette, M.G. (2006) Communication in bacteria: An ecological and evolutionary perspective. *Nat Rev Microbiol* 4, 249–58.

Knowles, M.R. and Boucher, R.C. (2002) Mucus clearance as a primary innate defense mechanism for mammalian airways. *J Clin Invest* 109, 571–7.

Lerat, E. and Moran, N.A. (2004) The evolutionary history of quorum-sensing systems in bacteria. *Mol Biol Evol* 21, 903–13.

Ley, R.E., Peterson, D.A. and Gordon, J.I. (2006) Ecological and evolutionary forces shaping microbial diversity in the human intestine. *Cell* 124, 837–48.

Ley, R.E., Turnbaugh, P.J., Klein, S. and Gordon, J.I. (2006) Microbial ecology: Human gut microbes associated with obesity. *Nature* 444, 1022–3.

Lievin-Le Moal, V. and Servin, A.L. (2006) The front line of enteric host defense against unwelcome intrusion of harmful microorganisms: Mucins, antimicrobial peptides, and microbiota. *Clin Microbiol Rev* 19, 315–37.

Macpherson, A.J., Geuking, M.B. and McCoy, K.D. (2005) Immune responses that adapt the intestinal mucosa to commensal intestinal bacteria. *Immunology* 115, 153–62

Magalhaes, J.G., Tattoli, I. and Girardin, S.E. (2007) The intestinal epithelial barrier: How to distinguish between the microbial flora and pathogens. *Semin Immunol* 19, 106–15.

McFarland, L.V. (2000) Normal flora: Diversity and functions. *Microb Ecol Health Dis* 12, 193–207.

McGlauchlen, K.S. and Vogel, L.A. (2003) Ineffective humoral immunity in the elderly. *Microbes Infect* 5, 1279–84.

Nobile, C.J. and Mitchell, A.P. (2007) Microbial biofilms: *E pluribus unum*. *Curr Biol* 17, R349–53.

Nochi, T. and Kiyono, H. (2006) Innate immunity in the mucosal immune system. *Curr Pharm Des* 12, 4203–13.

O'Hara, A.M. and Shanahan, F. (2006) The gut flora as a forgotten organ. *EMBO Rep* 7, 688–93.

O'Neill, L.A. (2006) How Toll-like receptors signal: What we know and what we don't know. *Curr Opin Immunol* 18, 3–9.

Otto, M. (2006) Bacterial evasion of antimicrobial peptides by biofilm formation. *Curr Top Microbiol Immunol* 306, 251–8.

Palmer, C., Bik, E.M., Eisen, M.B., Eckburg, P.B., Sana, T.R., Wolber, P.K., Relman, D.A. and Brown, P.O. (2006) Rapid quantitative profiling of complex microbial populations. *Nucleic Acids Res* 34(1), e5.

Palmer, J., Flint, S. and Brooks, J. (2007) Bacterial cell attachment, the beginning of a biofilm. *J Ind Microbiol Biotechnol* 34, 577–88

Pandey, S. and Agrawal, D.K. (2006) Immunobiology of Toll-like receptors: Emerging trends. *Immunol Cell Biol* 84, 333–41.

Parsek, M.R. and Greenberg, E.P. (2005) Sociomicrobiology: The connections between quorum sensing and biofilms. *Trends Microbiol* 13, 27–33.

Ponda, P.P. and Mayer, L. (2005) Mucosal epithelium in health and disease. *Curr Mol Med* 5, 549–56.

Preston-Mafham, J., Boddy, L. and Randerson, P.F. (2002) Analysis of microbial community functional diversity using sole-carbon-source utilisation profiles – A critique. *FEMS Microb Ecol* 42, 1–14.

Rakoff-Nahoum, S. and Medzhitov, R. (2006) Role of the innate immune system and host-commensal mutualism. *Curr Top Microbiol Immunol* 308, 1–18.

Reading, N.C. and Sperandio, V. (2006) Quorum sensing: The many languages of bacteria. *FEMS Microbiol Lett* 254, 1–11.

Relman, D.A. (2002) New technologies, human–microbe interactions, and the search for previously unrecognized pathogens. *J Infect Dis* 186 (Suppl 2), S254–8.

Riley, M.A. and Wertz, J.E. (2002) Bacteriocins: Evolution, ecology, and application. *Annu Rev Microbiol* 56, 117–37.

Rusch, V.C. (1989) The concept of symbiosis: A survey of terminology used in description of associations of dissimilarly named organisms. *Microecol Therapy* 19, 33–59.

Sansonetti, P.J. and Di Santo, J.P. (2007) Debugging how bacteria manipulate the immune response. *Immunity* 26, 149–61.

Schnare, M., Rollinghoff, M. and Qureshi, S. (2006) Toll-like receptors: Sentinels of host defence against bacterial infection. *Int Arch Allergy Immunol* 139, 75–85.

Seebah, S., Suresh, A., Zhuo, S., Choong, Y.H., Chua, H., Chuon, D., Beuerman, R. and Verma, C. (2007) Defensins knowledgebase: A manually curated database and information source focused on the defensins family of antimicrobial peptides. *Nucleic Acids Res* 35, D265–8.

Sheehan, J.K., Kesimer, M. and Pickles, R. (2006) Innate immunity and mucus structure and function. *Novartis Found Symp* 279, 155–66.

Sirard, J.C., Bayardo, M. and Didierlaurent, A. (2006) Pathogen-specific TLR signaling in mucosa: Mutual contribution of microbial TLR agonists and virulence factors. *Eur J Immunol* 36, 260–3.

Smith, V.H. (2002) Effects of resource supplies on the structure and function of microbial communities. *Antonie Van Leeuwenhoek* 81, 99–106.

Sonnenburg, J.L., Angenent, L.T. and Gordon, J.I. (2004) Getting a grip on things: How do communities of bacterial symbionts become established in our intestine? *Nat Immunol* 5, 569–73.

Thornton, D.J. and Sheehan, J.K. (2004) From mucins to mucus: Toward a more coherent understanding of this essential barrier. *Proc Am Thorac Soc* 1, 54–61.

Tjabringa, G.S., Vos, J.B., Olthuis, D., Ninaber, D.K., Rabe, K.F., Schalkwijk, J., Hiemstra, P.S. and Zeeuwen, P.L. (2005) Host defense effector molecules in mucosal secretions. *FEMS Immunol Med Microbiol* 45, 151–8.

Tlaskalova-Hogenova, H., Tuckova, L., Mestecky, J., Kolinska, J., Rossmann, P., Stepankova, R., Kozakova, H., Hudcovic, T., Hrncir, T., Frolova, L. and Kverka, M. (2005) Interaction of mucosal microbiota with the innate immune system. *Scand J Immunol* 62 (Suppl 1), 106–13.

Toivanen, P., Vaahtovuo, J. and Eerola, E. (2001) Influence of major histocompatibility complex on bacterial composition of fecal flora. *Infect Immun* 69, 2372–7.

Trinchieri, G. and Sher, A. (2007) Cooperation of Toll-like receptor signals in innate immune defence. *Nat Rev Immunol* 7, 179–90.

Turnbaugh, P.J., Ley, R.E., Mahowald, M.A., Magrini, V., Mardis, E.R. and Gordon, J.I. (2006) An obesity-associated gut microbiome with increased capacity for energy harvest. *Nature* 444, 1027–31

Vimr, E.R., Kalivoda, K.A., Deszo, E.L. and Steenbergen, S.M. (2004) Diversity of microbial sialic acid metabolism. *Microbiol Mol Biol Rev* 68, 132–53.

Waters, C.M. and Bassler, B.L. (2005) Quorum sensing: Cell-to-cell communication in bacteria. *Annu Rev Cell Dev Biol* 21, 319–46.

Weinberg, E.D. (2007) Antibiotic properties and applications of lactoferrin. *Curr Pharm Des* 13, 801–2.

Weng, L., Rubin, E.M. and Bristow, J. (2006) Application of sequence-based methods in human microbial ecology. *Genome Res* 16, 316–22.

West, A.P., Koblansky, A.A. and Ghosh, S. (2006) Recognition and signaling by toll-like receptors. *Annu Rev Cell Dev Biol* 22, 409–37.

Williams, O.W., Sharafkhaneh, A., Kim, V., Dickey, B.F. and Evans CM. (2006) Airway mucus: From production to secretion. *Am J Respir Cell Mol Biol* 34, 527–36.

Williams, P., Winzer, K., Chan, W.C. and Camara, M. (2007) Look who's talking: Communication and quorum sensing in the bacterial world. *Philos Trans R Soc Lond B Biol Sci* 362, 1119–34.

Winkler, P., Ghadimi, D., Schrezenmeir, J. and Kraehenbuhl, J.P. (2007) Molecular and cellular basis of microflora-host interactions. *J Nutr* 137, 756S–72S.

Woof, J.M. and Kerr, M.A. (2006) The function of immunoglobulin A in immunity. *J Pathol* 208, 270–82.

Xu, J. (2006) Microbial ecology in the age of genomics and metagenomics: Concepts, tools, and recent advances. *Mol Ecol* 15, 1713–31.

Xu, J. and Gordon, J.I. (2003) Honor thy symbionts. *Proc Natl Acad Sci U S A* 100, 10452–9.

Zhou, J. (2003) Microarrays for bacterial detection and microbial community analysis. *Curr Opin Microbiol* 6, 288–94.

Zoetendal, E.G., Akkermans, A.D.L., Akkermans-van Vliet, W.M., de Visser, J.A.G.M. and de Vos, W.M. (2001) The host genotype affects the bacterial community in the human gastrointestinal tract. *Microbial Ecology in Health and Disease* 13, 129–34.

Chapter 2

THE INDIGENOUS MICROBIOTA OF THE SKIN

2.1 ANATOMY AND PHYSIOLOGY OF HUMAN SKIN

The skin is one of the largest organs of the body in terms of its surface area (approximately 1.8 m^2) and weight (approximately 4.2 kg without blood). It is composed of epidermal, connective, nervous, and muscular tissues and has a variety of functions including protecting the underlying tissues from microbes. Basically, skin consists of two main layers – an outer epithelium, known as the epidermis, and an inner dermis (Fig. 2.1).

The epidermis is a keratinized, stratified, squamous epithelium within which five layers can be distinguished, the outermost layer of which (the stratum corneum [Fig. 2.2]) is approximately 0.02 mm thick and consists of dead, keratinized cells (known as corneocytes or squames [Fig. 2.3]), with lipids filling the intercellular spaces (Fig. 2.4). The squames are gradually sloughed off (a process known as desquamation), taking with them any attached microbes. As a result of this process, approximately 250 g of skin are lost each day, and the stratum corneum is renewed every 15 days.

The structure of the dermis is more complex as it contains hair follicles (approximately 5 million), sebaceous glands, and sudoriferous (sweat-producing) glands (Figs 2.1). Sebaceous glands secrete an oily substance, sebum, which consists of triglycerides (41%), wax monoesters (25%), fatty acids (16%), and squalene (12%). Sebaceous glands usually, but not always, open into a hair follicle and the hair, follicle, and gland are said to constitute a "pilosebaceous unit". The glands are also found in several hairless regions – the eyelids, penis, labia minora, and nipples – but not on the palms of the hands or soles of the feet (Fig. 2.5). There are two types of sudoriferous glands – apocrine and eccrine (Fig. 2.6). The fluid secreted by apocrine glands consists mainly of sialomucin and, as with sebaceous glands, this is discharged into a hair follicle. Apocrine glands are found at only a limited number of sites on the skin surface (Fig. 2.7). In contrast, eccrine glands have a wider distribution (Fig. 2.8) and are not associated with hair follicles. Their main function is in temperature regulation, which is accomplished by the cooling effect produced by the evaporation of sweat on the skin surface. Eccrine sweat has a complex composition and contains lactate, urea, serine, ornithine, citrulline, aspartic acid, Na$^+$, K$^+$, NH$_4^+$, glycogen, proteolytic enzymes, and effector molecules of the innate and acquired immune systems.

The presence of hair affects the temperature and moisture content of the skin surface, while the presence of glands, because of their secretions, alters the nutrient and moisture content, the pH, and osmolarity of a site. All of these factors affect the composition of the microbial community at a particular site; consequently, the distribution of the three types of glands on the skin surface (shown in Figs 2.5, 2.7, and 2.8) has a profound effect on the types of bacteria present and their population density.

2.2 CUTANEOUS ANTIMICROBIAL DEFENSE SYSTEMS

The antimicrobial defense mechanisms that operate on the skin surface are summarized in Fig. 2.9. Many of these (e.g. desquamation) prevent colonization of the

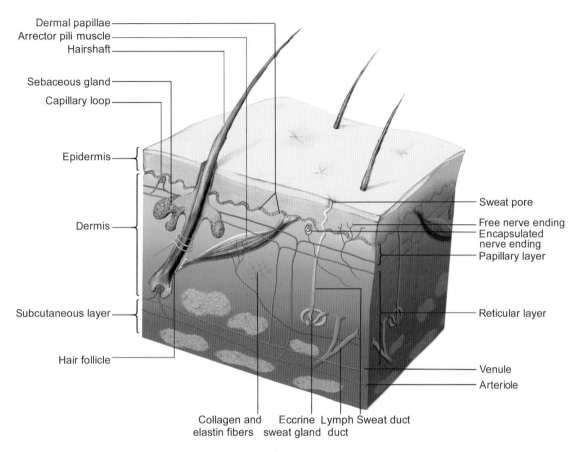

Dermal papillae
Arrector pili muscle
Hairshaft
Sebaceous gland
Capillary loop
Epidermis
Dermis
Subcutaneous layer
Hair follicle

Sweat pore
Free nerve ending
Encapsulated nerve ending
Papillary layer
Reticular layer
Venule
Arteriole

Collagen and elastin fibers
Eccrine sweat gland
Lymph duct
Sweat duct

Fig. 2.1 Diagram showing the major structures comprising human skin. Diagram courtesy of Roisin Mathews, © Department of Learning and Teaching Resources, Belfast Institute of Further and Higher Education, Belfast, UK.

Fig. 2.2 Section of human skin from the lower leg, showing various layers, including the stratum corneum. Scale bar, 200 μm. Reprinted with the permission of Macmillan Publishers Ltd, London, UK, from: Glaser, R., Harder, J., Lange, H., Bartels, J., Christophers, E. and Schroder, J.M. (2005) Antimicrobial psoriasin (S100A7) protects human skin from *Escherichia coli* infection. *Nat Immunol* 6, 57–64, © 2005.

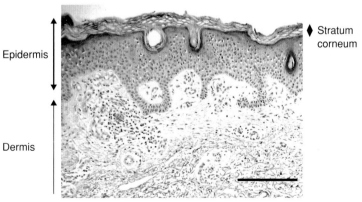

Epidermis

Dermis

Stratum corneum

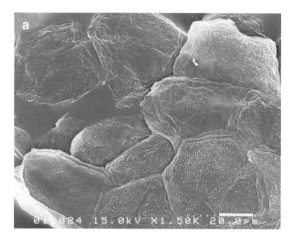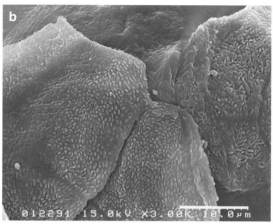

Fig. 2.3 Scanning electron micrographs showing squames on the surface of human skin. (a) Scale bar, 20 μm; (b) scale bar, 10 μm. Reproduced with the permission of The Japanese Society for Investigative Dermatology (© 2005) from: Katsuyama, M., Kobayashi, Y., Ichikawa, H., Mizuno, A., Miyachi, Y., Matsunaga, K. and Kawashima, M. (2005) A novel method to control the balance of skin microflora, Part 2. A study to assess the effect of a cream containing farnesol and xylitol on atopic dry skin. *J Dermatol Sci* 38, 207–13.

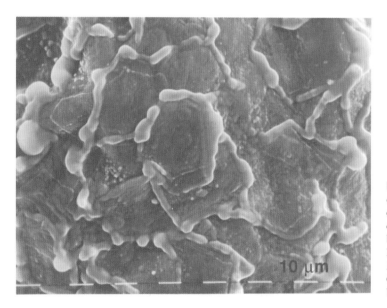

Fig. 2.4 Scanning electron micrograph of the surface of bovine skin after freeze-drying to preserve lipids. Keratinized cells can be seen with gaps between them filled by lipid-rich material. Reproduced from: Noble, W.C. (ed.). (1992) *The Skin Microflora and Microbial Skin Disease*. Cambridge University Press, Cambridge, UK.

skin by all microbes, while others (e.g. low pH, high salt content, antimicrobial peptides, and fatty acids), affect some species more than others, and therefore exert a selection pressure and so influence the types of microbes that can become established on the skin.

2.2.1 Innate defense systems

The continuous flow of air across body surfaces constitutes the first line of defense against microbial colonization as it prevents air-borne microbes and

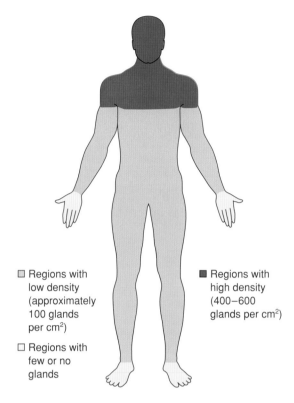

☐ Regions with
low density
(approximately
100 glands
per cm²)

☐ Regions with
few or no
glands

■ Regions with
high density
(400–600
glands per cm²)

Fig. 2.5 Distribution of sebaceous glands on the skin surface.

microbe-containing particles from settling on the skin. The stratum corneum (Figs 2.2 to 2.4) provides an effective barrier against further penetration of any microbes that do arrive at the skin surface. This layer of dead keratinocytes and lipids renders the surface of the skin very dry, thereby limiting the growth of a wide range of microbes (see section 1.2.2 and Table 1.4). Furthermore, the skin has a low pH (section 2.3) at most body sites as a result of the accumulation of acidic glandular secretions, acids produced during the keratinization process, and acids excreted by epithelial cells and microbes. Many microbes cannot survive at such low pHs, while the growth rate of many that do survive may be reduced considerably. As well as having a pH-lowering effect, many of the substances present on the skin surface are directly toxic to microbes. Free fatty acids, particularly lauric and myristic acids, are the most effective of these and have a wide antimicrobial spectrum. Linoleic and linolenic acids also have antimicrobial activity, particularly against transients such as *Staphylococcus aureus*. Sphingosine, a hydrolysis product of ceramide, is present at high concentrations on the skin surface and is very effective against *Staph. aureus*. Lauric acid and caproic, butyric, and myristic acids are inhibitory to *Propionibacterium acnes*, *Streptococcus pneumoniae*, *Streptococcus pyogenes*, corynebacteria, micrococci, *Candida* spp., and staphylococci.

Fig. 2.6 Scanning electron micrograph of a sweat duct in the skin from a human foot. Note the presence of bacteria (arrowed) – mainly as single cells, pairs, or small groups. Reprinted with the permission of Blackwell Publishing Ltd, from: Malcolm S.A, and Hughes T.C. (1980) The demonstration of bacteria on and within the stratum corneum using scanning electron microscopy. *British Journal of Dermatology* 102, 267–75.

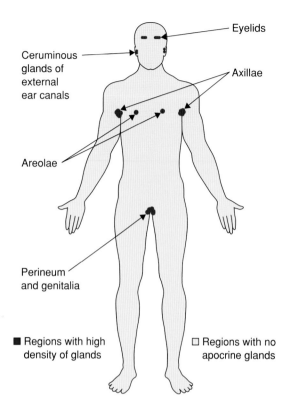

Fig. 2.7 Distribution of apocrine glands on the skin surface.

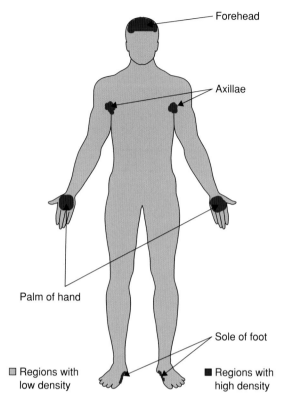

Fig. 2.8 Distribution of eccrine glands on the skin surface. In regions with a high density of eccrine glands, there may be as many as 600 glands/cm^2. An adult has a total of between 2 million and 5 million eccrine glands and, in temperate climates, produces 0.5–1.0 liters of sweat per day.

The evaporation of sweat from the skin surface leaves behind large quantities of NaCl and other solutes, and this results in any moisture present on the skin having a high osmolarity and a low water activity. Many organisms, particularly Gram-negative bacteria, are unable to grow at such low water activities (section 1.2.2). Finally, the constant shedding of squames, taking with them any adherent microbes, is a very effective means of limiting microbial colonization of the skin.

A number of Toll-like receptors (TLRs) are expressed by keratinocytes including TLR1–5 and TLR9 (Fig. 2.10.; see also section 1.2.4). Antimicrobial peptides produced by the skin include HBD-1, HBD-2, HBD-3, LL-37, adrenomedullin, dermcidin, RNase 7, and psoriasin. Figures 2.11–2.13 show the expression patterns of some of these in healthy skin.

Little is known about possible variations in antimicrobial peptide production at different skin sites. In the case of psoriasin, however, the levels of expression have been shown to vary markedly, with the highest

being found on the scalp and the lowest on the chest (Fig. 2.13).

2.2.2 Acquired immune defense systems

The skin-associated lymphoid tissue (SALT) is involved in the generation of humoral and cell-mediated immune responses and consists of: (1) Langerhans cells and dermal dendritic cells, both of which are antigen-presenting cells and circulate between the skin and the lymph nodes; (2) keratinocytes and endothelial cells, both of which produce cytokines; and (3) lymphocytes. Because of their long processes, Langerhans cells form a virtually continuous network over the skin surface, and this can capture almost any antigen entering the skin. Both IgA and IgG antibodies are secreted by the

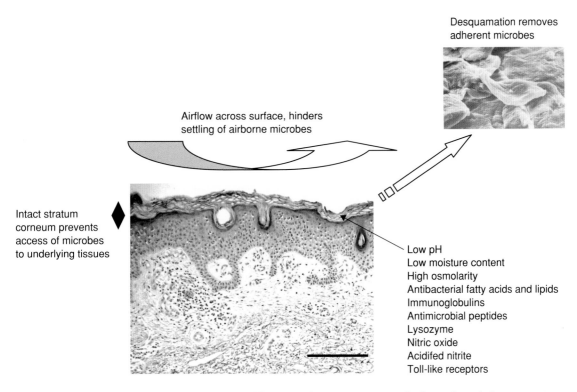

Desquamation removes adherent microbes

Airflow across surface, hinders settling of airborne microbes

Intact stratum corneum prevents access of microbes to underlying tissues

Low pH
Low moisture content
High osmolarity
Antibacterial fatty acids and lipids
Immunoglobulins
Antimicrobial peptides
Lysozyme
Nitric oxide
Acidifed nitrite
Toll-like receptors

Fig. 2.9 Diagram summarizing the main antimicrobial defense mechanisms operating at the skin surface. The large image in the center, showing the structure of the skin, is reprinted with the permission of Macmillan Publishers Ltd from: Glaser, R., Harder, J., Lange, H., Bartels, J., Christophers, E. and Schroder, J.M. (2005) Antimicrobial psoriasin (S100A7) protects human skin from *Escherichia coli* infection. *Nat Immunol* 6, 57–64, © 2005. The smaller image (top right) is reproduced with the permission of Elsevier, Amsterdam, The Netherlands, from: Noble, W.C. (ed.). (1974) *Microbiology of Human Skin*, Volume 2 in the series of *Major Problems in Dermatology*.

eccrine glands and are deposited on the skin surface, where they can help prevent microbial adhesion and exert an antimicrobial effect. Immunohistochemical studies have shown that many organisms present on the skin surface are coated with antibodies.

2.3 ENVIRONMENTAL DETERMINANTS OPERATING AT DIFFERENT SKIN REGIONS

The main factors affecting the growth and survival of microbes on the skin are summarized in Table 2.1.

The skin, in general, is a relatively dry environment, and any fluids present on its surface have a relatively high osmotic pressure. Gram-positive species (particularly staphylococci and micrococci) are better adapted to these conditions than Gram-negative species, and hence the microbial communities on the skin tend to be dominated by the former, while the latter are largely excluded.

The temperature of the skin varies widely with the anatomical location but is invariably lower than that of the internal organs (Fig. 2.14). The axillae and groin tend to have the highest temperatures, while the toes and fingers have the lowest.

The temperatures encountered (approximately 25–35°C) are ideal for the growth of mesophiles, whereas psychrophiles and thermophiles are largely excluded. Although different sites have different temperatures, these variations amount to only a few degrees centigrade, and therefore are unlikely to exert a dramatic selection pressure on the microbes colonizing a particular site. However, the temperature at a particular site

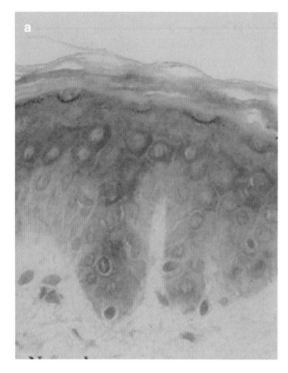 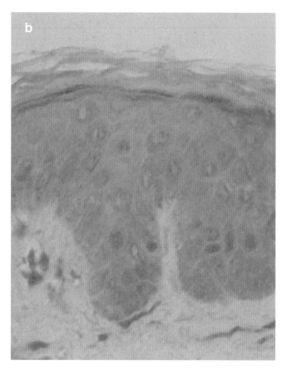

Fig. 2.10 Expression (brown-staining regions) of Toll-like receptors (a) TLR2 and (b) TLR4 in biopsies of healthy skin. Magnification ×500. Reproduced with the permission of Blackwell Publishing Ltd from: Jugeau, S., Tenaud, I., Knol, A.C., Jarrousse, V., Quereux, G., Khammari, A. and Dreno, B. (2005) Induction of toll-like receptors by *Propionibacterium acnes*. *Br J Dermatol* 153, 1105–13.

will affect the growth rate of the organisms present and will also influence the microbiota indirectly because of its effect on sweat production. Increased sweat production at a site with a high temperature results in an environment with a higher moisture content and alters the range (and concentration) of nutrients, the concentration of antimicrobial compounds, osmolarity, and pH.

One of the main factors governing the distribution of microbes on the skin is moisture availability. Because the water content of the stratum corneum is low (approximately 15% by weight), the skin surface is a relatively dry environment, thereby limiting microbial survival and growth. However, sweat production (approximately 200 ml per day) by the eccrine glands can increase the moisture content at the surface, particularly at sites from which sweat cannot easily evaporate (known as "occluded" regions), e.g. the axillae and regions between the toes (toe webs). Such occluded regions have relatively greater population densities

than dry areas (e.g. the palms of the hands) and support different microbial communities. Hence, corynebacteria, Gram-negative bacteria, and fungi are found at occluded sites, whereas they tend to be absent from dry regions. Staphylococci and micrococci, in contrast, are resistant to desiccation and so are not restricted to occluded sites.

Another consequence of sweat production is that the solutes present are left behind on the skin surface when sweat evaporates. NaCl is a major component of sweat; hence its concentration is greatly increased in regions with a high density of eccrine glands.

The stratum corneum is provided with oxygen directly from the atmosphere and also by diffusion from cutaneous capillaries. The oxygen concentration at the surface is known as the "transcutaneous oxygen pressure" and is highest on the chest (approximately 80% of that found in arterial blood), but is much lower on the forehead (approximately 27% of that found

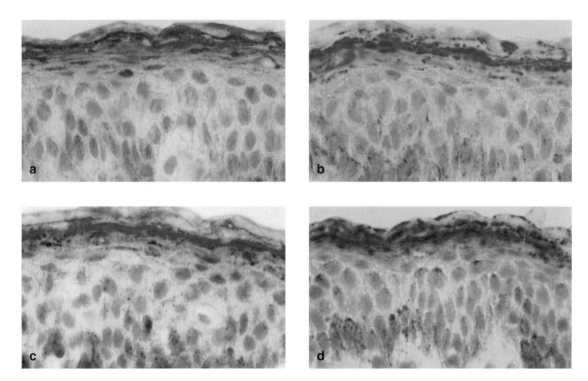

Fig. 2.11 Expression of (a) HBD-1, (b) HBD-2, (c), HBD-3, and (d) LL-37 in normal human skin. Magnification ×600. The red/brown staining denotes the presence of these antimicrobial peptides, which appear to be produced in the largest quantities in the epidermis. Reprinted with permission from: *European Journal of Immunology* (2005) 35, 1886–95, ©2005 Wiley-Vch Verlag GmbH & Co. KGaA, Weinheim, Germany.

in arterial blood). The surface of the skin, therefore, is predominantly an aerobic environment and is not conducive to the growth or survival of obligate anaerobes. However, owing to its consumption by host cells and resident microbes, oxygen levels are reduced within hair follicles and in the depths of the stratum corneum. This results in the creation of microaerophilic and/or anaerobic environments in these regions, thereby enabling the growth of microaerophiles and obligate anaerobes. Propionibacteria, which are microaerophiles or obligate anaerobes, are major colonizers of hair follicles.

Unlike internal tissues, the pH of the skin is not maintained at a constant, approximately neutral pH. The pH of the skin is usually acidic, but the exact pH value varies widely between sites (Fig. 2.15) and is related mainly to the density of sweat glands. The acidic pH results from the presence of lactic acid (produced by host cells and microbes), acidic amino acids (from sweat), fatty acids (from sebum), and acids produced during the keratinization process (urocanic acid and pyrrolidone carboxylic acid). Despite it being an acidic environment, the pHs encountered on the skin surface are generally too high for the growth of acidophiles but are suitable for neutrophiles. The pH of the skin is too low for alkaliphiles.

The stratum corneum consists mainly of protein (75–80%) and lipids (5–15%). A wide variety of other nutrients are also present, and these are derived from glandular secretions (sweat and sebum), interstitial fluid, and microbes (Table 2.2).

Lipids are an important carbon and energy source for many cutaneous microbes, and their concentration on the skin surface varies widely as shown in Fig. 2.16.

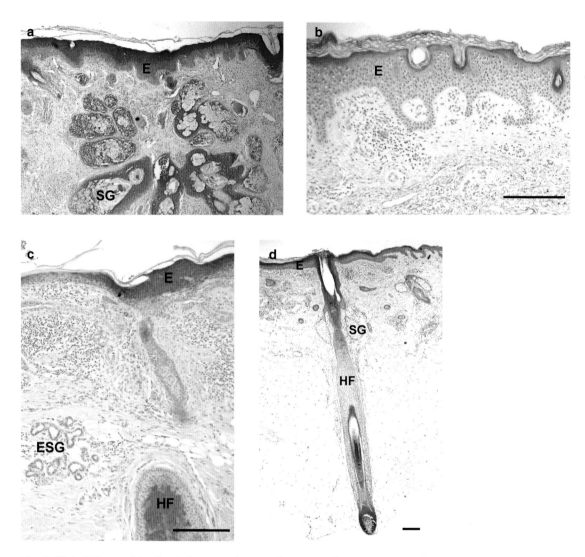

Fig. 2.12 (a–d) Immunohistochemical staining of various skin regions, showing expression of psoriasin (red staining): (a) in the suprabasal keratinocytes of the epidermis and the sebaceous glands of nose skin; (b) in the uppermost part of the epidermis of the lower leg; (c) in the uppermost part of the epidermis of the cheek; and (d) in the uppermost epidermal layers of the hair follicles and the surrounding epidermis. There is no expression in the eccrine sweat glands. Scale bars, 200 μm. SG, sebaceous glands; ESG, eccrine sweat glands; E, epidermis; HF, hair follicle. Reprinted with the permission of Macmillan Publishers Ltd from: *Nature Immunology* (2005) 6, 57–64, © 2005.

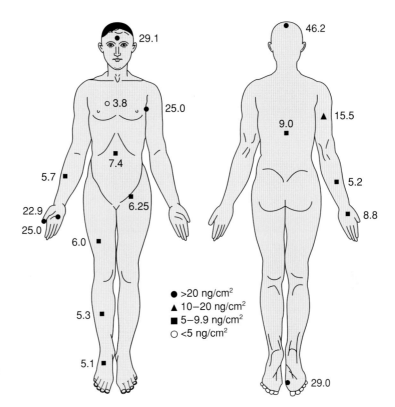

Fig. 2.13 Concentration of psoriasin present on various skin regions in eight healthy adults. Figures denote the median concentration of psoriasin/cm² for each region. Reprinted with the permission of Macmillan Publishers Ltd, London, UK, from: *Nature Immunology* (2005) 6, 57–64, © 2005.

Table 2.1 The main factors affecting microbial survival and growth on the skin.

Factor	Effect
Temperature	Allows growth of mesophiles, but prevents growth of thermophiles and psychrophiles
Low moisture content	Prevents survival or growth of many species, particularly Gram-negative bacteria
High osmolarity	Prevents survival or growth of many species, particularly Gram-negative bacteria
Low pH	Prevents survival or growth of many species
Oxygen concentration	Generally high and therefore prevents survival or growth of anaerobes. Oxygen concentration is low in hair follicles, however, thereby permitting growth of anaerobes and microaerophiles
Nutrient availability	Plentiful but consists mainly of host polymers
Interactions with other microbes	May be beneficial (see Fig. 2.41) or antagonistic (see Table 2.7)
Host defense systems	Prevent adhesion and/or survival of many types of microbes

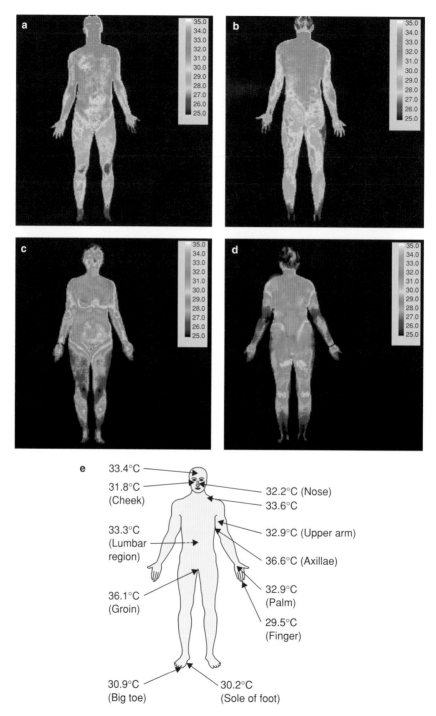

Fig. 2.14 The temperature of various regions of the skin in (a, b) adult males and (c, d) adult females. (e) Typical values of the temperature of various skin regions in adults. Images (a–d) kindly supplied by Professor Francis Ring, University of Glamorgan, Pontypridd, UK.

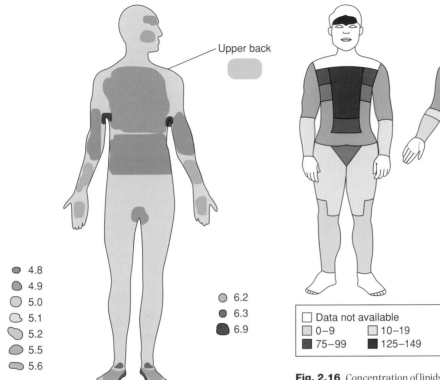

Fig. 2.15 Diagram showing the pH of the various regions of the skin.

Fig. 2.16 Concentration of lipids (in μg/cm²) on the skin surface of adults. Note that no skin region had a lipid content in the range 100–124 μg/cm².

2.4 THE INDIGENOUS MICROBIOTA OF THE SKIN

2.4.1 Members of the cutaneous microbiota

The organisms most frequently present on the skin surface belong to the genera *Corynebacterium*, *Staphylococcus*, *Propionibacterium*, *Micrococcus*, *Kocuria*, *Malassezia*, *Brevibacterium*, *Dermabacter*, *Acinetobacter*, and *Methylobacterium*.

2.4.1.1 *Corynebacterium* spp.

Corynebacterium is one of several genera of cutaneous microbes often described as being "coryneform" or "diphtheroid". The terms coryneform and diphtheroid refer to any non-acid-fast, nonbranching, non-sporing, pleomorphic, Gram-positive rod – whether aerobic or anaerobic. For many years, the aerobic coryneforms were split into two groups – those that required lipids for growth (known as "lipophilic" or "small colony" diphtheroids) and those that did not (known as "non-lipophilic" or "large colony" diphtheroids). The genus *Corynebacterium* contains members of both groups. The other coryneform genera found on the skin are *Propionibacterium*, *Brevibacterium*, *Dermabacter*, and *Turicella*. The main characteristics of *Corynebacterium* spp. can be summarized as follows:

- Pleomorphic Gram-positive bacilli;
- Produce metachromatic granules;
- Noncapsulate;
- Nonsporing;
- Nonmotile;
- G+C content of DNA is 46–74 mol%;
- Contain meso-diaminopimelic acid, short-chain (C_{22}–C_{36}) mycolic acids and dehydrogenated menaquinones with eight or nine isoprene units;
- Cell wall contains arabinogalactan;

Table 2.2 Nutrients available on the skin and their main sources (PMM, product of microbial metabolism).

Class of nutrient	Compound	Source	Microbes known to utilize or require the nutrient
Carbon/energy source	Glucose	Sweat; hydrolysis of glycoproteins	*Staphylococcus, Micrococcus, Propionibacterium, Corynebacterium, Brevibacterium*
	Ribose	Hydrolysis of nucleic acids	Some staphylococci, some micrococci, some propionibacteria, *Acinetobacter*
	Glycerol	Lipid hydrolysis	*Staphylococcus, Micrococcus, Propionibacterium*
	Amino acids	Hydrolysis of proteins; sweat	*Staphylococcus, Micrococcus,* lipophilic coryneforms, some propionibacteria, *Acinetobacter*
	Fatty acids	Sweat; hydrolysis of lipids; PMM	Some staphylococci, *Propionibacterium, Acinetobacter, Brevibacterium, Micrococcus,* lipophilic coryneforms, *Malassezia*
	Lactic acid	Sweat; PMM	*Staphylococcus, Micrococcus, Acinetobacter*
Nitrogen source	NH$_4^+$	Sweat; PMM	*Staphylococcus, Micrococcus,* aerobic coryneforms, *Malassezia, Acinetobacter*
	Amino acids	Hydrolysis of proteins; sweat	*Staphylococcus,* some micrococci, *Propionibacterium, Malassezia, Acinetobacter*
	Urea	Sweat	*Corynebacterium, Staphylococcus, Brevibacterium*
Essential amino acids	Amino acids	Hydrolysis of proteins; sweat; cerumen	*Staphylococcus, Propionibacterium, Malassezia*
Phosphorus source	Phosphate	Hydrolysis of nucleic acids; sweat	All organisms
Vitamins	Biotin	Sweat	Most staphylococci, *Micrococcus, Propionibacterium,* some aerobic coryneforms
	Thiamine	Sweat	*Staphylococcus, Micrococcus, Propionibacterium,* some aerobic coryneforms
Micronutrients	Na$^+$, K$^+$, Mg^{2+}, Cl$^-$, etc.	Sweat; dead keratinocytes; interstitial fluid	All organisms

• Catalase-positive;
• Facultative anaerobes or aerobes (*Corynebacterium jeikeium* and *Corynebacterium urealyticum*);
• Grow over a wide temperature range (15–40°C);
• Optimum growth at 37°C;
• Halotolerant (apart from *C. jeikeium*) – can grow at NaCl concentrations of up to 10%.
The genus contains 59 species, but only eight of these are regularly present on human skin. Their halotolerance enables them to colonize sites with a high salt content, e.g. occluded regions such as the axillae and toe webs. Apart from *C. urealyticum*, they are able to use carbohydrates and amino acids as sources of carbon and energy. Glucose and a range of amino acids are present in sweat, and proteases produced by the cutaneous microbiota liberate amino acids from skin proteins. Some amino acids are also essential growth requirements and are incorporated into cellular protein. Unlike the other cutaneous species, the lipophilic corynebacteria require a number of vitamins, including

riboflavin, nicotinamide, thiamine, pantothenate, and biotin. Many of these are present in sweat, while others are liberated by dying keratinocytes.

Corynebacteria adhere to human epidermal keratinocytes in vitro to the extent of between four and 25 bacteria per epithelial cell, depending on the particular species. However, little is known regarding the adhesins responsible for mediating attachment. Mannose-, galactose-, and *N*-acetylglucosamine-containing receptors are thought to be involved. Fibronectin and various lipids may also function as receptors.

Although many studies have investigated the prevalence of "*Corynebacterium* spp." on various skin regions, it is very difficult to give an overall view of their distribution and prevalence. This situation has arisen for a number of reasons including: (1) confusion over terminology – in different studies, authors refer to "coryneforms", "diphtheroids", "lipophilic coryneforms", "aerobic diphtheroids", "small-colony diph-

theroids", "aerobic corynebacteria", "corynebacteria", etc. without always defining what they mean; (2) difficulties in distinguishing between the numerous species; and (3) the many changes in nomenclature that have occurred during the past 20 years.

The distribution of lipophilic and nonlipophilic *Corynebacterium* spp. on the skin is summarized in Figs 2.17 and 2.18.

The genome of the lipophilic species, *C. jeikeium*, has been sequenced and has been shown to consist of a circular chromosome of 2 462 499 bp and a 14 323-bp bacteriocin-producing plasmid (http://cmr.tigr.org/tigr-scripts/CMR/GenomePage.cgi?org=ntcj02). The absence of a fatty acid synthase is responsible for its lipophilic phenotype. The organism is an important nosocomial pathogen and is frequently resistant to a number of antibiotics. In a recent study of the *Corynebacterium* spp. present on the back and forehead of healthy adults, *C. jeikeium* was found to be one of

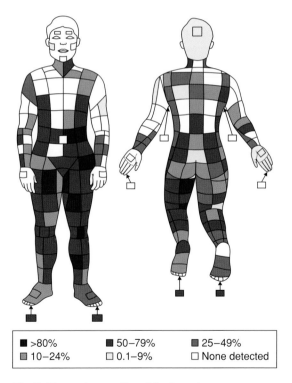

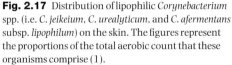

| ■ >80% | ■ 50–79% | ■ 25–49% |
| □ 10–24% | □ 0.1–9% | □ None detected |

Fig. 2.17 Distribution of lipophilic *Corynebacterium* spp. (i.e. *C. jeikeium*, *C. urealyticum*, and *C. afermentans* subsp. *lipophilum*) on the skin. The figures represent the proportions of the total aerobic count that these organisms comprise (1).

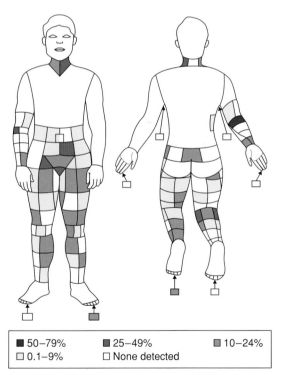

| ■ 50–79% | ■ 25–49% | ■ 10–24% |
| □ 0.1–9% | □ None detected | |

Fig. 2.18 Distribution of nonlipophilic *Corynebacterium* spp. (*C. xerosis*, *C. minutissimum*, *C. striatum*, *C. amycolatum*, and *C. afermentans* subsp. *afermentans*) on the skin. The figures represent the proportions of the total aerobic count that these organisms comprise (1).

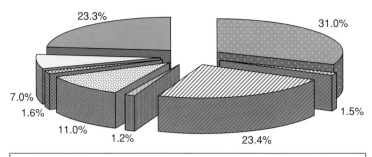

23.3%   31.0%

7.0%

1.6%   1.5%

11.0%   1.2%   23.4%

▣ *Corynebacterium jeikeium*	⊠ *Corynebacterium urealyticum*
▨ CDC Group G2	▥ CDC Group F1
▧ CDC Group G1	⊠ *Corynebacterium kroppenstedtii*
▤ *Corynebacterium afermentans* subsp. *lipophilum*	▦ Unidentified species

Fig. 2.19 Identity of *Corynebacterium* spp. isolated from the back and forehead of five healthy adults. The figures denote the proportions of the various species detected (2). CDC, Centers for Disease Control (Atlanta, Georgia, USA).

the most frequently isolated species and comprised almost one third of all isolates (Fig. 2.19).

2.4.1.2 *Propionibacterium* spp.

Propionibacteria are Gram-positive bacilli that often show bifurcations and/or branching, and their main characteristics can be summarized as follows:
- Gram-positive bacilli;
- G+C content of DNA is 57–68 mol%;
- Obligate anaerobes or microaerophiles;
- Catalase-positive (except for *Propionibacterium propionicum*);
- Growth over the pH range 4.5–8.0;
- Optimum pH for growth is 5.5–6.0;
- Nonmotile;
- Nonsporing;
- Propionic acid is a major end product of metabolism;
- Require biotin, nicotinamide, pantothenate, and thiamine for growth.

Four species are regularly found on human skin – *P. acnes*, *Propionibacterium avidum*, *Propionibacterium granulosum*, and *P. propionicum*.

Their anaerobic/microaerophilic nature enables them to colonize hair follicles, which have reduced oxygen levels. However, they are also found on the skin surface, where oxygen utilization by aerobes and facultative anaerobes provides an oxygen-depleted environment. Propionibacteria can utilize sugars, fatty acids, glycerol, amino acids, and RNA as carbon and energy sources. The main end products of glucose metabolism are propionic and acetic acids. Propionibacteria produce a range of proteases and lipases which enable

them to utilize host proteins and lipids on the skin surface. Propionibacteria require biotin, nicotinamide, pantothenate, and thiamine for growth. Biotin and thiamine are present in sweat, as are several other vitamins. Owing to their production of propionic acid, bacteriolytic enzymes, and bacteriocins, propionibacteria can inhibit the growth of a number of organisms.

Adhesion of *P. acnes* to human epidermal keratinocytes has been studied in vitro, and the number of bacteria adhering to each keratinocyte ranges from 0.1 to 5 depending on the particular strain. Unfortunately, little is known regarding the adhesins or receptors involved. The organism binds to fibronectin and oleic acid. Oleic acid is a constituent of sebum, and this fatty acid also promotes co-aggregation. Such co-aggregation, and the association of the resulting microcolony with oleic acid in a hair follicle, would help to maintain the organism within this habitat. Microcolony formation would also help to establish the microaerophilic environment needed by the organism.

P. acnes is the most prevalent member of the genus on human skin (Fig. 2.20) and inhabits predominantly the hair follicles. It accounts for approximately half of the total skin microbiota and, in most individuals, it outnumbers coagulase-negative staphylococci (CNS) by a factor of between ten and 100. However, as for any cutaneous organism, the population density at any site varies markedly between individuals (Fig. 2.21).

Two distinct phenotypes of *P. acnes*, known as types I and II, have been recognized for many years on the basis of serological agglutination tests and cell-wall sugar analysis. Recently, *recA*-based sequence analysis has revealed that *P. acnes* types I and II

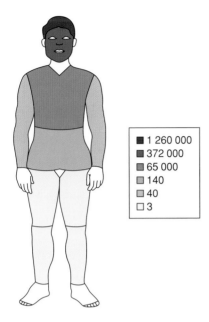

represent phylogenetically distinct groups and that type I strains can be further divided into two groups (1A and 1B). The 2.56-Mbp genome of *P. acnes* has been sequenced (http://cmr.tigr.org/tigr-scripts/CMR/GenomePage.cgi?org=ntpa02) and found to contain 2297 protein-coding genes, of which 66% have been assigned a role category. *P. acnes* is an opportunistic pathogen and plays an important role in acne, the most common skin infection of humans.

P. granulosum prefers a similar habitat to *P. acnes*, whereas *P. avidum* frequents regions that are rich in eccrine sweat glands, and *P. propionicum* is found mainly on the eyelids (Fig. 2.22).

2.4.1.3 *Staphylococcus* spp.

Staphylococci are Gram-positive cocci which occur singly, in pairs, or in clusters. Approximately half of the 35 species comprising the genus can be found on human skin. Three members of the genus (*Staph. aureus*, *Staphylococcus intermedius*, and *Staphylococcus delphni*) produce coagulase, an enzyme that converts fibrinogen to fibrin. Those that do not produce this enzyme, the CNS, are major inhabitants of the skin.

Fig. 2.20 Concentration of *P. acnes* at various sites on the skin of healthy adults. The figures denote the number of colony forming units (cfu) per cm² (3).

Legend:
- 1 260 000
- 372 000
- 65 000
- 140
- 40
- 3

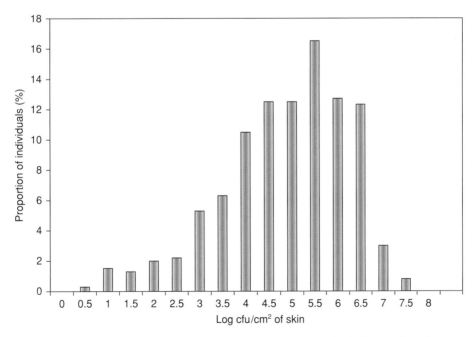

Fig. 2.21 Density of colonization of *Propionibacterium* spp. on the cheeks of 761 individuals, showing the wide variation between subjects (4).

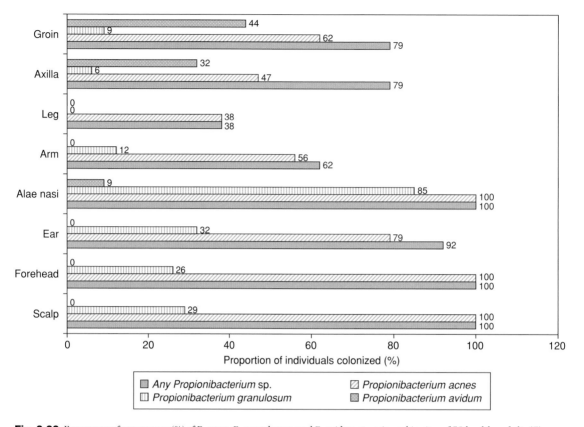

Fig. 2.22 Frequency of occurrence (%) of *P. acnes*, *P. granulosum*, and *P. avidum* at various skin sites of 50 healthy adults (5).

Staph. aureus, on the other hand, is a member of the respiratory microbiota, its primary habitat being the anterior nares (see Chapter 4). The main characteristics of CNS are as follows:
- Gram-positive cocci;
- Nonmotile;
- Nonsporing;
- G+C content of DNA is 30–39 mol%;
- Facultative anaerobes;
- Catalase-positive;
- Oxidase-negative (most species);
- Ferment a number of sugars to produce mainly lactic acid;
- Growth over the temperature range 10–45°C;
- Optimum temperature for growth is 30–37°C;
- Growth over pH range 4.0–9.0;
- Optimum pH for growth is 7.0–7.5;
- Halotolerant – all species grow at 10% NaCl, and many grow at 15% NaCl;

- Growth at low water activities (approximately 0.85);
- Production of a variety of hydrolytic enzymes – proteases, lipases, keratinases, and nucleases;
- Many species hydrolyze urea.

A number of factors contribute to their success as colonizers of human skin: their halotolerance, their ability to grow at low water activities and over a wide pH range, their capacity to degrade host polymers, and their ability to grow aerobically and anaerobically. Although CNS can be isolated from almost any skin region, individual species generally show a preference for a particular site. The results of one study that investigated the distribution of CNS on the skin is shown in Fig. 2.23. Unfortunately, when this study was carried out (in the early 1970s), only three species in the genus *Staphylococcus* were recognized – *Staph. aureus*, *Staphylococcus epidermidis*, and *Staphylococcus saprophyticus*. Since then, additional CNS have been identified, and *Staph. epidermidis* and *Staph. saprophyticus*

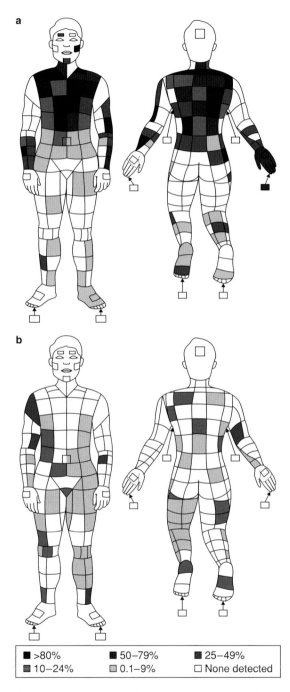

a

b

Legend		
■ >80%	■ 50–79%	■ 25–49%
■ 10–24%	▨ 0.1–9%	□ None detected

Fig. 2.23 Distribution on the skin of (a) *Staph. epidermidis* and (b) *Staph. saprophyticus* (see notes on the nomenclature of CNS in section 2.4.1.3). The figures represent the proportions of the total aerobic count that these organisms comprise (1).

have been further sub-divided, so that now at least 35 species of CNS are recognized.

The frequency of detection of a number of CNS species at different skin sites is shown in Fig. 2.24.

Although staphylococci possess a wide range of adhesins (Table 2.3), little is known regarding which are involved in the adhesion of these organisms to host cells in vivo.

The genomes of a number of CNS have been sequenced, including that of *Staph. epidermidis* (http://cmr.tigr.org/tigr-scripts/CMR/GenomePage.cgi?org=gse). CNS are important opportunistic pathogens and are responsible for most medical device-related infections. They can also cause infective endocarditis and urinary tract infections.

2.4.1.4 *Micrococcus* spp.

The genus *Micrococcus* originally contained ten species, but it has now been split into six genera: *Micrococcus* (containing *M. luteus*, *M. lylae*, and *M. antarcticus*), *Kocuria* (containing the former *M. roseus*, *M. varians*, and *M. kristinae*), *Kytococcus* (containing the former *M. sedentarius*), *Nesterenkonia* (containing the former *M. halobius*), *Dermacoccus* (containing the former *M. nishinomiyaensis*), and *Arthrobacter* (containing the former *M. agilis*). Of these ten species, *M. luteus* and *Kocuria varians* are the most frequently encountered on human skin (Fig. 2.25).

The main characteristics of *Micrococcus* spp. are as follows:
- Gram-positive cocci, usually in tetrads or clusters;
- G+C content of their DNA is 66–75 mol%;
- Obligate aerobes;
- Catalase-positive;
- Oxidase-positive;
- Nonmotile;
- Nonsporing;
- Halotolerant (up to 7.5% NaCl);
- Growth over the temperature range 25–37°C;
- Optimum growth at 37°C;
- Most species need amino acids as growth factors.

Micrococci are halotolerant, which contributes to their ability to survive on skin. They can use carbohydrates and amino acids as carbon and energy sources, and most strains produce proteases and keratinases and so can obtain amino acids from skin polymers. Other carbon sources utilized include glycerol, propionate, lactate, and acetate – all of which are metabolic end products of other organisms. Most micrococci require some amino acids as growth factors – commonly

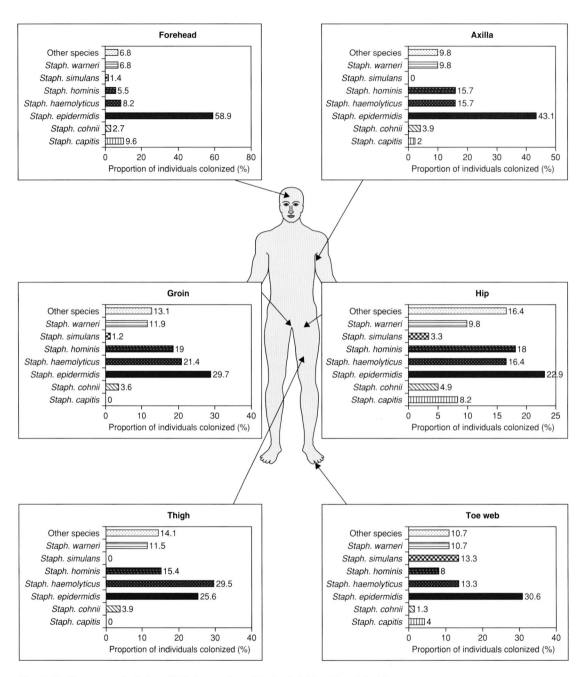

Fig. 2.24 Frequency of isolation of CNS from various skin sites in 16 healthy adults (6).

Table 2.3 Adhesins of *Staph. epidermidis* and their receptors.

Adhesin	Receptor
Teichoic acid	Fibronectin
Serine–aspartate dipeptide repeats protein G (SdrG)	Fibrinogen
Autolysin (AtlE)	Vitronectin
Fibrinogen-binding protein (Fbe)	Fibrinogen
Accumulation-associated protein (AAP)	Other *Staph. epidermidis* cells
Polysaccharide intercellular adhesion (PIA)	Other *Staph. epidermidis* cells; red blood cells
Lipase (GehD)	Collagen

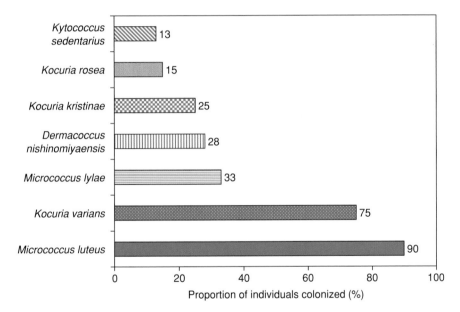

Fig. 2.25 Frequency of isolation of *Micrococcus* spp. and related organisms from various skin sites (including the forehead, cheek, chin, axilla, arm, and leg) of 19 healthy adults. The figures represent the proportion of individuals colonized by the organism at one or more of the sites surveyed (7).

arginine, cysteine, methionine, and tyrosine. Vitamin requirements vary with the species. *M. luteus* is the most frequently detected *Micrococcus* sp. on the skin and can be isolated from the forehead, legs, and arms of 60%, 70%, and 70% of individuals, respectively. It may comprise high proportions of the communities of these sites. It is less frequently found in the axillae. *M. lylae* is also frequently present on the forehead, arms, and legs.

The genus *Kocuria* belongs to the family *Micrococcaceae* and consists of nonmotile, nonsporing, aerobic Gram-positive cocci. They are catalase-positive, and

the G+C content of their DNA is 66–75 mol%. Two out of the five species belonging to this genus, *Koc. varians* and *Kocuria kristinae*, are frequently present on the forehead, legs, and arms. *Kocuria rosea* has also been detected on human skin and produces a keratinase.

2.4.1.5 *Malassezia* spp.

Malassezia spp. are lipophilic, dimorphic yeasts whose usual habitat is mammalian skin. The genus consists of 11 species, of which the following ten have been found

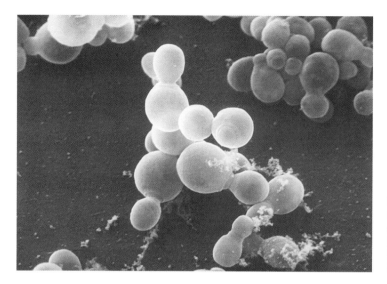

Fig. 2.26 Scanning electron micrograph of *Mal. furfur*. (Image courtesy of Janice Carr, Public Health Image Library, Centers for Disease Control and Prevention, Atlanta, Georgia, USA).

on the skin of healthy humans – *Mal. furfur, Mal. pachydermatis, Mal. sympodialis, Mal. globosa, Mal. slooffiae, Mal. restricta, Mal. obtuse, Mal. dermatis, Mal. japonica,* and *Mal. yamatoensis.* On healthy skin, *Malassezia* spp. are usually present in the yeast form, which consists of round, oval, or cylindrical cells (Fig. 2.26). In contrast, the pseudomycelial form is often present in diseased skin and can also be seen in culture. The organisms are difficult to grow, and this has hindered studies of their physiology.

Although they are usually cultured aerobically, *Malassezia* spp. can also grow under microaerophilic and anaerobic conditions. They can, therefore, colonize any region of the skin, provided that other conditions are suitable. They cannot ferment sugars, but can use lipids as the sole source of carbon and energy. All species, except *Mal. pachydermatis,* require preformed fatty acids (with a carbon-chain length greater than 10) for growth – they are incorporated directly into cellular lipids. These compounds are present in sebum and are also produced by the action of lipases on lipids present on the skin surface. The fatty acid and lipid requirements of these yeasts would explain, in part, their preferential colonization of skin regions that are rich in sebaceous glands. Ammonium ions and amino acids, both of which are present in sweat, are used as nitrogen sources. Ammonium ions are also available from the urea present in sweat by the action of ureases produced by cutaneous microbes such as *Corynebacterium* spp., staphylococci, and *Brevibacterium* spp. Amino

acids can be obtained from skin proteins by the action of proteases produced by *Malassezia* spp. and other cutaneous microbes. Methionine is the preferred source of sulfur for these organisms, although they can also utilize cysteine – both of these amino acids are present in sweat as well as in protein hydrolysates. The organisms do not appear to require any vitamins for growth.

Malassezia spp. are found on the skin of 75–98% of the population and in more than 90% of adults. Although they may be found at most body sites, they are present in higher numbers on the face, scalp, chest, and upper back because of the numerous sebaceous glands in these regions – from which they obtain the lipids that are essential for their growth (Fig. 2.27). Healthy individuals are usually colonized by between one and three *Malassezia* spp. – the most common being *Mal. globosa* and *Mal. restricta* (Fig. 2.28). In some studies, the proportions of *Mal. restricta* have been underestimated as the organism is difficult to grow.

Malassezia spp. are associated with a number of skin diseases, including pityriasis versicolor, seborrhoeic dermatitis, folliculitis, and atopic dermatitis.

2.4.1.6 *Acinetobacter* spp.

Acinetobacter spp. are aerobic Gram-negative coccobacilli occurring in pairs or small clusters, and their main characteristics are as follows:

• Gram-negative coccobacilli;
• Nonmotile;

a

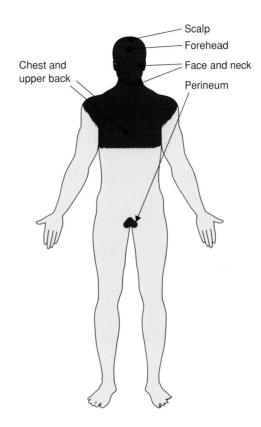

Scalp
Forehead
Chest and
upper back
Face and neck
Perineum

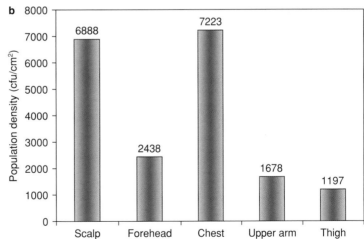

Fig. 2.27 (a) Skin sites harboring the highest proportions of *Malassezia* spp. (b) Population density of *Malassezia* spp. at various skin sites in 80 healthy adults (8).

- Obligate aerobes;
- Catalase-positive;
- Oxidase-negative;
- G+C content of their DNA is 39–47 mol%;

- Growth over the temperature range 20–42°C;
- Optimum temperature for growth is 33–35°C;
- Unable to metabolize glucose;
- Some species are halotolerant.

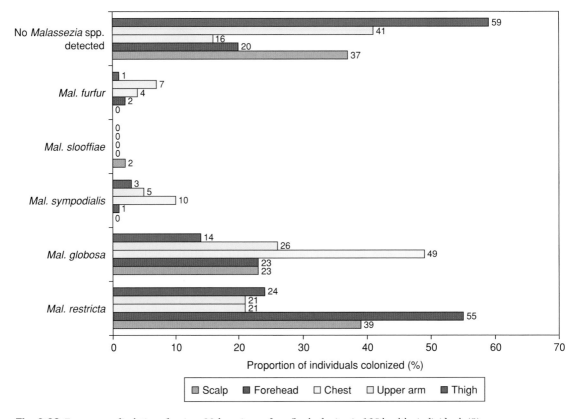

Fig. 2.28 Frequency of isolation of various *Malassezia* spp. from five body sites in 120 healthy individuals (8).

At least 19 species are recognized, of which *Acinetobacter lwoffii* is usually the most frequently detected on skin. Other species often found are genospecies 15BJ, *Acin. radioresistens*, and genospecies 3. Although they can utilize a wide range of organic compounds as carbon and energy sources, glucose cannot be metabolized. Organic acids (e.g. acetate, lactate, and pyruvate), amino acids, and alcohols appear to be their main carbon and energy sources. Lactate and pyruvate are present in sweat, while acetate and lactate are the metabolic end products of many cutaneous microbes. They can grow on ammonium ions as the sole nitrogen source, and these are plentiful in the cutaneous environment. They do not require any vitamins for growth. Their salt tolerance varies, with many strains being able to grow in 6% NaCl.

Because of their strict requirement for oxygen, they are found mainly on the skin surface – colonization of oxygen-depleted regions such as the deeper regions

of hair follicles is restricted. They are found on the skin of approximately 45% of the population and can be detected at numerous sites (Fig. 2.29). Several studies have shown that the isolation frequencies of *Acinetobacter* spp. are greater in summer than in winter, indicating the preference of these organisms for a moist environment. *Acin. baumannii* is an important, antibiotic-resistant nosocomial pathogen, but is rarely detected on the skin of nonhospitalized individuals.

2.4.1.7 *Brevibacterium* spp.

Brevibacterium spp. are aerobic Gram-positive bacilli, and their main characteristics are as follows:
- Gram-positive bacilli;
- Aerobic;
- Nonmotile;
- Catalase-positive;

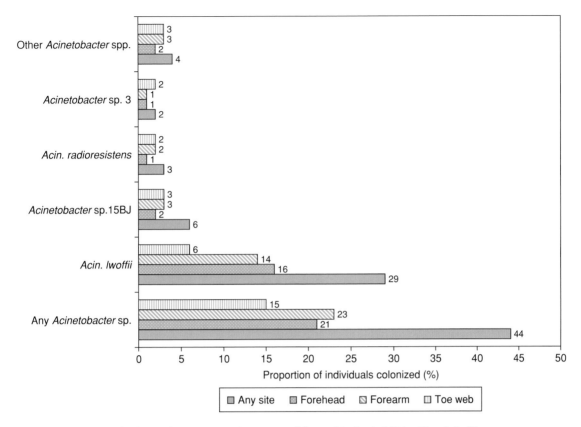

Fig. 2.29 Frequency of isolation of various *Acinetobacter* spp. at different skin sites in 192 healthy adults (9).

• Oxidative;
• Tolerate high salt concentrations (up to 15%);
• Optimum temperature for growth is 30–37°C;
• G+C content of their DNA is 64 mol%.

Four out of the seven species in the genus are regularly present on skin – *Brev. epidermidis*, *Brev. otitidis*, *Brev. mcbrellneri*, and *Brev. casei*. Because of their requirement for oxygen, they are found mainly on the skin surface rather than in hair follicles. Glucose, acetate, lactate, and amino acids can all be used as carbon and energy sources. Glucose and amino acids are present in sweat, while acetate and lactate are produced by many cutaneous microbes. All four cutaneous species produce a range of proteases including keratinases. Little is known about the prevalence of these organisms on human skin, although the preferred sites of colonization of *Brev. epidermidis* appear to be the perineum and the toe interspace (Fig. 2.30).

2.4.1.8 *Dermabacter hominis*

Dermabacter hominis is the sole member of the genus *Dermabacter*. It is a nonmotile, catalase-positive, oxidase-negative Gram-positive bacillus, and its DNA has a G+C content of 62 mol%. It ferments a range of sugars to produce mainly acetate and lactate. It produces a range of hydrolytic enzymes including proteases, nucleases, and an amylase, and the action of these enzymes on skin macromolecules would provide a range of potential nutrients for itself and other organisms.

2.4.1.9 *Methylobacterium* spp.

The main characteristics of *Methylobacterium* spp. are as follows:
• Gram-negative bacilli;
• Aerobic;

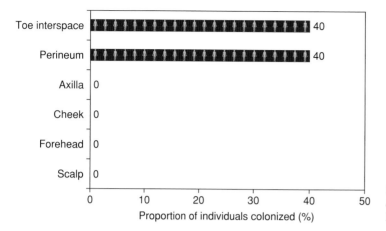

Fig. 2.30 Frequency of isolation of *Brev. epidermidis* from various skin sites on 30 healthy adult males (10).

- Nonmotile;
- Nonsporing;
- Oxidase-positive;
- Catalase-positive;
- Growth over the range 12–43°C;
- Optimum growth at 37°C;
- Grow over pH range 6–9;
- Optimum growth at pH 7.3.

Methylobacterium spp. can use methanol and methylamine as well as C_2, C_3, and C_4 compounds as carbon and energy sources. Colonies generally have a pink color and are often covered by a thin film. Species detected on the human foot include *Met. podarium*, *Met. extorquens*, *Met. rhodinum*, *Met. thiocyanatum*, *Met. zatmanii*, and *Methylobacterium* strain G296-15. They have also been detected in the oral cavity.

2.4.2 Community composition at different sites

Most of our knowledge of the composition of the microbial communities inhabiting the various regions of the skin is based on the results of culture-dependent studies and, therefore, is very incomplete. Very few culture-independent studies have been published, and at least one of these (the results of which are summarized in section 2.4.3) has reported that only 16% of the phylotypes detected on the forearm could be isolated from this site by standard cultivation techniques.

For reasons summarized in Fig. 2.9 and below, the skin is a relatively inhospitable environment for microbes. The characteristic features of the cutaneous environment are as follows:
- Relatively dry with little free water;
- High osmolarity;
- Low pH;
- Exposed to external environment;
- Subject to fluctuations in temperature, ion concentrations, osmolarity, and water content;
- Subjected to radiation and mechanical stress;
- A range of antimicrobial host defense mechanisms operates;
- Outer layers are continually being shed.

As has been described above in section 2.3, the skin is not a homogeneous system and, because of its particular anatomy and physiology, each of the many sites available on its surface provides a unique environment for microbial colonization. Some sites, such as the soles of the feet, axillae, and perineum are densely populated, while others, such as the forearm and leg, have sparser microbial communities. Such variations can be appreciated by considering the overall distribution of aerobic and facultative anaerobes on the skin surface, as shown in Fig. 2.31.

Furthermore, as well as displaying region-associated environmental variations on its surface, the hair follicles in the dermis (Fig. 2.1) offer yet another set of environmental conditions. There are, therefore, two main types of habitat available for microbial colonization – the external environment-contacting surface of the skin and, in the deeper layers of the skin, the hair follicles. Most organisms on the surface of the skin appear to be present as microcolonies, and these are usually no

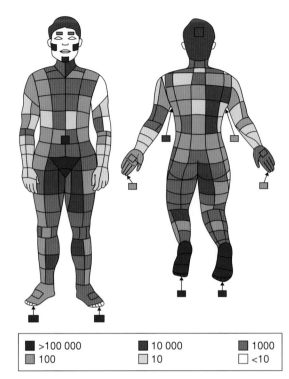

| ■ >100 000 | ■ 10 000 | ■ 1000 |
| ■ 100 | □ 10 | □ <10 |

Fig. 2.31 Distribution of aerobes and facultative anaerobes on the skin of a healthy adult male (1). Population density is shown in cfu/cm².

more than two cells thick. The number of cells in such colonies varies with the site and with the individual, and ranges from less than ten to 10^5 (Fig. 2.32). Not all skin squames are colonized by microbes, and less than 50% of follicles contain microbes. The sudoriferous glands appear not to be colonized by microbes.

Because of the wide variety of environments available for skin-colonizing microbes, it is difficult to generalize about the various microbial communities that constitute the "cutaneous microbiota". Nevertheless, some broad generalizations are worth making. Because Gram-positive species are better able to tolerate the conditions in the cutaneous environment, the various communities found on and in the skin are dominated by these organisms – Gram-negative species being found in appreciable numbers only at a limited number of sites. Of the Gram-positive species, *P. acnes* is generally present in the greatest proportions and accounts for approximately half of the cutaneous microbiota of most individuals.

Among the most frequently isolated bacteria from both the skin surface and the hair follicles are CNS. The most common isolate is *Staph. epidermidis*, which usually constitutes approximately 50% of the staphylococci present. However, at least 18 different species of CNS have been detected on the skin, the most frequently isolated species being *Staphylococcus hominis*, *Staphylococcus haemolyticus*, *Staphylococcus capitis*, and *Staphylococcus warneri*. The distribution of the various CNS on the skin, together with their relative proportions at various sites, are shown in Figs 2.24 and 2.33.

Staph. aureus cannot grow and replicate on the skin and is not regularly isolated from most skin regions – it is therefore regarded as a transient. Aerobic coryneforms (i.e. *Corynebacterium* spp. and *Brevibacterium* spp.) are major inhabitants of the skin surface and are particularly prevalent in moist intertriginous regions. Propionibacteria are found mainly in hair follicles, but are also present on the skin surface. They are the dominant organisms in communities inhabiting sebum-rich areas – e.g. on the head, chest, and back. Micrococci, particularly *M. luteus*, are found mainly on the skin surface. With the exception of *Acinetobacter* spp., Gram-negative bacteria do not generally comprise high proportions of most of the microbial communities found on the skin. As well as a variety of bacteria, fungi belonging to the genus *Malassezia* are also members of the cutaneous microbiota.

Although the microbiota at a particular site is dependent on a variety of environmental factors, its composition appears to be dictated mainly by the number and density of sebaceous and sudoriferous glands at that site (Figs 2.6 to 2.8). The reasons for this are as follows:
• They are important sources of a range of microbial nutrients.
• Sudoriferous glands are a major source of free water.
• Glandular secretions contain antimicrobial substances, each with a particular antimicrobial spectrum.
• Temperature regulation can result in the production of large quantities of sweat, which has a dramatic effect on the moisture content, pH, and nutrient concentration of the site.
• Sebum produced by sebaceous glands contributes to the "cement" between squames and therefore is involved in controlling water loss and hence water availability at the site.
Consequently, variations in the number and density of these glands affect many of the key environmental determinants at the skin surface, including temperature, water content, range and concentration of

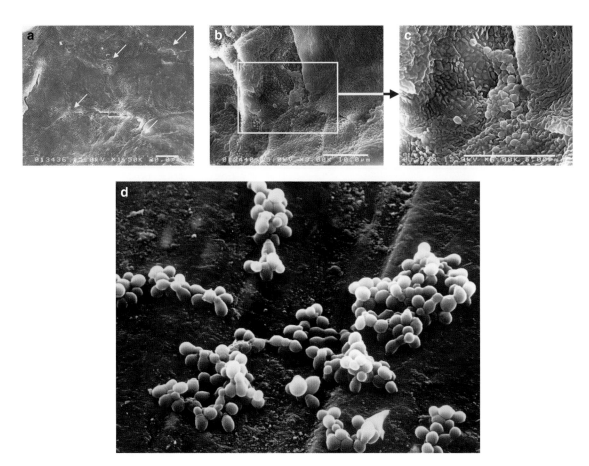

Fig. 2.32 Scanning electron micrograph showing microcolonies of bacteria on (a–c) the surface of the skin of the arm (scale bar, 10 μm) and (d) on a human foot. Microcolonies are denoted by arrows in (a) and are shown at greater magnification in (b) and (c). (a–c) Reprinted with the permission of The Japanese Society for Investigative Dermatology (© 2005) from: Katsuyama, M., Kobayashi, Y., Ichikawa, H., Mizuno, A., Miyachi, Y., Matsunaga, K. and Kawashima, M. (2005) A novel method to control the balance of skin microflora, Part 2. A study to assess the effect of a cream containing farnesol and xylitol on atopic dry skin. *J Dermatol Sci* 38, 207–13. (d) Reprinted with the permission of Blackwell Publishing Ltd, Oxford, UK, from: Malcolm, S.A. and Hughes, T.C. (1980) The demonstration of bacteria on and within the stratum corneum using scanning electron microscopy. *Br J Dermatol* 102, 267–75.

nutrients, osmolarity, pH, and the range and concentration of antimicrobial substances. Another major determinant of the composition of the microbiota of a site is whether or not the site is occluded (i.e. "covered") in some way – either as a result of the anatomy of the site (e.g. axillae, sub-mammary regions) or due to clothing. Occlusion hinders the evaporation of water, encourages the accumulation of secretions, and alters pH, etc. – it therefore has a major influence on microbial community composition. Some broad patterns

emerge that are related to these factors, and these are summarized in Table 2.4.

Hair follicles appear to be colonized by three main groups of organisms – propionibacteria, staphylococci, and *Malassezia* spp. – their relative proportions being 2500 : 50 : 1, respectively. In general, the propionibacteria tend to be localized in a narrow band within the follicle, whereas the staphylococci are more broadly distributed throughout the follicle. The propionibacterial "band" occurs at different depths within different follicles.

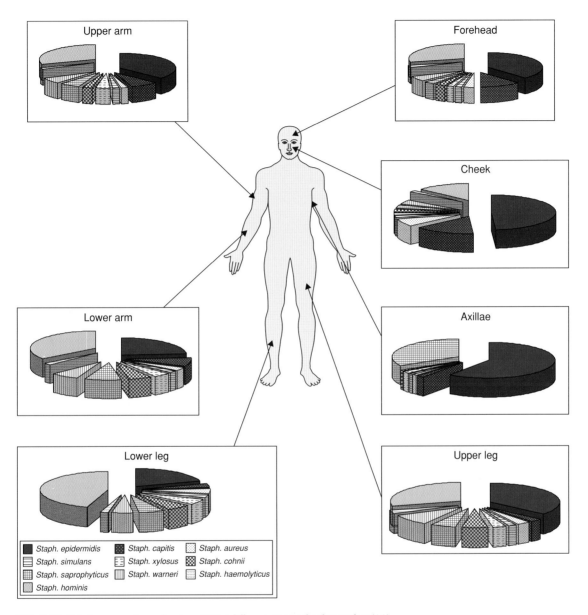

Fig. 2.33 Relative proportions of various CNS at different sites on the skin surface (11).

Because of the paucity of data, it is not possible to describe the microbiota of each of the many different regions of the skin. Those described below include regions from each of the main types of cutaneous environments present in humans – oily (head, neck, upper back, and trunk), moist (axillae, groin, perineum, and toe interspace), and dry (arms and legs). Unfortunately, few studies have involved a qualitative and quantitative analysis of all of the microbes present at a particular skin site – many studies have determined only the prevalence of a particular organism (or group of organisms) in a population. Furthermore, the absence

Table 2.4 General patterns of colonization by cutaneous microbes related to glandular distribution, moisture content, temperature, and occlusion of the region.

Region	Important environmental determinants	Effect on microbiota
Head	Many sebaceous and sudoriferous glands	High population density; dominated by propionibacteria; few corynebacteria
Axillae	Many sebaceous and sudoriferous glands; partially occluded therefore increased moisture, temperature, and pH	High microbial density; higher numbers of moisture-requiring corynebacteria, fungi, and *Acinetobacter* spp.
Perineum	Occluded therefore increased moisture and temperature	High microbial density; higher numbers of moisture-requiring corynebacteria, fungi, and *Acinetobacter* spp.
Toe webs	Occluded therefore increased moisture and temperature	High microbial density; higher numbers of moisture-requiring corynebacteria, fungi, brevibacteria, and *Acinetobacter* spp.
Arms and legs	Few sebaceous and eccrine glands; no apocrine glands; relatively dry regions	Low microbial density; mainly staphylococci and micrococci; few fungi
Hands	No sebaceous glands; exposed area therefore low water content	Mainly staphylococci; few fungi, corynebacteria, or propionibacteria

of a particular organism does not necessarily mean that the organism was not present at that site; it is more likely that it was not looked for in the investigations that were used to compile the figure. An additional complication is the enormous variation that exists between individuals with respect to the concentration of any species at a particular site (see Figs 2.21 and 2.34). The data shown in Fig. 2.35 and in Table 2.5 must be regarded, therefore, as indicating general trends rather than absolute values.

When attempting to define the microbiota of a particular site, it must also be remembered that the skin is breached by openings (e.g. mouth, nose, rectum, urethra, etc.) into anatomical regions with their own microbiotas. Such openings enable a variety of organisms indigenous to other regions to gain access to the skin surface. Many of these microbes cannot grow or reproduce on the skin surface because of the very different environmental factors operating there and so do not become members of the cutaneous microbiota.

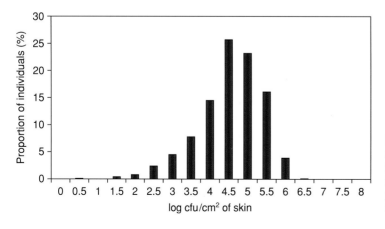

Fig. 2.34 Density of colonization of *Micrococcaceae* on the cheek of 761 individuals. The data demonstrate that the population density of organisms at a particular site can vary widely between individuals (4).

Table 2.5 Important environmental features of various regions of the skin together with details of the microbial communities present.

Region	Main environmental features	General composition of microbiota	Main species present
Scalp	(1) High density of sebaceous glands; (2) eccrine glands present; (3) abundant hair covering – traps moisture and a layer of air providing a warm and moist environment	(1) Dominated by propionibacteria and staphylococci; (2) lower, but substantial, proportion of coryneforms; (3) lower, but substantial, proportion of *Malassezia* spp.	*P. acnes, P. granulosum*; *Staph. capitis, Corynebacterium minutissimum*
Forehead	(1) High densities of sebaceous and eccrine glands; (2) very acidic; (3) exposed region, therefore variable temperature	(1) Dominated by propionibacteria; (2) lower, but substantial, proportion of staphylococci; (3) low proportions of coryneforms and *Malassezia* spp.	*P. acnes, P. granulosum*; *Staph. capitis, Staph. epidermidis*, and *Staph. hominis*; *Micrococcus luteus* and *Micrococcus lylae* frequently present; *Koc. varians* frequently present; cutaneous lipophillic corynebacteria (CLC) group are the most frequently isolated coryneforms, notably *C. minutissimum*
Toe interspace	(1) Occluded region, therefore relatively high moisture content, temperature, and pH; (2) absence of sebaceous glands, hair follicles, and apocrine glands; (3) eccrine glands present	(1) Dominated by aerobic coryneforms; (2) lower, but substantial, proportion of staphylococci; (3) low proportions of anaerobic coryneforms, micrococci, Gram-negative bacteria, *Malassezia* spp.	CLC group, *C. minutissimum*, and *Brev. epidermidis*; *Staph. epidermidis, Staph. haemolyticus, Staphylococcus simulans*
Axillae	(1) Occluded regions; (2) high densities of sebaceous, apocrine, and eccrine glands	(1) Dominated by coryneforms and staphylococci; (2) lower, but substantial, proportions of propionibacteria; (3) small numbers of a variety of other organisms present, including Gram-negative bacilli and cocci	*P. acnes* and *P. avidum*; large-colony diphtheroids (*C. minutissimum, Corynebacterium xerosis, Brevibacterium* spp.); *Staph. epidermidis, Staph. saprophyticus, Staph. aureus*; micrococci; *Escherichia coli, Klebsiella* spp., *Proteus* spp., *Enterobacter* spp., *Acinetobacter* spp.
Sole of foot	(1) Absence of hair follicles, sebaceous glands, and apocrine glands; (2) high density of eccrine glands; (3) occluded region because of socks and shoes – therefore high moisture content, temperature, pH, and CO_2 concentration	(1) Dominated by staphylococci; (2) lower, but substantial, proportion of aerobic coryneforms	*Staph. epidermidis, Staph. hominis, Staph. haemolyticus, Staph. warneri*; aerobic coryneforms; micrococci
Forearm and leg	(1) Relatively dry regions; (2) few eccrine and sebaceous glands; (3) no apocrine glands	(1) Low colonization density; (2) dominated by staphylococci; (3) much lower proportions of propionibacteria and coryneforms	*Staph. epidermidis, Staph. hominis*; *P. acnes*; *M. luteus*; *Koc. varians*

Table 2.5 (Cont'd)

Region	Main environmental features	General composition of microbiota	Main species present
Perineum	(1) Occluded region; (2) numerous sebaceous, apocrine, and eccrine glands	(1) High colonization density; (2) dominated by aerobic coryneforms; (3) lower, but substantial, proportion of CNS; (4) low proportions of a variety of other organisms	Lipophilic diphtheroids; *Staph. epidermidis*, *Staph. hominis*; wide variety of bacteria present, including many intestinal organisms; *Staph. aureus* frequently present; lactobacilli (from vagina) present in females
External auditory canal	(1) Occluded region; (2) numerous sebaceous glands; (3) numerous ceruminous glands, which secrete an acidic fluid with a high content of lipids and amino acids	(1) Dominated by staphylococci; (2) lower, but substantial, proportions of *Turicella otitidis* and *Alloiococcus otitidis*; (3) low proportions of *Corynebacterium* spp., Gram-negative species, and micrococci	*Staph. auricularis*, *Staph. epidermidis*, *Staph. capitis*, *Staph. warneri*; *Turicella otitidis*; *Alloiococcus otitidis* *Corynebacterium* spp.; *P. acnes*

These organisms are known as "transients" and include *Staph. aureus* (from the anterior nares) and a variety of intestinal microbes (from the rectum). Organisms from the environment (e.g. *Bacillus* spp. and *Pseudomonas* spp.) may also be found as transients on exposed regions of the skin surface.

2.4.3 Culture-independent studies of the cutaneous microbiota

At the time of writing (mid-2007), only five culture-independent studies of the cutaneous microbiota have been published. The first of these was a study of the microbiota of the external auditory canal in 24 individuals from whom 16S rRNA gene clone libraries were prepared. Although 35 different procaryotic phylotypes were found among the 2150 clones analyzed (Fig. 2.36), 90.1% of the clones had rRNA gene sequences corresponding to *Alloiococcus otitidis*, *Corynebacterium otitidis*, *Staphylococcus auricularis*, or *Corynebacterium auris*. These findings contrast with those derived from a cultural analysis which found that the predominant bacteria in the external auditory canal are CNS (Fig. 2.35). *All. otitidis* is a Gram-positive coccus which is difficult to culture.

In a study of the microbiota of the forehead, 16S rRNA gene clone libraries were prepared from five adults, and 87.5% of the 416 clones analyzed had 16S rRNA gene sequences corresponding to *P. acnes*, *Methylophilus methylotrophus*, *Staph. epidermidis*, and *Acinetobacter johnsonii* (Fig. 2.37a). A total of 32 phylotypes were detected, of which only ten corresponded to species previously reported to have been found in the cutaneous microbiota. A culture-based analysis of the same samples reported a much higher proportion of *P. acnes* and failed to detect either *Methyl. methylotrophus* or *Acin. johnsonii* (Fig. 2.37b).

The fungi present on the skin of healthy adults have been investigated using 18S rRNA gene clone libraries (Fig. 2.38). The dominant fungal species were found to be *Mal. restricta*, *Mal. globosa*, and *Mal. sympodialis*.

The presence of *Methylobacterium* spp. on the feet has been investigated by PCR amplification of DNA using primers specific for the 16S rRNA gene of these organisms. The organisms were detected on the soles and in the toe webs of all five healthy adults investigated (Table 2.6). *Methylobacterium* spp. were also isolated from the feet of nine out of ten adults sampled.

Finally, in a study of the microbial communities residing on the forearms of six healthy adults, each was found to have a high diversity – the mean number of

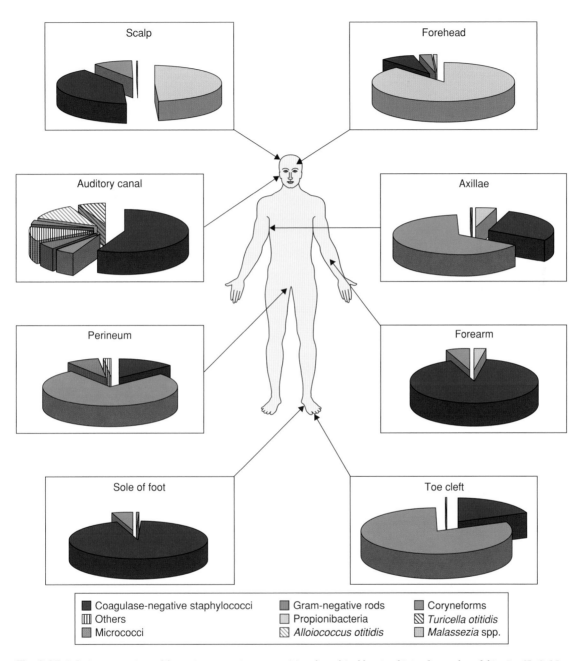

Fig. 2.35 Relative proportions of the various organisms comprising the cultivable microbiota of a number of skin sites (5–8, 10, 12–19).

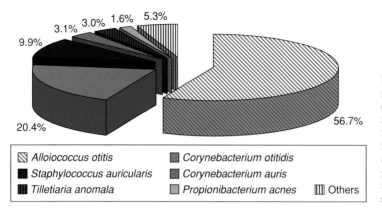

Fig. 2.36 Relative proportions of the predominant phylotypes detected in a culture-independent analysis of the microbiota of the external auditory canal in 24 individuals. DNA was extracted from samples taken from the external auditory canal, the bacterial 16S rDNA genes present were amplified by PCR, and the resulting products were cloned and sequenced (20).

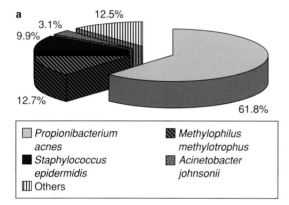

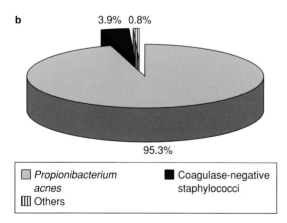

phylotypes detected in each subject was 48, of which only four (*P. acnes, Corynebacterium tuberculostearicum, Streptococcus mitis,* and *Finegoldia* AB109769) were present in all subjects (Fig. 2.39). A total of 182 phylotypes were detected, and 95% of these belonged to three phyla – *Actinobacteria, Firmicutes,* and *Proteobacteria* (Fig. 2.40a). Most (85%) of the clones had sequences corresponding to known species, but the sequences of 30 phylotypes were novel. More than half (58%) of the clones detected belonged to the genera *Propionibacterium, Corynebacterium, Staphylococcus,* and *Streptococcus* (Fig. 2.40b) and, in subjects re-investigated 8–10 months later, members of these genera were the only ones to be found in each subject on both occasions. In each individual, the microbiota of the left and right forearms had a high degree of similarity.

2.4.4 Interactions among members of the cutaneous microbiota

Although the skin is an aerobic environment, anaerobic/microaerophilic propionibacteria constitute one of the dominant groups of cutaneous microbes. Their ability to dominate the cutaneous microbiota is dependent, in part, on the presence of aerobic (micrococci, *C. jeikeium* and *C. urealyticum, Brevibacterium* spp., *Acinetobacter* spp.) and facultative (staphylococci, *D. hominis, Corynebacterium* spp., *Malassezia* spp.) organisms, which consume oxygen and create the atmospheric conditions necessary for the growth of propionibacteria. With regard to food chains and webs, a number of beneficial interactions are possible between members of the cutaneous microbiotas (Fig. 2.41).

Fig. 2.37 (a) Relative proportions of the predominant phylotypes detected in a culture-independent analysis of the microbiota of the foreheads of five adults. DNA was extracted from samples taken from the foreheads, the bacterial 16S rRNA genes present were amplified by PCR, and the resulting products were cloned and sequenced. (b) The results of a culture-based analysis of the same samples (21).

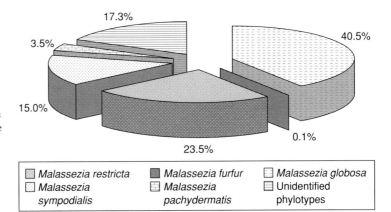

Fig. 2.38 Culture-independent analysis of the fungi present on the forearms of five healthy adults. DNA was extracted from samples taken from the forearms, the fungal 18S rRNA genes present were amplified by PCR, and the resulting products were cloned and sequenced (22).

Table 2.6 *Methylobacterium* spp. detected on the feet of healthy adults (23).

Species detected by a culture-independent technique	Species detected by culture
Met. extorquens	*Met. extorquens*
Met. rhodinum	*Methylobacterium* strain G296-15
Met. thiocyanatum	*Met. podarium*
Met. zatmanii	—
Methylobacterium strain G296-15	—

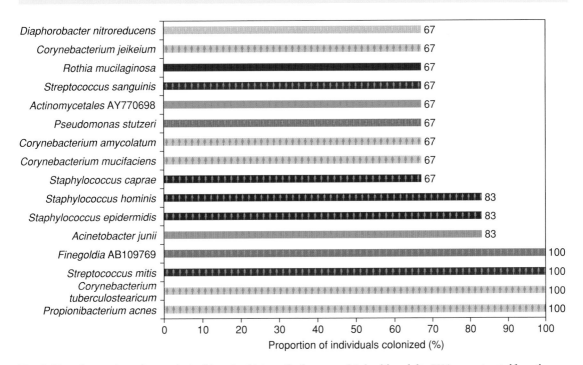

Fig. 2.39 Culture-independent analysis of the microbiota on the forearms of six healthy adults. DNA was extracted from the samples, the bacterial 16S rDNA genes present were amplified by PCR, and the resulting products were cloned and sequenced. The bars denote the frequency of detection of those phylotypes present in at least four of the six individuals (24).

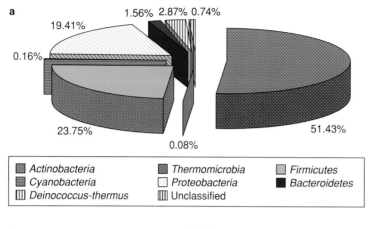

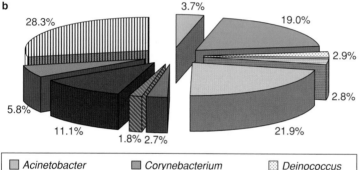

Fig. 2.40 Culture-independent analysis of the microbiota on the forearms of six healthy adults. DNA was extracted from the samples, the bacterial 16S rRNA genes present were amplified by PCR, and the resulting products were cloned and sequenced. A total of 1221 clones were analyzed. (a) Relative proportions of clones from the various phyla detected. (b) Relative proportions of clones from the most frequently detected genera (24).

In the competition for space and nutrients that takes place between microbes trying to establish themselves in a particular habitat, a number of strategies have been developed to enable one microbe to out-compete others. These include: (1) the production of antimicrobial agents; (2) interference with adhesion mechanisms; (3) alteration of the environment; and (4) depletion of essential nutrients. Unfortunately, only the first of these has been studied in any detail with regard to the cutaneous microbiota, and examples of antagonistic substances produced by cutaneous microbes are given in Table 2.7.

2.5 OVERVIEW OF THE CUTANEOUS MICROBIOTA

The skin, in general, provides a dry, acidic environment in which the main nutrients available to microbes are lipids and proteins. The main sites of microbial colonization are the skin surface and hair follicles. The skin is colonized by a number of microbial communities, each with a distinct, site-dependent composition that is strongly influenced by the density of sebaceous and sudoriferous glands at the particular site.

Most of our knowledge of the composition of the various cutaneous communities has been derived from culture-based studies – very few culture-independent investigations have been carried out. Unfortunately, many studies have focused on detecting the presence or absence of species belonging to a particular genus (e.g. staphylococci, corynebacteria, etc.) at a site rather than attempting to identify all of the species present. Gram-positive species (belonging to one or more of the genera *Propionibacterium*, *Staphylococcus*, and *Corynebacterium*) are usually the numerically dominant organisms at any site. In general, propionibacteria

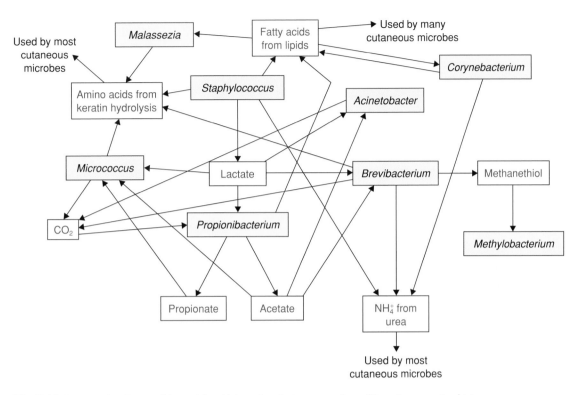

Fig. 2.41 Diagram showing possible nutritional interactions between members of the cutaneous microbiota.

Table 2.7 Antagonistic substances produced by cutaneous microbes and their possible effects in vivo.

Antimicrobial compound	Examples
Carbon dioxide	Produced by many bacteria – can inhibit growth of dermatophytes
Lysozyme	Produced by staphylococci – kills micrococci, *Brevibacterium* spp., and *Corynebacterium* spp.
Proteases	Produced by *P. acnes* – kill other *Propionibacterium* spp. and some staphylococci
Propionic acid	Produced by propionibacteria – inhibits many other species, particularly at the low pHs found on skin surface
Acetic acid	Produced by propionibacteria and *Dermabacter hominis* – inhibits many other species, particularly at the low pHs found on skin surface
Lactic acid	Produced by staphylococci and *D. hominis* – inhibits many other species
Fatty acids liberated from lipids	Lipases are produced by many skin organisms – the fatty acids generated kill streptococci and Gram-negative species
Bacteriocins	Produced by staphylococci, *Corynebacterium* spp., *Propionibacterium* spp., *Micrococcus* spp., and *Brevibacterium* spp. – inhibit or kill many cutaneous organisms

are the predominant organisms of sebum-rich regions (e.g. scalp, forehead) and staphylococci dominate dry regions (e.g. arms, legs), while corynebacteria comprise the highest proportions of microbes in communities inhabiting moist regions (e.g. axillae, perineum). Apart from *Acinetobacter* spp., few Gram-negative species are present on the skin surface. As well as bacteria, fungi (*Mallassezia* spp.) are found at many sites. Transients are often present on the skin surface, and these are derived from the environment and from other body sites that have openings onto the skin surface, e.g. the rectum, vagina, etc. The results of recent culture-independent studies imply that the composition of the resident microbial community of a site is far more complex that originally believed.

2.6 SOURCES OF DATA USED TO COMPILE FIGURES

1 Bibel, D.J. and Lovell, D.J. (1976) *J Invest Dermatol* 67, 265–9.
2 Kazmierczak, A.K., Szarapinska-Kwaszewska, J.K. and Szewczyk, E.M. (2005) *Pol J Microbiol* 54, 27–35.
3 Leyden, J.J., McGinley, K.J. and Vowels, B. (1998) *Dermatol* 196, 55–8.
4 Bojar, R.A. and Holland, K.T. (2002) *World J Microbiol Biotechnol* 18, 889–903.
5 McGinley, K.J., Webster, G.F. and Leyden, J.J. (1978) *Appl Environ Microbiol* 35, 62–6.
6 Marples, R. (1982) *Curr Med Res Opin* 7, 67–70.
7 Kloos, W.E. and Musselwhite, M.S. (1975) *Appl Microbiol* 30, 381–5.
8 Lee, Y.W., Yim, S.M., Lim, S.H., Choe, Y.B. and Ahn, K.J. (2006) *Mycoses* 49, 405–10.
9 Berlau, J., Aucken, H., Malnick, H. and Pitt, T. (1999) *Eur J Clin Microbiol Infect Dis* 18, 179–83.
10 Leyden, J.J. and McGinley, K.J. (1992) Coryneform bacteria. In: Noble, W.C (ed.). *The Skin Microflora and Microbial Skin Disease*. Cambridge University Press, Cambridge, UK, pp. 102–17.
11 Kloos, W.E., Musselwhite, M.S. and Zimmerman, R.J. (1975) A comparison of the distribution of *Staphylococcus* species on human and animal skin. In: Jeljaszewicz, J. (ed.). *Staphylococci and Staphylococcal Diseases*. Proceedings of III International Symposium on Staphylococci and Staphylococcal Infections. Warsaw, Poland, September 1975. Published by Fischer, Stuttgart, Germany, 1976, pp. 967–73.
12 McGinley, K.J., Webster, G.F., Ruggieri, M.R. and Leyden, J.J. (1980) *J Clin Microbiol* 12, 672–5.
13 Larson, E.L., McGinley, K.J., Foglia, A.R., Talbot, G.H. and Leyden, J.J. (1986) *Clin Microbiol* 23, 604–8.
14 Aly, R. and Maibach, H.I. (1977) *Appl Environ Microbiol* 33, 97–100.
15 Kates, S.G., Nordstrom, K.M., McGinley, K.J. and Leyden, J.J. (1990) *J Am Acad Dermatol* 22, 578–82.
16 Noble, W.C. (1992) Staphylococci on the skin. In: Noble, W.C. (ed.). *The Skin Microflora and Microbial Skin Disease*. Cambridge University Press, Cambridge, UK, pp. 135–52.
17 Noble, W.C. (1992) Other cutaneous bacteria. In: Noble, W.C. (ed.). *The Skin Microflora and Microbial Skin Disease*. Cambridge University Press, Cambridge, UK, pp. 210–31.
18 Leeming, J.P., Notman, F.H. and Holland, K.T. (1989) *J Appl Bacteriol* 67, 47–52.
19 Marshall, J., Leeming, J.P. and Holland, K.T. (1987) *J Appl Bacteriol* 62, 139–46.
20 Frank, D.N., Spiegelman, G.B., Davis, W., Wagner, E., Lyons, E. and Pace, N.R. (2003) *J Clin Microbiol* 41, 295–303.
21 Dekio, I., Hayashi, H., Sakamoto, M., Kitahara, M., Nishikawa, T., Suematsu, M. and Benno, Y. (2005) *J Med Microbiol* 54, 1231–8.
22 Paulino, L.C., Tseng, C.H., Strober, B.E. and Blaser, M.J. (2006) *J Clin Microbiol* 44, 2933–41.
23 Anesti, V., Vohra, J., Goonetilleka, S., McDonald, I.R., Straubler, B., Stackebrandt, E., Kelly, D.P. and Wood, A.P. (2004) *Environ Microbiol* 6, 820–30.
24 Gao, Z, Tseng, C.H., Pei, Z. and Blaser, M.J. (2007) *Proc Natl Acad Sci U S A* 104, 2927–32.

2.7 FURTHER READING

2.7.1 Books

Agache, P. and Humbert, P. (eds). (2004) *Measuring the Skin*. Springer-Verlag, Berlin, Germany.
Lesher, J.L., Aly, R., Babel, D.E., Cohen, P.R., Elston, D.M. and Tomecki, K.J. (eds). (2000) *An Atlas of Microbiology of the Skin*. CRC Press, Boca Raton, Florida, USA.
Noble, W.C. (ed.). (1992) *The Skin Microflora and Microbial Skin Disease*. Cambridge University Press, Cambridge, UK.
Wilson, M. (2005) *Microbial Inhabitants of Humans: Their Ecology and Role in Health and Disease*. Cambridge University Press, Cambridge, UK.

2.7.2 Reviews and papers

Ashbee, H.R. (2006) Recent developments in the immunology and biology of *Malassezia* species. *FEMS Immunol Med Microbiol* 47, 14–23.
Ashbee, H.R. (2007) Update on the genus *Malassezia*. *Med Mycol* 45, 287–303.

Barak, O., Treat, J.R. and James, W.D. (2005) Antimicrobial peptides: Effectors of innate immunity in the skin. *Adv Dermatol* 21, 357–74.

Berlau, J., Aucken, H., Malnick, H. and Pitt, T. (1999) Distribution of *Acinetobacter* species on skin of healthy humans. *Eur J Clin Microbiol Infect Dis* 18, 179–83.

Bojar, R.A. and Holland, K.T. (2002) The human cutaneous microbiota and factors controlling colonization. *World J Microbiol Biotechnol* 18, 889–903.

Braff, M.H. and Gallo, R.L. (2006) Antimicrobial peptides: An essential component of the skin defensive barrier. *Curr Top Microbiol Immunol* 306, 91–110.

Bruggemann, H., Henne, A., Hoster, F., Liesegang, H., Wiezer, A., Strittmatter, A., Hujer, S., Durre, P. and Gottschalk, G. (2004) The complete genome sequence of *Propionibacterium acnes*, a commensal of human skin. *Science* 305, 671–3.

Crespo-Erchiga, V. and Florencio, V.D. (2006) *Malassezia* yeasts and pityriasis versicolor. *Curr Opin Infect Dis* 19, 139–47.

Dekio, I., Hayashi, H., Sakamoto, M., Kitahara, M., Nishikawa, T., Suematsu, M. and Benno, Y. (2005) Detection of potentially novel bacterial components of the human skin microbiota using culture-independent molecular profiling. *J Med Microbiol* 54, 1231–8.

Elias, P.M. (2007) The skin barrier as an innate immune element. *Semin Immunopathol* 29, 3–14.

Elsner, P. (2006) Antimicrobials and the skin physiological and pathological flora. *Curr Probl Dermatol* 33, 35–41.

Frank, D.N., Spiegelman, G.B., Davis, W., Wagner, E., Lyons, E., and Pace, N.R. (2003) Culture-independent molecular analysis of microbial constituents of the healthy human outer ear. *J Clin Microbiol* 41, 295–303.

Fredricks, D.N. (2001) Microbial ecology of human skin in health and disease. *J Investig Dermatol Symp Proc* 6, 167–9.

Funke, G., von Graevenitz, A., Clarridge III, J.E. and Bernard, K.A. (1997) Clinical microbiology of coryneform bacteria. *Clin Microbiol Rev* 10, 125–59.

Gao, Z., Tseng, C.H., Pei, Z. and Blaser, M.J. (2007) Molecular analysis of human forearm superficial skin bacterial biota. *Proc Natl Acad Sci U S A* 104, 2927–32.

Harder, J. and Schroder, J.M. (2005) Antimicrobial peptides in human skin. *Chem Immunol Allergy* 86, 22–41.

Hendolin, P.H., Karkkainen, U., Himi, T., Markkanen, A., Ylikoski, J., Stroman, D.W., Roland, P.S., Dohar, J. and Burt, W. (2001) Microbiology of normal external auditory canal. *Laryngoscope* 111, 2054–9.

Hopwood, D., Farrar, M.D., Bojar, R.A. and Holland, K.T. (2005) Microbial colonization dynamics of the axillae of an individual over an extended period. *Acta Derm Venereol* 85, 363–4.

Izadpanah, A. and Gallo, R.L. (2005) Antimicrobial peptides. *J Am Acad Dermatol* 52, 381–90.

Joly-Guillou, M.L. (2005) Clinical impact and pathogenicity of *Acinetobacter*. *Clin Microbiol Infect* 11, 868–73.

Jung, K., Brauner, A., Kuhn, I., Flock, J.I. and Mollby, R.

(1998) Variation of coagulase-negative staphylococci in the skin flora of healthy individuals during one year. *Microbial Ecol Health Dis* 10, 85–90.

Kazmierczak, A.K., Szarapinska-Kwaszewska, J.K. and Szewczyk, E.M. (2005) Opportunistic coryneform organisms – Residents of human skin. *Pol J Microbiol* 54, 27–35.

Mack, D., Davies, A.P., Harris, L.G., Rohde, H., Horstkotte, M.A. and Knobloch, J.K. (2007) Microbial interactions in *Staphylococcus epidermidis* biofilms. *Anal Bioanal Chem* 387, 399–408.

McInturff, J.E., Modlin, R.L. and Kim, J. (2005) The role of Toll-like receptors in the pathogenesis and treatment of dermatological disease. *J Invest Dermatol* 125, 1–8.

Miller, L.S. and Modlin, R.L. (2007) Human keratinocyte Toll-like receptors promote distinct immune responses. *J Invest Dermatol* 127, 262–3.

Miller, L.S. and Modlin, R.L. (2007) Toll-like receptors in the skin. *Semin Immunopathol* 29, 15–26.

Morishita, N. and Sei, Y. (2006) Microreview of pityriasis versicolor and *Malassezia* species. *Mycopathologia* 162, 373–6.

Nagase, N., Sasaki, A., Yamashita, K., Shimizu, A., Wakita, Y., Kitai, S. and Kawano, J. (2002) Isolation and species distribution of staphylococci from animal and human skin. *J Vet Med Sci* 64, 245–50.

Niyonsaba, F. and Ogawa, H. (2005) Protective roles of the skin against infection: Implication of naturally occurring human antimicrobial agents beta-defensins, cathelicidin LL-37 and lysozyme. *J Dermatol Sci* 40, 157–68.

Paulino, L.C., Tseng, C.H., Strober, B.E., and Blaser, M.J. (2006) Molecular analysis of fungal microbiota in samples from healthy human skin and psoriatic lesions. *J Clin Microbiol* 44, 2933–41.

Perry, A.L. and Lambert, P.A. (2006) *Propionibacterium acnes*. *Lett Appl Microbiol* 42, 185–8.

Raz, R., Colodner, R. and Kunin, C.M. (2005) Who are you – *Staphylococcus saprophyticus*? *Clin Infect Dis* 40, 896–8.

Rieg, S., Seeber, S., Steffen, H., Humeny, A., Kalbacher, H., Stevanovic, S., Kimura, A., Garbe, C. and Schittek, B. (2006) Generation of multiple stable dermcidin-derived antimicrobial peptides in sweat of different body sites. *J Invest Dermatol* 126, 354–65.

Rippke, F., Schreiner, V. and Schwanitz, H.J. (2002) The acidic milieu of the horny layer: New findings on the physiology and pathophysiology of skin pH. *Am J Clin Dermatol* 3, 261–72.

Rosen, T. (2007) The *Propionibacterium acnes* genome: From the laboratory to the clinic. *J Drugs Dermatol* 6, 582–6.

Runeman, B., Rybo, G., Forsgren-Brusk, U., Larko, O., Larsson, P. and Faergemann, J. (2005) The vulvar skin microenvironment: Impact of tight-fitting underwear on microclimate, pH and microflora. *Acta Derm Venereol* 85, 118–22.

Schauber, J. and Gallo, R.L. (2007) Expanding the roles of antimicrobial peptides in skin: Alarming and arming keratinocytes. *J Invest Dermatol* 127, 510–12.

Schittek, B., Hipfel, R., Sauer, B., Bauer, J., Kalbacher, H., Stevanovic, S., Schirle, M., Schroeder, K., Blin, N., Meier, F., Rassner, G. and Garbe, C. (2001) Dermcidin: A novel human antibiotic peptide secreted by sweat glands. *Nat Immunol* 2, 1133–7.

Schroder, J.M. and Harder, J. (2006) Antimicrobial skin peptides and proteins. *Cell Mol Life Sci* 63, 469–86.

Sugita, T., Suto, H., Unno, T., Tsuboi, R., Ogawa, H., Shinoda, T. and Nishikawa, A. (2001) Molecular analysis of *Malassezia* microflora on the skin of atopic dermatitis patients and healthy subjects. *J Clin Microbiol* 39, 3486–90.

Tauch, A., Kaiser, O., Hain, T., Goesmann, A., Weisshaar, B., Albersmeier, A., Bekel, T., Bischoff, N., Brune, I., Chakraborty, T., Kalinowski, J., Meyer, F., Rupp, O., Schneiker. S., Viehoever, P. and Puhler, A. (2005) Complete genome sequence and analysis of the multiresistant nosocomial pathogen *Corynebacterium jeikeium* K411, a lipid-requiring bacterium of the human skin flora. *J Bacteriol* 187, 4671–82.

Webster, G.F. (2007) Skin microecology: The old and the new. *Arch Dermatol* 143, 105–6.

Chapter 3

THE INDIGENOUS MICROBIOTA OF THE EYE

3.1 ANATOMY AND PHYSIOLOGY OF THE EYE

The main anatomical features of the eye are shown in Fig. 3.1. The outermost layer of the eyeball consists of a transparent fibrous layer (the cornea) at the front and a protective layer of connective tissue (the sclera) at the back. The cornea is covered by the conjunctiva (a transparent layer of modified skin), and this also lines the eyelid and the eyelid margin. Nutrients are supplied to the sclera by a layer of vascular tissue known as the choroid. The retina covers the rear three-quarters of

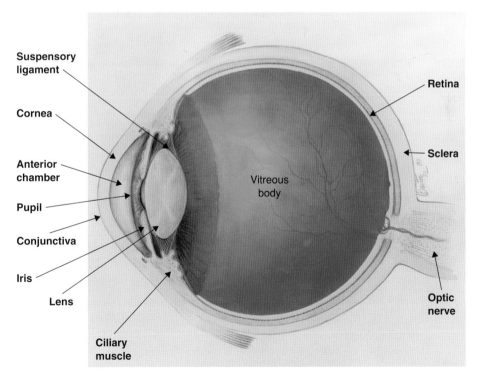

Fig. 3.1 The main anatomical features of the eye and its associated structures (diagram courtesy of National Eye Institute, National Institutes of Health, Bethesda, Maryland, USA).

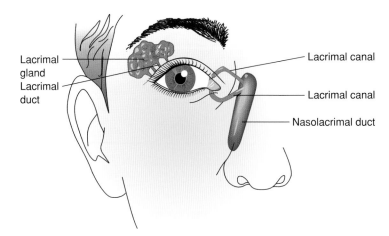

Lacrimal gland

Lacrimal duct

Lacrimal canal

Lacrimal canal

Nasolacrimal duct

Fig. 3.2 Anterior view of the lacrimal apparatus. Tears are produced by the lacrimal gland and flow through the lacrimal duct onto the conjunctiva. Excess tears drain into the nasal cavity via the lacrimal canals and nasolacrimal duct.

the inner surface of the eyeball and contains the light-sensitive cells onto which light is focused by the lens. The region in front of the lens is known as the anterior cavity, and this is filled with a fluid (the aqueous humor), which supplies nutrients and oxygen to the cornea, iris,

and lens. The region behind the lens is known as the vitreous chamber, which is filled with a transparent gel known as the vitreous humor.

The eyebrows and eyelashes protect the eyeballs from the direct rays of the sun, foreign objects, and

Epithelium

Fig. 3.3 Cross-section through the normal human conjunctiva. The arrows point to two goblet cells. Image kindly supplied by Marcia M. Jumblatt, Department of Ophthalmology and Visual Science, University of Louisville School of Medicine, Louisville, Kentucky, USA.

perspiration. A very important additional function of the eyelids is to spread tears over the eyeballs during blinking. Tears are produced, and delivered to the eyeballs, by a collection of glands, ducts, and canals known as the lacrimal apparatus (Fig. 3.2). Excess fluid drains into the nasal cavity via the nasolacrimal duct. Tears have lubricating, moistening, cleaning, and protective functions.

3.2 ANTIMICROBIAL DEFENSE SYSTEMS OF THE EYE

The eyelids protect the conjunctiva from microbes in three important ways. Firstly, blinking (which occurs at a rate of one blink every 5 s) protects against microbe-laden foreign objects. Secondly, the eyelids remove foreign debris and desquamated cells as they move over the cornea. Finally, they distribute tears (with a total volume of approximately 7 µl per eye) as a thin film (known as the "tear film") over the surface of the eye. As will be described later, tears have a wide range of antimicrobial properties.

The conjunctival epithelium is a stratified squamous epithelium consisting of between five and seven layers of cells, and is an effective barrier against microbial invasion as long as it remains intact. It contains numerous mucin-secreting goblet cells as well as intra-epithelial lymphocytes and dendritic cells (Fig. 3.3).

The cytoplasmic membranes of the epithelial cells of the outermost layer have numerous folds (known as microplicae) which project a distance of 0.5–0.75 µm. These serve to increase the surface area, and thereby aid the movement of nutrients and waste products across the cell membranes, as well as helping to stabilize the tear film which covers the conjunctiva.

The tear film protects the conjunctiva from dehydration, microbes, and particulate matter (Fig. 3.4). Its thickness is dependent on a number of factors including age, gender, time since last blink, and the temperature and relative humidity of the environment. Estimates range from 2.7 to 46 µm, and this wide variation arises as a result of not only the factors just mentioned, but also because of the different assay methods used. There is some evidence to suggest that the tear film consists of three layers – lipid, aqueous, and mucous – although some investigators propose that the latter two layers are not clearly distinguishable but merge into one another (Fig. 3.5).

The lipid layer is between 0.6 and 2.0 µm thick and consists of phospholipids and neutral lipids – the former

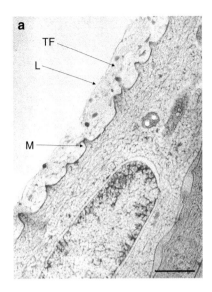

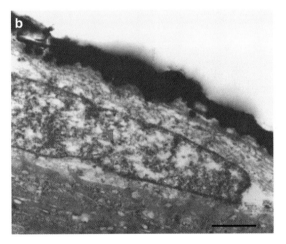

Fig. 3.4 (a) Transmission electron micrograph of the mouse corneal epithelium showing the tear film (TF), including the lipid layer (L) on its outer surface. Microplicae (M) are also clearly visible. (b) Another sample after fixation in glutaraldehyde and tannic acid showing the tear film as a dark layer on the surface of the corneal epithelium. Scale bar, 1.0 µm. From: Tran, C.H., Routledge, C., Miller, J., Miller, F. and Hodson, S.A. (2003) *Investigative Ophthalmology and Visual Sciences*, 44, 3520–5. Copyright 2003 by the Association for Research in Vision and Ophthalmology, Rockville, Maryland, USA. Reprinted by permission of the Association for Research in Vision and Ophthalmology via the Copyright Clearance Center, Danvers, Massachusetts, USA.

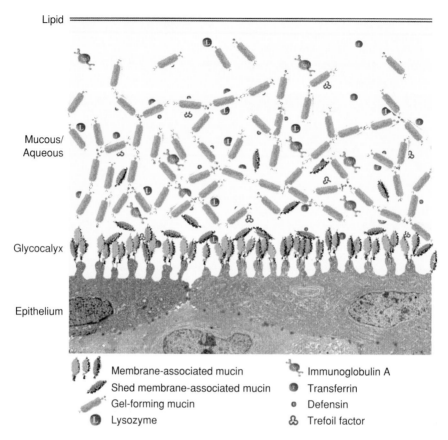

Lipid

Mucous/
Aqueous

Glycocalyx

Epithelium

Membrane-associated mucin

Shed membrane-associated mucin

Gel-forming mucin

Lysozyme

Immunoglobulin A

Transferrin

Defensin

Trefoil factor

Fig. 3.5 The outermost layer of the tear film consists of lipids, and this covers an aqueous layer containing mainly soluble mucins, proteins, antimicrobial defense components (e.g. lysozyme, IgA, defensins, transferrin, etc.), and other molecules. The aqueous layer merges with the mucous layer, which contains a higher concentration of the gel-forming mucin MUC5AC, as well as of trefoil factors (proteins involved in repair and defense of the mucosa). Underlying this layer is the glycocalyx of the conjunctival epithelial cells. Reprinted with the permission of Elsevier, Amsterdam, The Netherlands (©2004), from Gipson, I.K. (2004) *Exp Eye Res* 78, 379–88.

accumulate at the air interface, while the latter are in contact with the underlying aqueous layer. The lipids are secreted by the meibomian glands of the upper eye-lid. This layer has three main functions: (1) it prevents evaporation of the underlying aqueous layers; (2) it facilitates spreading of the aqueous layers because of its surfactant properties; and (3) it traps dust particles, thereby protecting the epithelium from their abrasive properties. Beneath the lipid layer is the aqueous layer (between 4 and 7 μm thick), which is produced by the lacrimal glands and the conjunctival epithelium. As well as nutrients and waste products from epithelial cells, the aqueous layer also contains effector molecules

of the innate and adaptive immune defense systems, i.e. antibodies, enzymes, and antimicrobial peptides. Its main functions are to hydrate, nourish, and oxygenate the conjunctival epithelium and also to protect it from microbes and from physical damage. This aqueous layer merges with the underlying mucoid layer, which is the innermost and thickest of the three layers and has a depth of between 3 and 30 μm.

The mucoid layer is produced by the goblet and squamous epithelial cells of the conjunctiva, and contains a high proportion of mucins which bind microbes, thereby preventing them from adhering to the underlying ocular structures. Two types of mucins

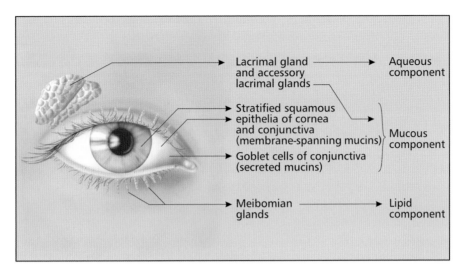

Fig. 3.6 Constituents of the tear film and their sources. Reprinted with the permission of Elsevier, Amsterdam, The Netherlands (©2006) from Paulsen, F.P. and Berry, M.S. (2006) Mucins and TFF peptides of the tear film and lacrimal apparatus. *Prog Histochem Cytochem* 41, 1–53.

– membrane-bound and soluble – are present in the mucoid layer. The membrane-bound mucins (produced by the squamous epithelial cells) have been identified as MUC1, MUC4, and MUC16, while the main soluble, gel-forming mucin (secreted by the goblet cells) has been shown to be MUC5AC. Small quantities of other soluble mucins (MUC2 and MUC7) may also be present. The gel-forming mucins (MUC5AC and MUC2) form a network covering the conjunctiva, and this traps microbes, foreign particles, and exfoliated epithelial cells. Blinking causes the network to collapse into a single strand which is pushed towards the inner corner of the eye. It then becomes compacted into a small clump and is pushed onto the skin where it dries and eventually either falls off or is removed by rubbing. Hence, microbes, foreign particles, and debris are regularly removed from the conjunctival surface. The origins of the various constituents of the tear film are shown in Fig. 3.6.

The tear film contains a variety of antimicrobial compounds (Table 3.1, Fig. 3.7), and tears have been shown to rapidly kill a wide range of organisms.

Tears contain high levels of secretory immunoglobulin A (sIgA), while IgG and IgE are also present but at lower concentrations. IgM and IgD are rarely detected. IgA constitutes approximately 17% of the total protein content and has a number of protective functions (Table 3.1). Toll-like receptors (TLRs) are important components of the innate immune response (see section 1.2.4), and have been detected in both the conjunctiva (TLR2, 4, and 9; Fig. 3.8) and the cornea (TLR1, 2, 3, 4, 5, 6, 7, 9, and 10).

3.3 ENVIRONMENTAL DETERMINANTS ON THE CONJUNCTIVAL SURFACE

The only region of the eye colonized by microbes is the conjunctiva. The main source of host-derived nutrients for microbes colonizing this region is tear fluid, and its composition is shown in Table 3.2. Tears have a high water content and are isotonic – their osmolarity is approximately 320 mOsm/kg. They are produced at a rate of approximately 1.2 μl/min, but this rate can increase greatly in response to physical and emotional stimuli. A single tear produced by someone who is crying has a volume of 50 μl, and large numbers of tears are produced each minute.

The main constituents of tear fluid are proteins, and as many as 491 proteins have been identified in tear fluid from a single healthy individual. Glucose and lactate are present in tear fluid, and these can be used as carbon and energy sources by a wide range of organisms; however, their low concentrations mean that the main sources of nutrients for conjunctival microbes are likely to be the proteins, mucins, and lipids present.

Table 3.1 Antimicrobial compounds that may be present in the tear film.

Component	Antimicrobial activities
Lysozyme (0.6–2.6 g/100 ml)	Kills some Gram-positive species; synergic with lactoferrin, SLPI, HBD-2, HBD-4, and LL-37; agglutinates bacteria
Lactoferrin (0.2 g/100 ml)	Iron-binding, therefore limits microbial growth; bactericidal – synergic interaction with lysozyme, IgA, LL-37, and human β-defensins; prevents bacterial adhesion to host cells; enhances the activity of natural killer cells
Lactoferricin	Bactericidal
Transferrin	Iron-binding, therefore limits microbial growth
Secretory phospholipase A$_2$ (5.5 mg/100 ml)	Bactericidal against many Gram-positive species
Secretory leukocyte protease inhibitor	Microbicidal; synergic with lysozyme; inhibits pro-inflammatory activities of bacterial components, e.g. LPS; inhibits serine proteases released by neutrophils, thereby protecting tissues against these enzymes
Human β-defensins (HBD-1, HBD-2, HBD-3)	Microbicidal – active against bacteria, fungi, and viruses
LL-37	Bactericidal; neutralizes activities of LPS and lipoteichoic acid; synergic with lysozyme; chemotactic for neutrophils and monocytes
LEAP-1 (liver-expressed antimicrobial peptide-1; hepcidin)	Microbicidal
LEAP-2	Microbicidal
Lactoperoxidase	Catalyses the reaction between hydrogen peroxide and thiocyanate (both of which are present in tears), resulting in the production of hypothiocyanite – which is active against a wide range of microbes
Bactericidal/permeability-inducing protein (2.8 μg/100 ml)	Bactericidal against Gram-negative bacteria
Caeruloplasmin	Chelates Cu ions; acts as a superoxide dismutase
Prealbumin	Enhances lysozyme activity
Fibronectin	Facilitates phagocytosis
Sialin	Binds microbes
Plasminogen activator	Chemoattractant for leukocytes
β-Lysin	Antibacterial – particularly effective against micrococci; acts synergistically with lysozyme
Fatty acids	Kill streptococci and fungi
Antibodies (mainly IgA)	Prevent adhesion of microbes to host cells; induce aggregation of microbes, thereby facilitating their removal; involved in opsonization; neutralize toxins and viruses
MUC7	Peptides derived from the N-terminal region of this mucin have antifungal and antibacterial properties
Glycoprotein 340	Agglutinates some bacteria

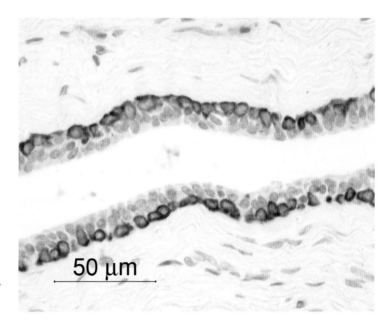

Fig. 3.7 Immunohistochemical detection (brown staining) of the expression of bactericidal/permeability-inducing protein in epithelial cells of the excretory lacrimal gland duct. Reproduced with kind permission of Springer Science and Business Media, Berlin, Germany, from Peuravuori, H., Aho, V.V., Aho, H.J., Collan, Y. and Saari, K.M. (2006) *Graefes Arch Clin Exp Ophthalmol* 244, 143–8.

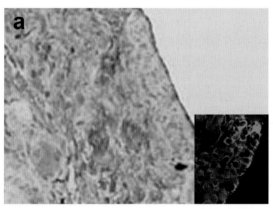

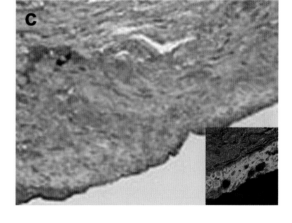

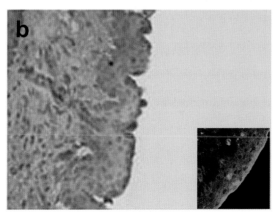

Fig. 3.8 Immunohistochemical detection of TLRs in the conjunctival epithelium of healthy individuals. Brown staining indicates the expression of (a) TLR2, (b) TLR4, and (c) TLR9. Nuclei were counterstained with hematoxylin (original magnification, ×40). Frames inside the pictures are confocal scanning micrographs (original magnification, ×60) of the epithelium. Reprinted with the permission of the American Academy of Ophthalmology, San Francisco, California, USA (©2005), from Bonini, S., Micera, A., Iovieno, A., Lambiase, A. and Bonini, S. (2005) Expression of Toll-like receptors in healthy and allergic conjunctiva. *Ophthalmology* 112, 1528–34.

Table 3.2 Composition of the tear film. See Table 3.1 for details of the antimicrobial peptides and proteins that may be present.

Component	Concentration
Total protein	0.3–2.6 g/100 ml
Albumin	0.1–0.2 g/100 ml
Immunoglobulins	0.07–0.4 g/100 ml
Lysozyme	0.6–2.6 g/100 ml
Lactoferrin	0.22 g/100 ml
MUC5AC	6.7 mg/100 ml
Epidermal growth factor	0.5 µg/100 ml
Amino acids	8 mg/100 ml
Anti-proteinases	6 mg/100 ml
Glucose	0.2–0.5 mmol/l
Lactate	1–5 mmol/l
Urea	54 mg/100 ml
Cholesterol	0.02–0.2 g/100 ml
Other lipids (mainly wax monoesters and sterol esters)	NDA
Fatty acids	NDA
Enzymes (lactate dehydrogenase, amylase)	NDA
Na^+	120–160 mmol/l
Cl^-	118–135 mmol/l
HCO_3^-	20–25 mmol/l
Mg^{2+}	0.7–0.9 mmol/l
K^+	20–42 mmol/l
Ca^{2+}	0.5–1.1 mmol/l

NDA, no data available.

Table 3.3 Hydrolysis of macromolecules by members of the ocular microbiota. In the case of mucins, a single species can usually only partially hydrolyze the molecule, and a microbial consortium is needed for complete hydrolysis (section 1.5.3).

Macromolecule	Organisms able to hydrolyze the macromolecule
Mucin	Viridans streptococci (e.g. *Streptococcus oralis*, *Streptococcus sanguinis*), *Strep. pneumoniae*, *Propionibacterium acnes*
Protein	*Staph. epidermidis*, *Staph. aureus*, *P. acnes*, *M. luteus*, *Strep. pneumoniae*, viridans streptococci, lactobacilli, *H. influenzae*
Lipid	*Staph. epidermidis*, *Staph. aureus*, *P. acnes*, *Corynebacterium* spp.

Table 3.3 lists members of the ocular microbiota that are able to degrade these macromolecules (sometimes only partially in the case of mucins), and thereby provide amino acids, sugars, and fatty acids as carbon, nitrogen, and energy sources. A number of bacteria isolated from the conjunctivae of humans, mainly viridans streptococci, are able to grow on ocular mucins. The sialidases that they secrete cleave sugar residues from the mucin molecule, and the residue may then be degraded further by glycosidases and/or proteases (see section 1.5.3). Urea can be hydrolyzed by a number of bacteria including *Staphylococcus epidermidis*, *Staphylococcus aureus*, *Micrococcus luteus*, some *Corynebacterium* spp., and some viridans streptococci, resulting in the liberation of ammonia and the formation of ammonium ions, which can be used as nitrogen sources.

The tear film has a pH of between 7.14 and 7.82, with a mean value of approximately 7.5, although this increases slightly with increasing tear flow rate. This pH is suitable for the growth of most ocular microbes. During prolonged eyelid closure, the pH decreases to approximately 7.25 due to the dissolution of trapped CO_2 produced by the cornea.

While the temperature of most body sites remains at a fairly constant value of 37°C, the temperature of the conjunctiva is usually several degrees lower than this – except in very hot and humid conditions. The main causes of heat loss from the eye are evaporation of the tear film, convection, and radiation. As these processes are markedly affected by variations in the temperature, humidity, and velocity of the surrounding air, the ocular surface, more than any other region of the body, is subject to marked temperature changes. An ambient temperature of 20°C in still air results in a conjunctival temperature of 32°C. Evaporation is increased with increasing air movement, so that even lower temperatures will be reached in a windy environment. When the air temperature falls to 6°C the temperature of the conjunctivae can drop to approximately 27°C in still air and to 17°C when the air velocity is 4 m/s. Given that the microbial growth rate halves for each 10°C fall in temperature, the effect of such temperature changes would have a profound effect on the growth of ocular microbes. As will be discussed later (in section 3.4.2), the temperature of the environment (probably in combination with humidity) has been shown to affect the composition of the ocular microbiota. Furthermore, certain organisms, such as *Mycobacterium tuberculosis*, have a narrow temperature range for growth (30–39°C), while some streptococci (e.g. *Streptococcus mutans* and *Streptococcus sobrinus*) and Gram-negative species (e.g. *Pseudomonas cepacia*, *Alcaligenes faecalis*) do not grow well below 30°C, and therefore the lower conjunctival temperature could select against the survival of such organisms.

The very large surface area-to-volume ratio of the tear film means that it is highly aerobic – its dissolved oxygen content is usually at least 75% of that of oxygen-saturated water. The survival of anaerobes and microaerophiles (e.g. *Propionibacterium* spp. and *Peptostreptococcus* spp.), therefore, will be limited to microhabitats with a low oxygen content created by oxygen-consuming aerobes and facultative anaerobes.

As described in section 3.2, the conjunctival surface is equipped with a wide range of antimicrobial mechanisms that will have to be overcome before microbial colonization can occur.

Most of the accessory structures of the eye (i.e. eyebrows, eyelids, and eyelashes) are colonized by microbes, but the environmental factors operating at these sites are similar to those that operate either at the skin (in the case of the eyelids) or hair (in the case of eyelashes and eyebrows), and so will not be discussed further. The microbiotas of these sites are also what would be expected of skin and its associated structures.

3.4 THE INDIGENOUS MICROBIOTA OF THE EYE

The conjunctivae of many individuals appear either to harbor only sparse microbial populations or are sterile. This is likely to be a consequence of the low nutrient content of tear fluid, the low temperature of the conjunctiva, and the presence of a wide range of antimicrobial mechanisms. However, as virtually all studies of the ocular microbiota have been culture-based, these findings may not reflect the real situation. The finding of only small numbers of cultivable microbes on the outer surface of the eye has prompted debate as to

whether the organisms that can be detected there constitute a resident microbiota or are merely transients derived from neighboring skin.

Evidence in favor of the latter hypothesis is that: (1) in some studies, no cultivable microbes have been detected on the conjunctivae of as many as 65% of healthy individuals, and (2) the organisms that can be cultivated are usually, but not always, typical of those found on the skin. However, there are a number of possible explanations for the failure to detect microbes in high proportions of ocular samples: (1) the amount of sample obtained (a few microliters of tear fluid) from the conjunctiva is extremely small, thereby making detection difficult by traditional culture methods; (2) the culture techniques employed in many studies have been inadequate, e.g. many have not involved anaerobic incubation of the samples and few have employed a variety of media; and (3) remarkably few studies have used molecular detection methods to investigate the microbiota of healthy eyes (although such studies have been carried out on infected eyes) and, therefore, would fail to detect not-yet-cultivated or difficult-to-grow species. For example, in a recent study of the conjunctival microbiota of patients with conjunctivitis, no bacteria were cultivated from 55% of those samples in which bacteria were detected by PCR amplification of 16S rRNA genes.

Findings that imply that the conjunctival surface does have a resident microbiota include:
• While many (but not all) of the organisms that are cultured from the conjunctiva can also be found on the skin, not all of the organisms found on the skin can be detected on the conjunctiva. Consequently, while many of the organisms detected may originate from the skin, the ocular environment selects for the survival of only some of these species, and these are regularly and consistently detectable on the conjunctiva.
• Some of the organisms often present on the conjunctiva are not members of the cutaneous microbiota, e.g. viridans streptococci, *Streptococcus pneumoniae*, *Peptostreptococcus* spp., *Lactobacillus* spp., and *Haemophilus influenzae*.

3.4.1 Members of the ocular microbiota

Many of the organisms comprising the ocular microbiota are also present on the skin and are described in Chapter 2. Although *Strep. pneumoniae* and viridans streptococci are also found on the conjunctiva, they are best known as members of the respiratory microbiota and therefore are described in detail in Chapter 4. The genera *Peptostreptococcus* and *Lactobacillus* are described in Chapters 5 and 6, respectively.

The major adhesins of members of the ocular microbiota that are relevant to colonization of the conjunctiva are shown in Fig. 3.9.

3.4.2 Composition of the ocular microbiota

As described in section 3.1, the only region of the eye exposed to the external environment is the conjunctiva. Associated structures, such as the eyelids, support a microbial community identical to that found on

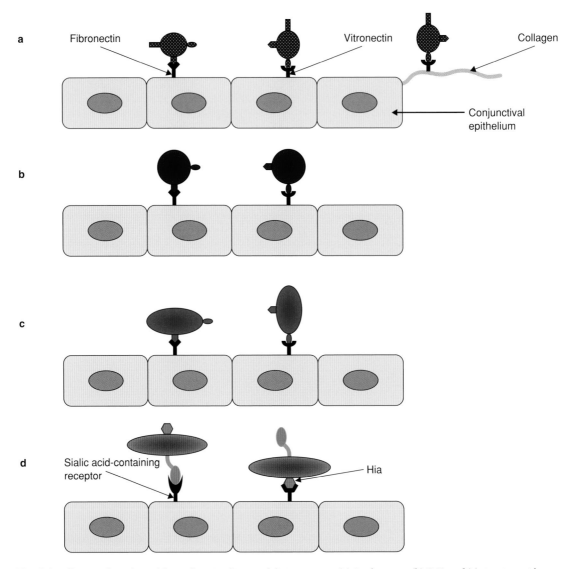

Fig. 3.9 Adhesins of members of the ocular microbiota and their receptors. (a) *Staph. aureus*, (b) CNS, and (c) streptococci have adhesins that recognize fibronectin and vitronectin on the surface of conjunctival epithelial cells. *Staph. aureus* can also bind to any collagen that is exposed when the epithelium is damaged. (d) *H. influenzae* has fimbrial adhesins that mediate binding to sialic acid-containing molecules and to sulfated glycosaminoglycans. It also has an autotransporter protein (Hia), which binds to an as-yet unknown receptor.

adjacent skin regions. This section, therefore, will describe only the microbiotas of the conjunctival surface and the eyelid margin which, as pointed out in section 3.1, is also covered by the conjunctiva. Unfortunately, at the time of writing (mid-2007), there are no reports of the use of molecular techniques to characterize the indigenous microbiota of disease-free eyes – although they have been employed to investigate the conjunctival microbiota of patients with conjunctivitis and other diseases. Our knowledge of the ocular microbiota, therefore, is derived entirely from culture-based studies. Microbes can usually be cultivated from the conjunctiva, although the proportion of samples from which no organisms can be grown ranges from 0 to 65% in different studies. The ocular microbiota is usually very simple and generally consists of no more than two species (Fig. 3.10). The population density is also usually low, although the number of bacteria cultivated varies enormously in different studies, ranging from less than 100 cfu to more than 5×10^4 cfu.

The organisms most frequently isolated are coagulase-negative staphylococci (CNS) and *Propionibacterium acnes* and, to a lesser extent, *Peptostreptococcus* spp. and *Corynebacterium* spp. (Fig. 3.11). Other organisms

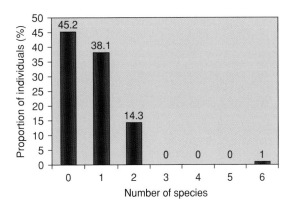

Fig. 3.10 The number of different bacterial species cultured from the conjunctival surface. The figures denote the proportions (%) of individuals in a group of 42 healthy adults who harbored the indicated number of species (1).

frequently detected, but generally accounting for lower proportions of the microbiota, include *Staph. aureus*, streptococci, micrococci, *Lactobacillus* spp., and facultative Gram-negative rods.

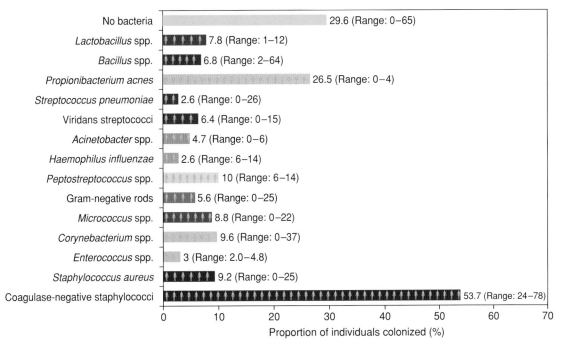

Fig. 3.11 Frequency of detection of various microbes on the conjunctivae of healthy adults. The data shown are mean values (and ranges) derived from the results of 17 culture-based studies involving 4623 individuals from a number of countries (1–17).

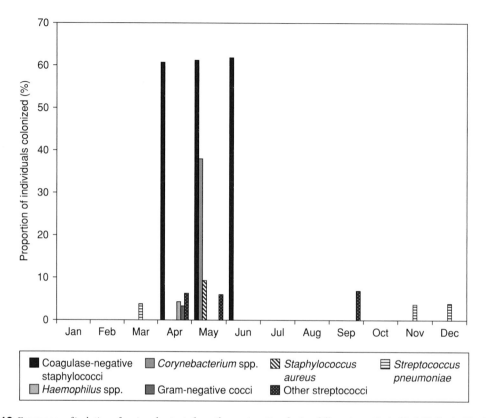

Fig. 3.12 Frequency of isolation of various bacteria from the conjunctiva during different months in Madrid, Spain. The bars represent the proportion (%) of individuals colonized by the indicated species during the months when there was a statistically significant increase in the frequency of isolation of those species. All of the organisms were also isolated on all other sampling occasions, but the data are omitted for the sake of clarity. The study involved a total of 4432 individuals (18).

The frequency of detection of various bacterial species on the conjunctivae has been shown to be affected by climatic factors – particularly temperature and relative humidity (Fig. 3.12). In a study carried out in Spain (Madrid), the frequencies of isolation of a number of organisms (CNS, *Corynebacterium* spp., streptococci, and *H. influenzae*) were found to be significantly higher during the period from April to June than during January to March, and this corresponded to an increase in the average temperature from 8.2 to 16.5°C during these periods. The corresponding humidities during these periods decreased from 63% to 52%.

Although the ocular microbiota is generally dominated by CNS, few studies have undertaken speciation of the isolates. In one study where this was carried out, *Staph. epidermidis* was found to predominate, and comprised 57% of the CNS that were isolated (Fig. 3.13).

Studies of the *Staph. epidermidis* isolates obtained on different occasions from the same individual over a 6-month period have shown that they are not identical as they have different pulsed-field gel electrophoresis patterns

Few studies have identified the coryneforms isolated from conjunctivae. In one recent study, all of the *Corynebacterium* spp. isolated were found to be lipophilic species with CDC (Centers for Disease Control) Group G predominating (Fig. 3.14). These lipophilic species are not frequently isolated from any other site on the human body, supporting the view that the conjunctiva does have its own distinctive, indigenous microbiota.

During periods of eye closure, the composition of tears alters and the pH decreases, and this has been found to increase the total number of microbes present

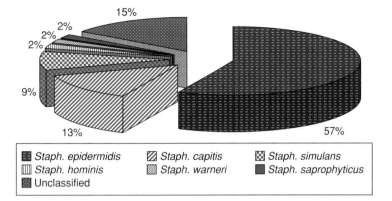

Fig. 3.13 Identity of those species of CNS that are members of the ocular microbiota. The figures denote the proportions of the various species isolated from the eyes of 34 adults (19).

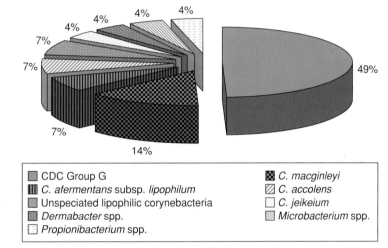

Fig. 3.14 Identity of those coryneform bacteria that are members of the ocular microbiota. The figures denote the proportion of the various species isolated from the conjunctivae of 92 adults (20). CDC, Centers for Disease Control (Atlanta, Georgia, USA).

without altering their relative proportions. The conjunctival microbiota alters with age, with a trend towards a greater frequency of isolation of coryneforms and Gram-negative species and a decreased frequency of *Strep. pneumoniae* with increasing age (Fig. 3.15).

The indigenous microbiotas of the left and right eye are remarkably similar in individuals free of any eye disease. Hence, in a study of 410 adults, the chances of finding a particular bacterial species on one conjunctiva if it was present on the other were generally very high (e.g. for CNS and diphtheroids), whereas if the organism was not on one conjunctiva then the chances of it being found on the other conjunctiva were very low (Fig. 3.16).

The microbiota of the eyelid margin is very similar

to that of the conjunctiva although, in general, greater numbers of microbes are cultivated from the lid margin (Fig. 3.17).

In a comparative study of the bacteria present on the conjunctivae and eyelid margins of healthy individuals, the chances of finding a particular bacterial species on the left eyelid margin if it was present on the left conjunctiva were very high (Fig. 3.18).

3.4.3 Interactions among members of the ocular microbiota

The dominant genera of the ocular microbiota, the staphylococci, coryneforms, and propionibacteria, are

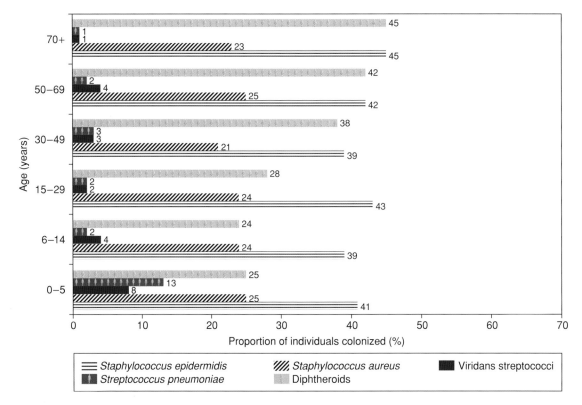

Fig. 3.15 Changes in the frequency of isolation of various organisms from the conjunctivae with age (21).

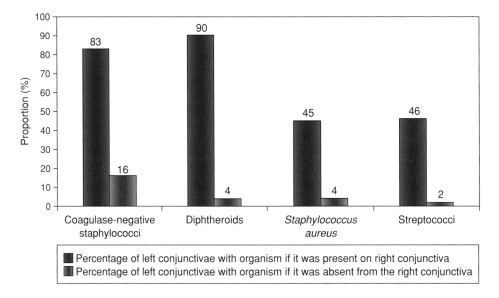

Fig. 3.16 Comparative microbiology of the left and right conjunctivae. Bacteria were cultured from the left and right conjunctivae of 410 adults, and the results were used to determine whether the presence of an organism on one conjunctiva of an individual could predict the likelihood of it being found on the other conjunctiva of that individual (22).

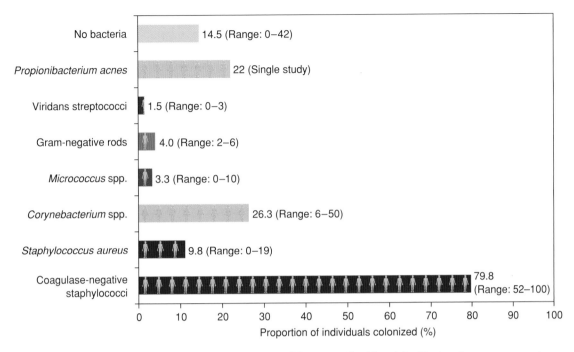

Fig. 3.17 Frequency of detection of various microbes on the eyelid margins of healthy adults. The data shown are mean values (and ranges) derived from the results of four studies involving a total of 496 individuals from a number of countries (1, 9, 13, 22).

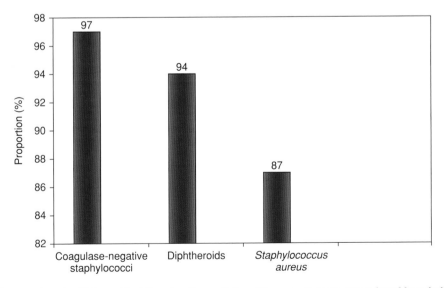

Fig. 3.18 Comparative microbiology of the left conjunctiva and left eyelid margin. Bacteria were cultured from the left conjunctiva and left eyelid margin of 410 adults, and the results were used to determine whether the presence of an organism on the left conjunctiva of an individual could predict the likelihood of it being found on the left eyelid margin of that individual. The bars denote the proportion of left eyelids that were positive for a particular species detected on the left conjunctiva (22).

all known to produce bacteriocins. However, the role of these molecules in maintaining the composition of the ocular microbiota has not been established. Similarly, metabolic end products (e.g. acetic and lactic acids) of ocular microbes may be responsible for excluding certain organisms from the conjunctival surface. The results of a number of investigations imply that antagonistic effects do operate in vivo. For example, a study of organisms isolated from the conjunctivae of healthy adults revealed that when *Staph. aureus* was present on the conjunctiva, the chances of finding either CNS or coryneforms was significantly reduced. This may be attributable to various bacteriocins produced by *Staph. aureus*. Hence, the bacteriocin Bac1829 is inhibitory to a number of *Corynebacterium* spp., while BacR1 is inhibitory to not only *Corynebacterium* spp. but also to several *Staphylococcus* spp. The study also found a synergistic relationship between CNS and corynebacteria – when either of these groups of organisms was present, the chance of finding members of the other group was significantly enhanced. This has also been observed in a number of other studies.

Although the basis of this synergy has not been established, it is known that some *Corynebacterium* spp. are dependent on certain fatty acids for growth and that some staphylococci have lipases, which release fatty acids from lipids. Ocular strains of micrococci have been shown to secrete a number of antimicrobial peptides. These are active mainly against Gram-positive species and show very high activity against micrococci, corynebacteria, *Sarcina* spp., and *Bacillus* spp. and moderate activity against streptococci, clostridia, and mycobacteria. Their effect, if any, on the composition of the ocular microbiota has not been established.

The production of metabolic end products and the degradation of polymers by ocular microbes may enable a number of nutritional inter-relationships to become established between members of the ocular microbiota. Examples of such possibilities are summarized in Fig. 3.19.

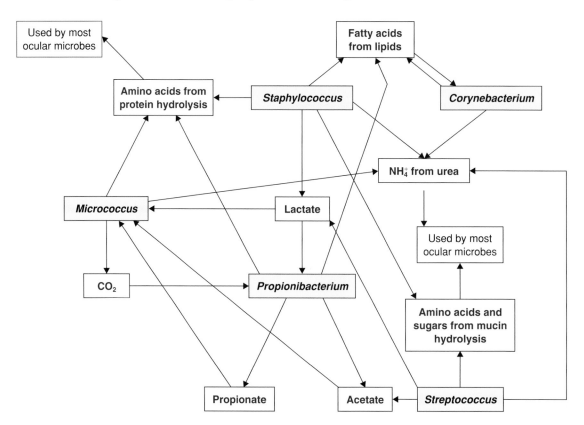

Fig. 3.19 Nutritional interactions that may occur between members of the ocular microbiota.

3.5 OVERVIEW OF THE OCULAR MICROBIOTA

The conjunctival surface is equipped with a variety of mechanical and chemical antimicrobial defense mechanisms which are highly effective at minimizing microbial colonization. The conjunctival surfaces of a large proportion of individuals appear to be free of cultivable microbes and, when a microbial community is found to be present, it tends to have a low population density and a simple composition – usually no more than two species. CNS and, to a lesser extent, *P. acnes* are the most frequently isolated organisms. Other organisms occasionally found include *Staph. aureus* and species belonging to the genera *Corynebacterium, Streptococcus, Lactobacillus, Peptostreptococcus, Bacillus,* and *Micrococcus.* Gram-negative species are infrequently isolated. The eyelid margins have a similar cultivable microbiota to that found on the conjunctiva, but the population density tends to be higher. Our knowledge of the ocular microbiota is based on culture-dependent analysis as, unfortunately, no culture-independent studies of the microbiota of disease-free eyes have been published.

3.6 SOURCES OF DATA USED TO COMPILE FIGURES

1 Gritz, D.C., Scott, T.J., Sedo, S.F., Cevallos, A.V., Margolis, T.P. and Whitcher, J.P. (1997) *Cornea* 16, 400–5.
2 Ta, C.N., He, L., Nguyen, E. and De Kaspar, H.M. (2006) *Eur J Ophthalmol* 16, 359–64.
3 Arantes, T.E., Cavalcanti, R.F., Diniz, M. de F., Severo, M.S., Lins Neto, J. and Castro, C.M. (2006) *Arq Bras Oftalmol* 69, 33–6.
4 Ta, C.N., Chang, R.T., Singh, K., Egbert, P.R., Shriver, E.M., Blumenkranz, M.S., et al. (2003) *Ophthalmology* 110, 1946–51.
5 Locatelli, C.I., Kwitko, S. and Simonetti, A.B. (2003) *Braz J Microbiol* 34, 203–8.
6 Yamauchi, Y., Minoda, H., Yokoi, K., Maruyama, K., Kumakura, S., Usui, M., Cruz, J.M.C. and Fukutake, K. (2005) *Ocular Immun Inflam* 13, 301–4.
7 Campos, M.S., Campos e Silva, L.Q., Rehder, J.R., Lee, M.B., O'Brien, T. and McDonnell, P.J. (1994) *Acta Ophthalmol (Copenh)* 72, 241–5.
8 Iskeleli, G., Bahar, H., Eroglu, E., Torun, M.M. and Ozkan, S. (2005) *Eye & Contact Lens* 31, 124–6.
9 Albietz, J.M. and Lenton, L.M. (2006) *Cornea* 25, 1012–19.
10 Chisari, G., Cavallaro, G., Reibaldi, M. and Biondi, S. (2004) *Int J Clin Pharmacol Ther* 42, 35–8.
11 Srinivasan, R., Reddy, R.A., Rene, S., Kanungo, R. and Natarajan, M.K. (1999) *Indian J Ophthalmol* 47, 185–9.
12 Kato, T. and Hayasaka, S. (1998) *Jpn J Ophthalmol* 42, 461–5.
13 Stapleton, F., Willcox, M.D., Fleming, C.M., Hickson, S., Sweeney, D.F. and Holden, B.A. (1995) *Infect Immun* 63, 4501–5.
14 Soudakoff, P.S. (1954) *Am J Ophthalmol* 38, 374–6.
15 Oguz, H., Oguz, E., Karadede, S. and Aslan, G. (1999) *Int Ophthalmol* 23, 117–20.
16 Perkins, R.E., Kundsin, R.B., Pratt, M.V., Abrahamsen, I. and Leibowitz, H.M. (1975) *J Clin Microbiol* 1, 147–9.
17 Singer, T.R., Isenberg, S.J. and Apt, L. (1988) *Br J Ophthalmol* 72, 448–51.
18 Rubio, E.F. (2004) *Eye* 18, 778–84.
19 Larkin, D.F. and Leeming, J.P. (1991) *Eye* 5, 70–4.
20 von Graevenitz, A., Schumacher, U. and Bernauer, W. (2001) *Curr Microbiol* 42, 372–4.
21 Osato, M. (1996) Normal ocular flora. In: Pepose, J.S., Holland, G.N., and Wilhelmus, K.R. (eds). *Ocular Infection and Immunity.* Mosby-Year Book Inc., St. Louis, Missouri, USA, pp. 191–9.
22 Allansmith, M.R., Ostler, H.B. and Butterworth, M. (1969) *Arch Ophthalmol* 82, 37–42.

3.7 FURTHER READING

Albietz, J.M. and Lenton, L.M. (2006) Effect of antibacterial honey on the ocular flora in tear deficiency and meibomian gland disease. *Cornea* 25, 1012–19.
Arantes, T.E.F., Cavalcanti, R.F., Diniz, M. de F.A., Severo, M.S., Neto J.L. and de Castro, C.M.M.B. (2006) Conjunctival bacterial flora and antibiotic resistance pattern in patients undergoing cataract surgery. *Arq Bras Oftalmol* 69, 33–6.
Barkana, Y., Almer, Z., Segal, O., Lazarovitch, Z., Avni, I. and Zadok, D. (2005) Reduction of conjunctival bacterial flora by povidone/iodine, ofloxacin and chlorhexidine in an outpatient setting. *Acta Ophthalmol Scand* 83, 360–63.
Berry, M., Harris, A., Lumb, R. and Powell, K. (2002) Commensal ocular bacteria degrade mucins. *Br J Ophthalmol* 86, 1412–16.
Bilen, H., Ates, O., Astam, N., Uslu, H., Akcay, G. and Baykal, O. (2007) Conjunctival flora in patients with type 1 or type 2 diabetes mellitus. *Adv Ther* 24, 1028–35.
Chaidaroon, W., Ausayakhun, S., Pruksakorn, S., Jewsakul, S.O. and Kanjanaratanakorn, K. (2006) Ocular bacterial flora in HIV-positive patients and their sensitivity to gentamicin. *Jpn J Ophthalmol* 50, 62–80.
Chang, J.H., McCluskey, P.J. and Wakefield, D. (2006) Toll-like receptors in ocular immunity and the immunopathogenesis of inflammatory eye disease. *Br J Ophthalmol* 90, 103–8.
Chisari, G., Cavallaro, G., Reibaldi, M. and Biondi, S. (2004) Presurgical antimicrobial prophylaxis: Effect on ocular flora in healthy patients. *Int J Clin Pharmacol Ther* 42, 35–8.

Dartt, D.A. (2004) Control of mucin production by ocular surface epithelial cells. *Exp Eye Res* 78, 173–85.

de Aguiar Moeller, C.T., Branco, B.C., Yu, M.C.Z., Farah, M.E., Santos, M.A.A. and Höfling-Lima, A.L. (2005) Evaluation of normal ocular bacterial flora with two different culture media. *Can J Ophthalmol* 40, 448–53.

de Souza, G.A., Godoy, L.M. and Mann, M. (2006) Identification of 491 proteins in the tear fluid proteome reveals a large number of proteases and protease inhibitors. *Genome Biol* 7, R72.

Ermis, S.S., Aktepe, O.C., Inan, U.U., Ozturk, F. and Altindis, M. (2004) Effect of topical dexamethasone and ciprofloxacin on bacterial flora of healthy conjunctiva. *Eye* 18, 249–52.

Evans, D.J., McNamara, N.A. and Fleiszig, S.M. (2007) Life at the front: Dissecting bacterial-host interactions at the ocular surface. *Ocul Surf* 5, 213–27.

Fung, K., Morris, C. and Duncan, M. (2002) Mass spectrometric techniques applied to the analysis of human tears: A focus on the peptide and protein constituents. *Adv Exp Med Biol* 506, 601–5.

Gipson, I.K. (2004) Distribution of mucins at the ocular surface. *Exp Eye Res* 78, 379–88.

Huang, L.C., Jean, D., Proske, R.J., Reins, R.Y. and McDermott, A.M. (2007) Ocular surface expression and in vitro activity of antimicrobial peptides. *Curr Eye Res* 32, 595–609.

Iskeleli, G., Bahar, H., Eroglu, E., Torun, M.M. and Ozkan, Ş. (2005) Microbial changes in conjunctival flora with 30-day continuous-wear silicone hydrogel contact lenses. *Eye & Contact Lens* 31, 124–6.

Jumblatt, M.M., Imbert, Y., Young, W.W. Jr., Foulks, G.N., Steele, P.S. and Demuth, D.R. (2006) Glycoprotein 340 in normal human ocular surface tissues and tear film. *Infect Immun* 74, 4058–63.

King-Smith, P.E., Fink, B.A., Hill, R.M., Koelling, K.W. and Tiffany, J.M. (2004) The thickness of the tear film. *Curr Eye Res* 29, 357–68.

Kirkwood, B.J. (2007) Normal flora of the external eye. *Insight* 32, 12–13.

Kumar, A. and Yu, F.S. (2006) Toll-like receptors and corneal innate immunity. *Curr Mol Med* 6, 327–37.

Kumar, A., Zhang, J. and Yu, F.X. (2006) Toll-like receptor 2-mediated expression of β-defensin-2 in human corneal epithelial cells. *Microbes Infect* 8, 380–9.

Li, J., Shen, J. and Beuerman, R.W. (2007) Expression of toll-like receptors in human limbal and conjunctival epithelial cells. *Mol Vis* 13, 813–22

McCulley, J.P. and Shine, W.E. (2004) The lipid layer of tears: Dependent on meibomian gland function. *Exp Eye Res* 78, 361–5

McIntosh, R.S., Cade, J.E., Al-Abed, M., Shanmuganathan, V., Gupta, R., Bhan, A., Tighe, P.J. and Dua, H.S. (2005) The spectrum of antimicrobial peptide expression at the ocular surface. *Invest Ophthalmol Vis Sci* 46, 1379–85.

Micera. A., Stampachiacchiere, B., Aronni, S., dos Santos, M.S. and Lambiase, A. (2005) Toll-like receptors and the eye. *Curr Opin Allergy Clin Immunol* 5, 451–8.

Moeller, C.T., Branco, B.C., Yu, M.C., Farah, M.E., Santos, M.A. and Hofling-Lima, A.L. (2005) Evaluation of normal ocular bacterial flora with two different culture media. *Can J Ophthalmol* 40, 448–53.

Ohashi, Y., Dogru, M. and Tsubota, K. (2006) Laboratory findings in tear fluid analysis. *Clin Chim Acta* 369, 17–28.

Paulsen, F. (2006) Cell and molecular biology of human lacrimal gland and nasolacrimal duct mucins. *Int Rev Cytol* 249, 229–79.

Paulsen, F.P. and Berry, M.S. (2006) Mucins and TFF peptides of the tear film and lacrimal apparatus. *Prog Histochem Cytochem* 41, 1–53.

Paulsen, F.P., Pufe, T., Schaudig, U., Held-Feindt, J., Lehmann, J., Schröder, J.M. and Tillmann, B.N. (2001) Detection of natural peptide antibiotics in human nasolacrimal ducts. *Invest Ophthalmol Vis Sci* 42, 2157–63.

Peuravuori, H., Aho, V.V., Aho, H.J., Collan, Y. and Saari, K.M. (2006) Bactericidal/permeability – increasing protein in lacrimal gland and in tears of healthy subjects. *Graefe's Arch Clin Exp Ophthalmol* 244, 143–8.

Rubio, E.F. (2004) Climatic influence on conjunctival bacteria of patients undergoing cataract surgery. *Eye* 18, 778–84.

Spurr-Michaud, S., Argueso, P. and Gipson, I. (2007) Assay of mucins in human tear fluid. *Exp Eye Res* 84, 939–50.

Ueta, M., Iida, T., Sakamoto, M., Sotozono, C., Takahashi, J., Kojima, K., Okada, K., Chen, X., Kinoshita, S. and Honda, T. (2007) Polyclonality of *Staphylococcus epidermidis* residing on the healthy ocular surface. *J Med Microbiol* 56, 77–82.

von Graevenitz, A., Schumacher, U. and Bernauer, W. (2001) The corynebacterial flora of the normal human conjunctiva is lipophilic. *Curr Microbiol* 42, 372–4.

Yu, F.S. and Hazlett, L.D. (2006) Toll-like receptors and the eye. *Invest Ophthalmol Vis Sci* 47, 1255–63.

Zhao, H., Jumblatt, J.E., Wood, T.O. and Jumblatt, M.M. (2001) Quantification of MUC5AC protein in human tears. *Cornea* 20, 873–7.

Chapter 4

THE INDIGENOUS MICROBIOTA OF THE RESPIRATORY TRACT

The principal function of the respiratory system is to supply oxygen to, and remove carbon dioxide from, the blood, which is transported throughout the body by the cardiovascular system. Exchange of gases between the two systems occurs at the respiratory membrane which is housed in the lung. Essentially, the respiratory system consists of: (1) a collection of tubes (known as the respiratory tract) for air conduction; (2) a respiratory membrane for gaseous exchange; and (3) a ventilatory mechanism (the lungs, diaphragm, and associated muscles) for delivering gases to and from the respiratory membrane. The total surface area of the respiratory mucosa amounts to approximately 80 m^2 – this constitutes more than 25% of the total area of the mucosal surfaces in an adult. The only section of the respiratory system colonized by microbes is the respiratory tract.

4.1 ANATOMY AND PHYSIOLOGY OF THE RESPIRATORY TRACT

The respiratory tract consists of two main regions (Fig. 4.1): (1) the upper respiratory tract (nose and pharynx) and (2) the lower respiratory tract (larynx, trachea, bronchi, bronchioles, alveolar ducts, alveolar sacs, and alveoli).

The upper respiratory tract is heavily colonized by microbes, whereas the lower respiratory tract does not have an indigenous microbiota. However, microbes can be detected in the latter (particularly in the trachea and larynx) due to the aspiration of microbe-containing secretions from the upper respiratory tract.

4.1.1 Nose

The nose consists of an external, visible portion and a large internal cavity within the skull (Fig. 4.2). Air is taken in via the external nares (nostrils) which contain hairs (vibrissae) to filter out large particles. It then passes over three bony protrusions (conchae) which

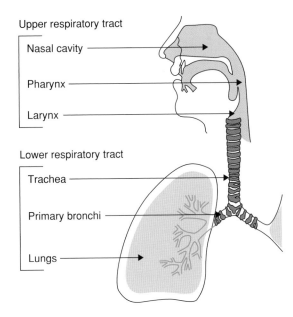

Fig. 4.1 The respiratory tract of humans. (From http://www.training.seer.cancer.gov/module_anatomy/anatomy_physiology_home.html; funded by the US National Cancer Institute's Surveillance, Epidemiology and End Results [SEER] Program with Emory University, Atlanta, Georgia, USA.)

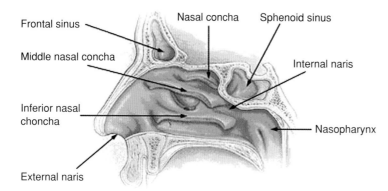

Frontal sinus

Middle nasal concha

Inferior nasal choncha

External naris

Nasal concha   Sphenoid sinus

Internal naris

Nasopharynx

Fig. 4.2 The nose and nasal cavity of humans. (From http://www.training.seer.cancer.gov/module_anatomy/anatomy_physiology_home.html funded by the US National Cancer Institute's Surveillance, Epidemiology and End Results [SEER] Program with Emory University, Atlanta, Georgia, USA.)

divide the nasal cavity into three air passages known as meati. The conchae are covered with mucus which traps any small particles and warms and moisturizes the air.

Each naris opens into a vestibule which constitutes one of the anterior chambers of the nasal cavity. The anterior zone of the vestibule is lined by a keratinized, stratified, squamous epithelium which has hairs, sebaceous glands, and sweat glands. This extends inwards for 1–2 cm and joins the nonkeratinized squamous epithelium of the posterior region. This forms a transitional zone between the keratinized epithelium of the anterior region and the respiratory mucosa of the rest of the nasal cavity. The respiratory mucosa is a pseudo-stratified, ciliated, columnar epithelium with many mucus-secreting goblet cells (Fig. 4.3). The cilia move the mucus and entrapped particles backwards through the internal nares and into the pharynx, where they are swallowed or expelled by coughing. Approximately 1 liter of fluid is secreted by the nasal epithelium (and swallowed) each day.

4.1.2 Pharynx

The pharynx is a tube that extends from the internal nares down to the larynx and consists of three main regions – the nasopharynx, the oropharynx, and the laryngopharynx (Fig. 4.4).

The nasopharynx is lined with a pseudo-stratified, ciliated, columnar epithelium, the cilia of which propel mucus towards the oral cavity (Fig. 4.3). At the back of the nasopharynx is the pharyngeal tonsil. The oropharynx, as well as having connections to the nasopharynx and laryngopharynx, has an opening to

the oral cavity (the fauces) through which food, drink, and air pass. It also contains the palatine and lingual tonsils. The oropharynx and laryngopharynx are lined with a nonkeratinized, stratified, squamous epithelium. The pharynx acts as a passageway for food, drink, and air, but also harbors the tonsils which are important lymphoid tissues.

4.1.3 Larynx

The larynx connects the pharynx with the trachea and acts as a sphincter to prevent the entrance into the lungs of anything other than air (Fig. 4.5). It is lined mainly with a ciliated, columnar epithelium.

4.1.4 Trachea

The trachea connects the larynx to the lungs and branches at its lower end to form the two bronchi. It is lined with a pseudo-stratified, ciliated, columnar epithelium containing numerous goblet cells (Fig. 4.3). The cilia move mucus and trapped particles upwards towards the pharynx.

4.1.5 Bronchi and bronchioles

Each of the first two bronchial branches from the base of the trachea (the left and right primary bronchi) divides successively to form secondary bronchi, tertiary bronchi, bronchioles, and finally terminal bronchioles. As these structures get narrower, the lining epithelium changes: hence, the bronchi are lined with a

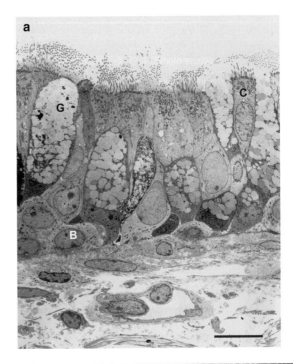

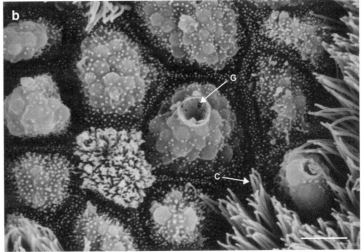

Fig. 4.3 (a) Transmission electron micrograph of a cross-section through the respiratory epithelium showing (G) goblet cells, (C) ciliated cells, and (B) basal cells. Scale bar, 10 μm. (b) Scanning electron micrograph of the respiratory epithelium showing the apices of (G) goblet cells and (C) cilia. Secretory granules can be seen within the goblet cells beneath their cytoplasmic membranes. Scale bar, 30 μm. Reproduced with permission from: Jeffery, P.K. and Li, D (1997) Airway mucosa: Secretory cells, mucus and mucin genes. *European Respiratory Journal* 10, 1655–62. © European Respiratory Society Journals Ltd, Sheffield, UK.

pseudo-stratified, ciliated, columnar epithelium containing goblet cells. Larger bronchioles have a ciliated, simple, columnar epithelium with a few goblet cells, while smaller bronchioles have a ciliated, simple, cuboidal epithelium with fewer goblet cells. The terminal bronchioles, on the other hand, have a nonciliated, simple, cuboidal epithelium without goblet cells.

4.1.6 Alveolus

The terminal bronchioles divide further to form respiratory bronchioles, each of which subdivides into several (between two and 11) alveolar ducts. Both of these structures are lined with a nonciliated, simple, cuboidal epithelium. Each alveolar duct opens into

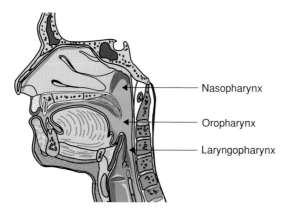

Fig. 4.4 The major regions of the pharynx in humans. (From http://www.training.seer.cancer.gov/ module_anatomy/anatomy_physiology_home.html; funded by the US National Cancer Institute's Surveillance, Epidemiology and End Results (SEER) Program with Emory University, Atlanta, Georgia, USA.)

an alveolar sac, which consists of a number of alveoli (Figs 4.5 and 4.6).

An alveolus is a cup-shaped structure lined mainly by type I alveolar cells, which are simple squamous epithelial cells that form an almost continuous lining – this is the site of gaseous exchange and constitutes the respiratory membrane. Interspersed among the type I cells are type II alveolar cells, which are cuboidal epithelial cells with numerous microvilli – these secrete alveolar fluid, and are also stem cells, and therefore can replace damaged type I cells. The alveolar fluid provides a warm, moist interface between the air and the type I cells, thereby facilitating gaseous exchange. The fluid also contains surfactants (phospholipids and proteins), which prevent the alveoli from collapsing in on themselves. The alveoli are surrounded by capillary networks, and the air in the alveoli is separated from the blood in the capillaries by the respiratory membrane, which is only approximately 0.5 μm thick, thereby facilitating gaseous diffusion. Approximately 300 million alveoli are present in the lungs, and these have a total surface area of approximately 70 m^2, thereby

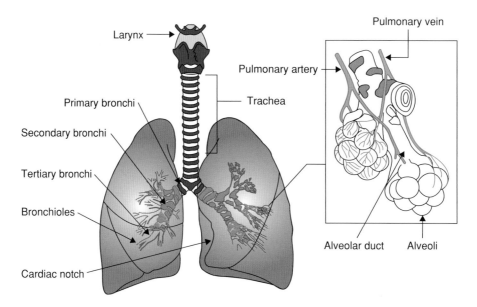

Fig. 4.5 The lower respiratory tract of humans. (From http://www.training.seer.cancer.gov/module_anatomy/ anatomy_physiology_home.html; funded by the US National Cancer Institute's Surveillance, Epidemiology and End Results (SEER) Program with Emory University, Atlanta, Georgia, USA.)

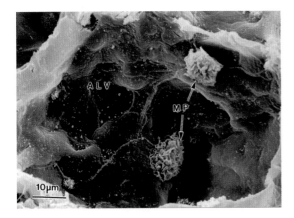

Fig. 4.6 Scanning electron micrograph of a rat lung showing the outlines of erythrocytes in the surrounding capillaries and two macrophages (MP) in the alveolar space (ALV). Alveolar macrophages constitute approximately 95% of the leukocytes in the airspaces of human lungs. Reproduced with permission from: Martin, T.R. and Frevert, C.W. (2005) Innate immunity in the lungs. *Proc Am Thorac Soc* 2, 403–11. Official journal of the American Thoracic Society © American Thoracic Society, New York, New York, USA.

enabling the rapid exchange of oxygen and carbon dioxide between blood and the atmosphere.

4.2 ANTIMICROBIAL DEFENSE SYSTEMS OF THE RESPIRATORY TRACT

Elaborate mechanisms have evolved to protect the respiratory tract from infection by the large numbers of microbes present in the considerable volumes of air (approximately 10 000 liters) inhaled each day. One of the most important of these is mucociliary clearance, which is largely responsible for maintaining the respiratory portion virtually free of microbes and particulate matter. This system involves trapping microbes and particulate matter in a layer of mucus covering the epithelial surface. The mucus is then propelled by ciliated epithelial cells towards the oropharynx, where it is either swallowed or expectorated. The system (often termed the "mucociliary escalator") is found in the posterior two thirds of the nasal cavity and the nasopharynx, and from the larynx down to, but not including,

the terminal bronchioles. Ciliated cells are the most numerous cells in these regions, and each has approximately 200 cilia on its outer surface (Figs 4.3 and 4.7). The ciliated epithelium is covered by a film (50–60 μm thick) of airway surface liquid (ASL), which consists of two layers. The lower layer (the periciliary fluid) is watery and is in the form of a sol, while the upper layer (mucous gel layer) is more viscous and forms a gel; the layers are separated by a thin layer of surfactant (Fig. 4.7).

ASL is also referred to as airway lining fluid (ALF) or epithelial lining fluid (ELF). Interspersed among the ciliated cells are a variety of secretory cells – goblet, Clara, and serous cells – as well as the openings of submucosal glands. Collectively, these cells and glands produce the periciliary fluid and mucus of the ASL. The Clara cells (Fig. 4.8) are present only in the terminal bronchioles and produce surfactant (which will be described later), and are stem cells, which can give rise to both ciliated and mucous cells.

ASL contains a variety of compounds, many of which can serve as nutrients for microbes while others have antimicrobial activities (Table 4.1) – its composition is described in section 4.3.3.

All ten Toll-like receptors (TLRs) are expressed by airway epithelial cells (Fig. 4.9), and these are thought to play an important role in regulating the composition of the indigenous microbiota, although the means by which they do so remains largely unresolved (see section 1.2.4). Most studies of TLRs have focused on their role in the recognition of pathogenic microbes rather than members of the indigenous microbiota.

4.2.1 Nasal cavity

Large microbe-laden particles in inhaled air are removed by hairs in the nostrils while smaller particles and microbes become trapped in the mucus covering the nasal mucosa. In the posterior two thirds of the nasal mucosa the mucociliary escalator propels the mucus-entrapped particles into the pharynx. However, in the anterior region of the nasal mucosa (where the mucociliary escalator does not operate) entrapped microbes are killed mainly by microbicidal compounds produced by the epithelium (Fig. 4.10) which are secreted into the nasal fluid (Table 4.1). Sneezing also provides an effective means of expelling mucus-entrapped microbes from this region.

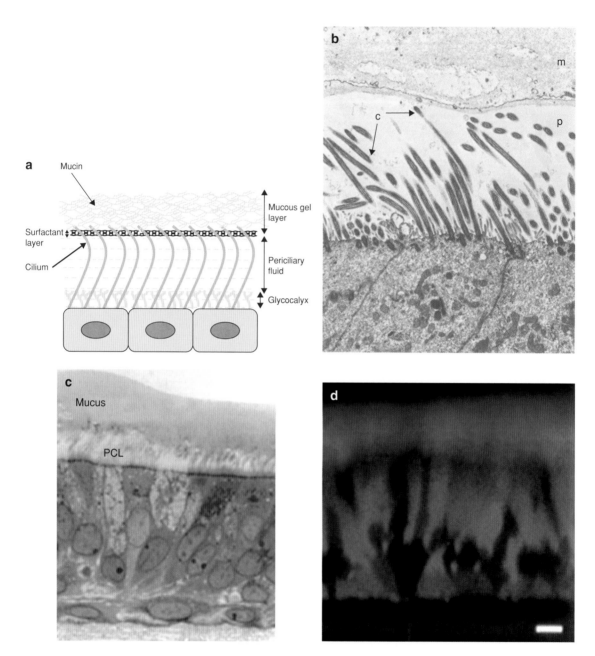

Fig. 4.7 (a) Diagrammatic representation of the main components of the mucociliary escalator. (b) Transmission electron micrograph of a cross-section through the trachea of a rat showing (p) the periciliary layer, (m) the mucus layer, and (C) the cilia. Reprinted from Yoneda, K. (1976) Mucous blanket of rat bronchus: Ultrastructural study. *Am Rev Respir Dis* 114, 837–42, © American Thoracic Society, New York, New York, USA. (c) Cross-section through a culture of human airway epithelium – the two layers of ASL can be clearly seen. PCL, periciliary layer. (d) Confocal image of living human airway epithelial culture. The cells are stained green and the ASL was visualized with Texas red dextran. Scale bar, 10 μm. (c) and (d) Reproduced with the permission of the American Society for Clinical Investigation, Ann Arbor, Michigan, USA, from: Knowles, M.R. and Boucher R.C. (2002) Mucus clearance as a primary innate defense mechanism for mammalian airways. *J Clin Invest* 109, 571–7.

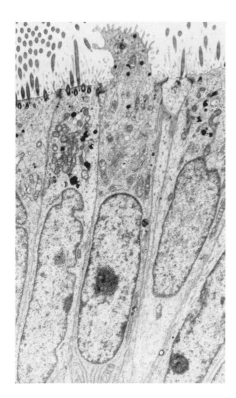

Fig. 4.8 Transmission electron micrograph of a Clara cell with its apex protruding into the airway lumen. Scale bar, 2.0 μm. Reproduced with permission from: Jeffery, P.K. and Li, D (1997) Airway mucosa: Secretory cells, mucus and mucin genes. *European Respiratory Journal* 10, 1655–62. © European Respiratory Society Journals Ltd, Sheffield, UK.

4.2.2 Other regions of the conducting portion

In the rest of the conducting portion of the respiratory tract, mucociliary clearance is one of the main defense mechanisms against microbes. In the bronchi and bronchioles, this is helped by the branching nature of the system which results in microbes and particles in inhaled air impacting onto the mucous layer. Mucus-entrapped microbes can also be expelled by sneezing and coughing. These physical methods of expelling microbes are supplemented by the presence of antimicrobial compounds in the ASL (Table 4.1). The effects of some of the antimicrobial peptides

found in ASL on respiratory microbes are shown in Fig. 4.11.

4.2.3 Respiratory portion

The mucociliary escalator does not operate in the respiratory bronchioles or in the alveoli. These epithelial surfaces are covered by alveolar lining fluid which is a plasma ultrafiltrate, together with secretions produced mainly by type II alveolar cells. The alveoli also have a substantial population of macrophages (approximately 7 macrophages per alveolus) together with smaller numbers of lymphocytes and polymorphonuclear leukocytes (PMNs). Alveolar lining fluid contains a number of antimicrobial components including lysozyme, free fatty acids, immunoglobulins (mainly IgG), iron-binding proteins, and surfactant proteins (SPs). Four SPs have been detected in the fluid (SP-A, SP-B, SP-C, and SP-D) of which SP-A and SP-D are important in defending this section of the respiratory tract from microbes. These two surfactant proteins function as opsonins, enabling phagocytosis of microbes by macrophages which are plentiful in the alveoli (Fig. 4.6). They also inhibit the growth of some Gram-negative bacteria. SP-A enhances PMN migration and stimulates the oxidative burst in PMNs.

4.3 ENVIRONMENTAL DETERMINANTS WITHIN THE RESPIRATORY TRACT

4.3.1 Atmospheric composition

In the nares, the air contains approximately 21% oxygen and 0.04% carbon dioxide. In the alveoli, the oxygen content decreases to 14%, while the carbon dioxide content increases to 5% as a result of gaseous exchange with the bloodstream. The respiratory tract, therefore, is predominantly an aerobic environment and provides atmospheric conditions suitable for the growth of obligate aerobes and facultative anaerobes. Nevertheless, as obligate anaerobes can be isolated from some regions of the respiratory tract, anaerobic microhabitats must be present. These arise as a result of oxygen consumption by aerobes and facultative species, and the growth of obligate anaerobes will be exacerbated by local anatomical features that hinder oxygen replenishment, e.g. the convoluted surfaces of some epithelial cells and the crypts of the tonsils.

Table 4.1 Antimicrobial compounds present in nasal fluid and in ASL. Details of the antimicrobial activities of the various compounds are given in Table 1.12.

Antimicrobial compound	Presence (and concentration when known)	
	Nasal fluid	Airway surface liquid
Lysozyme	250–500 µg/ml	0.1–1.0 mg/ml
Lactoferrin	80–200 µg/ml	0.1–1.0 mg/ml
Secretory leukocyte proteinase inhibitor	10–80 µg/ml	0.01–0.1 mg/ml
Secretory phospholipase A$_2$	Yes	Yes
Human β-defensin-1	Yes	1 µg/ml
Human β-defensin-2	0.3–0.4 µg/ml	1 µg/ml
Human neutrophil peptide-1	Yes	10 µg/ml
LL-37	Yes	Yes
Elafin	70 ng/ml	Yes
Statherin	Yes	NDA
Lactoperoxidase	Yes	0.65 mg/mg of secreted protein
Bactericidal/permeability-increasing protein (BPI)	Yes	Yes
Glandulin	Yes	NDA
Anionic peptide	NDA	0.8–1.3 mM
Nitric oxide and reactive nitrogen species	Yes	Yes
SPLUNC1	Yes	Yes
IgA	49–218 mg/100 ml	5–200 mg/100 ml
IgG	14–136 mg/100 ml	9–200 mg/100 ml
IgM	<10 mg/100 ml	Yes

NDA, no data available.

4.3.2 pH

The pH of ASL is generally slightly acidic throughout the respiratory tract, with a mean value of 6.78. In the nasal cavity, the pH of the mucosa gradually increases from the anterior nares (pH = 5.5) to almost neutral (pH = 6.95) at a distance of 6 cm from the tip of the nose.

4.3.3 Nutrients

The main types of host-derived nutrients available to the respiratory microbiota depend on the anatomical region (Fig. 4.12).

In all regions, the fluid present on the mucosal surface contains a plasma transudate. In addition, food and saliva passing through the pharynx may serve as additional sources of nutrients for microbes colonizing this region. However, the transit time of food and saliva is very rapid (between 1 and 2 s per swallow), making it unlikely that significant quantities of dietary constituents would be transferred to the ASL.

4.3.3.1 Composition of nasal fluid, ASL, and alveolar lining fluid

Nasal fluid contains secretions from the nasal mucosa and submucosal glands, the transudate from nasal blood vessels, products of cells that are resident in the mucosa (plasma cells, lymphocytes, etc.), and tears, which enter via the nasolacrimal ducts. More than 1000 different proteins have been detected, but fewer than half of these have been identified. The osmolarity of nasal fluid is 277 mOsm/liter, which is very similar to that of plasma (285 mOsm/L), and its pH ranges from 5.5 to 7.0. Approximately 1 liter of fluid is produced by the nasal epithelium each day, and its main constituents are listed in Table 4.2.

A total of 20–100 ml of ASL is produced each day. It consists mainly of water (90–95%), mucins, and proteins (Table 4.2).

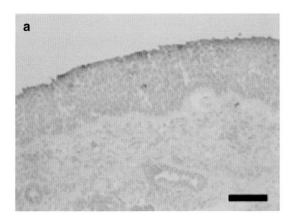

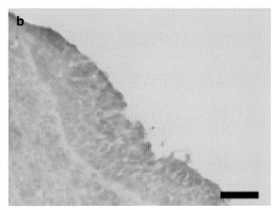

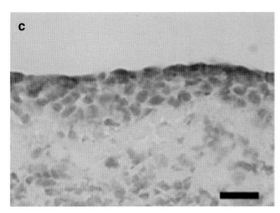

Fig. 4.9 Immunohistochemical localization (bright red staining regions) of (a) TLR 3, (b) TLR 4, and (c) TLR 2 in biopsies of nasal mucosa from healthy individuals. The most intensive immunoreactivity for all three TLRs was within the airway epithelium, particularly in epithelial cells in the apical region of the epithelium. Staining was also apparent in a few scattered intraepithelial leukocytes. (a) and (c) scale bars, 85 μm; (b) scale bar, 30 μm. Reprinted from: Fransson, M., Adner, M., Erjefält, J., Jansson, L., Uddman, R. and Cardell, L.O. (2005) Up-regulation of Toll-like receptors 2, 3 and 4 in allergic rhinitis. *Respiratory Research* 6, 100 doi:10.1186/1465-9921-6-100. © 2005 Fransson et al; licensee BioMed Central Ltd, London, UK.

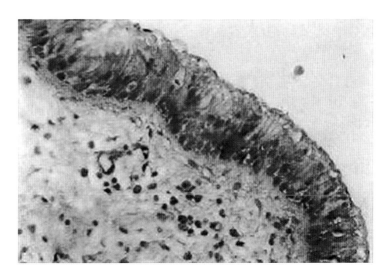

Fig. 4.10 Immunostaining of HBD-2 in the nasal mucosa of a healthy adult. Positive expression of HBD-2 (dark red staining regions) can be seen in the epithelial cells, especially in the basal layer (magnification ×400). Reprinted from: Chen, P.H. and Fang, S.Y. (2004) Expression of human beta-defensin 2 in human nasal mucosa. *Eur Arch Otorhinolaryngol* 261, 238–41 with kind permission of Springer Science and Business Media, Berlin, Germany.

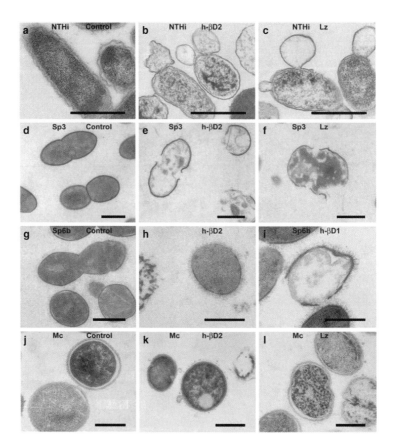

Fig. 4.11 Ultrastructural changes in (a) to (c) *H. influenzae*; (d) to (f) *Strep. pneumoniae* strain 3; (g) to (i) *Strep. pneumoniae* strain 6B, and (j) to (l) *Mor. catarrhalis* exposed to lysozyme, human β-defensin-2, or human β-defensin-1. Untreated *H. influenzae* is shown in (a). Bacteria were treated for 30 min with (b) β-defensin-2 (10 μg/ml), or with (c) 1 mg/ml human lysozyme. Scale bar, 0.5 μm. Untreated *Strep. pneumoniae* strain 3 is shown in (d). Bacteria were treated for 30 min with (e) β-defensin-2 (10 μg/ml), or with (f) 1 mg/ml human lysozyme. Scale bar, 0.5 μm. Untreated *Strep. pneumoniae* strain 6B is shown in (g). Bacteria were treated (h) for 30 min with β-defensin-2 (10 μg/ml), or (i) for 3 h with β-defensin-1 (10 μg/ml). Scale bar, 0.5 μm. Untreated *Mor. catarrhalis* is shown in (j). Bacteria were treated for 30 min with (k) β-defensin-2 (10 μg/ml), or with (l) 1 mg/ml human lysozyme. Scale bar, 0.5 μm. Reproduced from: Lee, H.Y., Andalibi, A., Webster, P. et al. (2004) Antimicrobial activity of innate immune molecules against *Streptococcus pneumoniae*, *Moraxella catarrhalis* and nontypeable *Haemophilus influenzae*. *BMC Infect Dis* 4, 12. Published online May 5, 2004. doi: 10.1186/1471-2334-4-12. Copyright © 2004 Lee et al.; licensee BioMed Central Ltd, London, UK. This is an Open Access article: verbatim copying and redistribution of this article are permitted in all media for any purpose, provided this notice is preserved along with the article's original URL.

Alveolar lining fluid is a complex mixture of proteins and lipids. The total protein content is in the region of 9.0 mg/ml, of which approximately half is albumin. It also contains a mixture of surface-active compounds, known as pulmonary surfactant, which lowers the surface tension of the fluid, thereby preventing the alveoli from collapsing during expiration. Pulmonary surfactant consists mainly of phospholipids (approximately

92%), the rest being protein together with small quantities of vitamins C and E. Its pH is approximately 6.9

4.3.3.2 Contribution of microbial residents of the respiratory tract to nutrient availability

In all regions of the respiratory tract, the pool of available nutrients in respiratory secretions is increased by

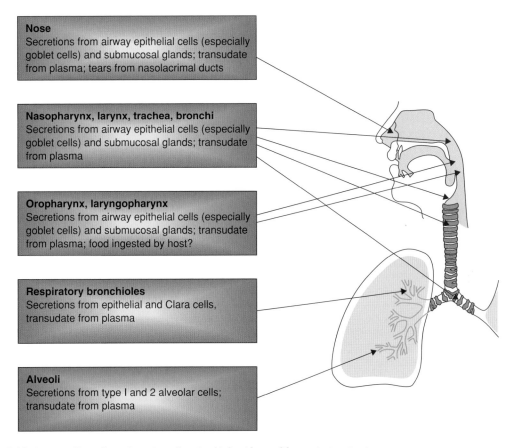

Nose
Secretions from airway epithelial cells (especially goblet cells) and submucosal glands; transudate from plasma; tears from nasolacrimal ducts

Nasopharynx, larynx, trachea, bronchi
Secretions from airway epithelial cells (especially goblet cells) and submucosal glands; transudate from plasma

Oropharynx, laryngopharynx
Secretions from airway epithelial cells (especially goblet cells) and submucosal glands; transudate from plasma; food ingested by host?

Respiratory bronchioles
Secretions from epithelial and Clara cells, transudate from plasma

Alveoli
Secretions from type I and 2 alveolar cells; transudate from plasma

Fig. 4.12 Sources of host-derived nutrients for microbial residents of the respiratory tract.

the activities of some members of the indigenous microbiota. Hence, macromolecules in the respiratory secretions will be converted by proteases, sialidases, glycosidases, lipases, etc. to carbohydrates, amino acids, and fatty acids for use as carbon, nitrogen, and energy sources (Fig. 4.13).

4.4 INDIGENOUS MICROBIOTA OF THE RESPIRATORY TRACT

4.4.1 Members of the respiratory microbiota

Each of the various regions of the upper respiratory tract is colonized by a wide range of microbes. Those most frequently detected include viridans streptococci, *Streptococcus pyogenes*, *Streptococcus pneumoniae*, *Neisseria* spp., *Haemophilus* spp., *Moraxella* spp., *Staphylococcus*

aureus, coagulase-negative staphylococci (CNS), *Corynebacterium* spp., *Propionibacterium* spp., *Prevotella* spp., *Porphyromonas* spp., *Mollicutes* (*Mycoplasma* spp. and *Ureaplasma* spp.), and *Kingella kingae*. *Corynebacterium* spp., CNS, and *Propionibacterium* spp. are described in Chapter 2, while *Prevotella* spp. and *Porphyromonas* spp. are major constituents of the oral microbiota and are described in Chapter 8. The other residents of the respiratory tract are described below.

4.4.1.1 *Streptococcus* spp.

Streptococci are Gram-positive spherical or ovoid cocci which usually occur in pairs or chains. The genus consists of at least 39 species, all of which are nutritionally fastidious and are incapable of respiratory metabolism. Some species require high CO_2 levels (5%) for growth, and the growth of most species is stimulated by

Table 4.2 Main constituents of nasal fluid and ASL. Constituents with antimicrobial properties are shown in Table 4.1.

Constituent	Presence and concentration	
	Nasal fluid	Airway surface liquid
Proteins	414–895 mg/100 ml	3 g/100 ml
Mucins	52–112 mg/100 ml	0.5–1.0 g/100 ml
Albumin	31–105 mg/100 ml	48–73 mg/100 ml
Lipocalin-1	8–18 mg/100 ml	NDA
DNA	40 µg/ml	280 µg/ml
Uric acid	5–16 µmol/l	NDA
Hyaluronic acid	NDA	3 mg/100 ml
Heparin	NDA	Yes
Chondroitin sulfate	NDA	Yes
IgA	49–218 mg/100 ml	5–200 mg/100 ml
IgG	14–136 mg/100 ml	9–200 mg/100 ml
Carbohydrates	NDA	950 mg/100 ml
Lipids	NDA	840 mg/100 ml (mainly phospholipids)
Urea	3.3 mM	NDA
Glutathione	NDA	429 µmol/l
Na^+	98–225 mM	80–85 mM
Cl^-	158–188 mM	75–80 mM
K^+	23–68 mM	15 mM
Ca^{2+}	3–14 mM	NDA
HPO_4^{2-}	3–7 mM	NDA

NDA, no data available.

increased concentrations of the gas. Some species are aciduric and can survive at pHs as low as 4.1. The older classification systems based on the hemolytic reactions (α-, β-, and nonhemolytic) and immunochemical properties (Lancefield groups A, B, C, D, etc.) of the different species are still widely used to divide the genus into a number of major groups. The main characteristics of *Streptococcus* spp. are as follows:

• Gram-positive cocci in pairs or chains;
• Facultative anaerobes;
• Nonsporing;
• Catalase-negative;
• Nonmotile;
• G+C content of DNA is 34–46 mol%;
• Growth over the temperature range 20–42°C;
• Optimum growth at approximately 37°C;
• Fermentation of carbohydrates to produce mainly lactate.

4.4.1.1.1 Strep. pyogenes
Strep. pyogenes is a β-hemolytic streptococcus (i.e. its colonies are surrounded by clear zones of hemolysis on blood agar) and has a Lancefield group A cell-wall antigen – it is also frequently known as the "Group A streptococcus". The genomes of several strains of the organism have been sequenced (http://cmr.tigr.org/tigr-scripts/CMR/GenomePage.cgi?org=ntsp16). *Strep. pyogenes* can bind to mucin, one of the main components of ASL, by means of two proteinaceous adhesins. One of these is the M protein, while the identity of the other has not been determined. α2-6-Linked sialic acid residues of the mucin molecule are the complementary receptors for these bacterial adhesins. Sialic acid residues of membrane proteins of pharyngeal cells are also the receptors for M protein-mediated adhesion to these epithelial cells. On the basis of the antigenicity of the N-terminal region of its M protein, more than 80 serotypes of *Strep. pyogenes* are recognized. Another typing scheme (based on the genes encoding M and M-like proteins – *emm* genes) has resulted in the recognition of 5 *emm* patterns – A to E. Other adhesins that may be involved in the adhesion of the organism to respiratory epithelial cells and other substrates in the respiratory tract are listed in Table 4.3.

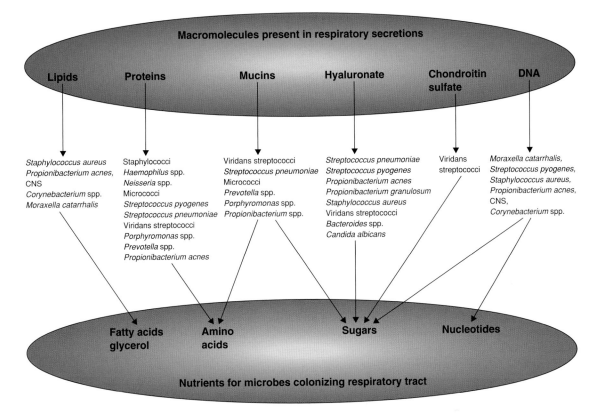

Fig. 4.13 Host-derived macromolecules as nutrient sources for the respiratory microbiota.

Table 4.3 Adhesins of *Strep. pyogenes* and their complementary receptors.

Adhesin	Receptor
Lipoteichoic acid	Fibronectin
M protein	Fibrinogen, fibronectin, laminin, sialic acid-containing molecules, heparan sulfate, heparin, galactose
Sfb1 (streptococcal fibronectin-binding protein 1)	Fibronectin
Protein F2	Fibronectin
Serum opacity factor	Fibronectin
Fbp54 (fibronectin-binding protein 54)	Fibronectin
Fba (fibronectin-binding protein a)	Fibronectin
Glyceraldehyde-3-phosphate dehydrogenase	Fibronectin, fibrinogen
Hyaluronic acid	CD44 ligands
Cpa (collagen-binding protein of Group A streptococci)	Type I collagen
SpeB cysteine protease	Integrins, laminin
Vitronectin-binding protein	Vitronectin
Lbp (laminin-binding protein)	Laminin
Pullulanase	Not known

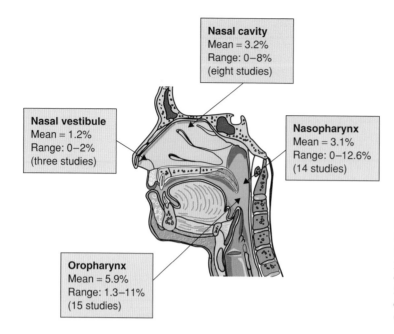

Nasal cavity
Mean = 3.2%
Range: 0–8%
(eight studies)

Nasal vestibule
Mean = 1.2%
Range: 0–2%
(three studies)

Nasopharynx
Mean = 3.1%
Range: 0–12.6%
(14 studies)

Oropharynx
Mean = 5.9%
Range: 1.3–11%
(15 studies)

Fig. 4.14 Prevalence of *Strep. pyogenes* at various sites within the respiratory tract. The figures represent the proportion of individuals (adults and children) harboring the organism at the indicated site and are based on studies carried out in a number of countries (1–26).

Strep. pyogenes produces a number of extracellular enzymes that can break down macromolecules present in airway fluids to generate molecules that can act as carbon, nitrogen, and energy sources. These include a hyaluronidase (which releases glucuronic acid and N-acetylglucosamine), several proteinases, deoxyribonucleases (at least three have been identified), and a pullulanase (which degrades pullulan and starch). M-protein type 57 strains of the organism produce a bacteriocin, streptococcin A-57, that is inhibitory for a number of Gram-positive species.

The main sites within the respiratory tract colonized by *Strep. pyogenes* are the oropharynx, nasopharynx, and nasal cavity, although the organism is also encountered, but less frequently, in the nasal vestibule (Fig. 4.14).

The rates of carriage of the organism vary widely between different studies, and such differences are related to the age of the population investigated, genetic background variables, and socioeconomic conditions such as housing, occupation, access to health care, hygiene levels, family size, overcrowded living conditions, school attendance, etc.

Strep. pyogenes is responsible for a wide range of infections including pharyngitis, tonsillitis, impetigo, erysipelas, bacteremia, meningitis, pneumonia, necrotizing fasciitis, puerperal sepsis, myositis, cellulitis, pericarditis,

septic arthritis, streptococcal toxic shock syndrome, acute rheumatic fever, acute post-streptococcal glomerulonephritis, and reactive arthritis.

4.4.1.1.2 Strep. pneumoniae
Strep. pneumoniae is an α-hemolytic streptococcus (i.e. on blood agar, its colonies are surrounded by a greenish zone of partial hemolysis), which is distinguished from other members of this group by its susceptibility to ethylhydrocupreine (optochin). The genomes of a number of strains of the organism have been sequenced (http://cmr.tigr.org/tigr-scripts/CMR/GenomePage.cgi?org=bsp). The cocci are usually arranged in pairs which are enclosed within a polysaccharide capsule. More than 90 serotypes can be distinguished on the basis of the antigenicity of the capsule.

A number of adhesins are involved in maintaining the organism in the respiratory tract, and these are listed in Table 4.4. In some cases, the receptors for these adhesins remain to be identified. However, glycosaminoglycans are considered to be important in adhesion (Fig. 4.15) and contain receptors for some of the organism's adhesins.

The organism produces a number of macromolecule-degrading exoenzymes (a hyaluronidase, three sialidases, a glycosidase, and several proteases) that could

Table 4.4 Adhesins of *Strep. pneumoniae* and their complementary receptors.

Adhesin	Receptor
Choline binding protein A (CbpA) (also known as SpsA and PspC)	Polymeric immunoglobulin receptor (pIgR)
6-Phosphogluconate dehydrogenase	NDA
Pneumococcal adherence and virulence factor A (PavA)	Fibronectin
Glyceraldehyde-3-phosphate dehydrogenase	Plasmin, plasminogen
Alpha-enolase	Plasmin, plasminogen
Pneumococcal surface adhesin A	NDA

NDA, no data available.

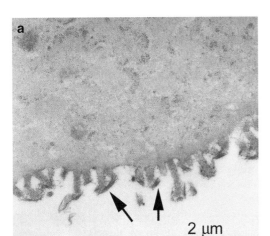 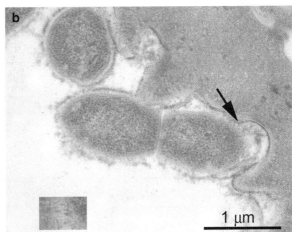

Fig. 4.15 Adhesion of *Strep. pneumoniae* to nasopharyngeal epithelial cells and the role of glycosaminoglycans as receptors. (a) Transmission electron micrograph of a secretory nasopharyngeal epithelial cell. Glycosaminoglycans (GAG) are present on the microvilli as bristle structures of varying length (arrows). (b) Electron micrograph showing attachment of *Strep. pneumoniae* to the cells. Intermingled GAG structures are present (arrow) at one of the adhesion sites and are shown in the insert at higher magnification. Reprinted from: Tonnaer, E.L., Hafmans, T.G., Van Kuppevelt, T.H., Sanders, E.A., Verweij, P.E. and Curfs, J.H. (2006) Involvement of glycosaminoglycans in the attachment of pneumococci to nasopharyngeal epithelial cells. *Microbes and Infection* 8, 316–22, ©2006, with permission from Elsevier, Amsterdam, The Netherlands.

generate a range of microbial nutrients from host polymers.

The main sites within the respiratory tract colonized by *Strep. pneumoniae* are the oropharynx, nasopharynx, and nasal cavity, although the organism may also be present in the nasal vestibule (Fig. 4.16).

The frequency of carriage is much greater in children than in adults, and virtually all children are colonized with *Strep. pneumoniae* sometime during the first 2 years of life – acquisition of the organism occurs either during the first few months of life or else after the age of 6 months. The highest carriage rates are among pre-school-age children (Fig. 4.17). Carriage is affected by a variety of factors including genetic background variables, vaccination status, and socioeconomic conditions, and is generally much higher in children in developing countries.

Strep. pneumoniae is an important human pathogen, being responsible for a range of diseases including pneumonia, meningitis, otitis media, and sinusitis.

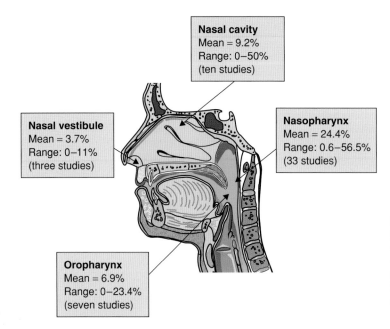

Nasal cavity
Mean = 9.2%
Range: 0–50%
(ten studies)

Nasal vestibule
Mean = 3.7%
Range: 0–11%
(three studies)

Nasopharynx
Mean = 24.4%
Range: 0.6–56.5%
(33 studies)

Oropharynx
Mean = 6.9%
Range: 0–23.4%
(seven studies)

Fig. 4.16 Prevalence of *Strep. pneumoniae* at various sites within the respiratory tract. The figures represent the proportion of individuals (adults and children) harboring the organism at the indicated site and are based on studies carried out in a number of countries (1, 3, 5–10, 12, 17, 19, 20, 23, 27–45)

4.4.1.1.3 Viridans group streptococci

The viridans group streptococci are comprised of α-hemolytic streptococci other than *Strep. pneumoniae*. Some viridans streptococci produce a sialidase, and many produce a range of glycosidases and are, therefore, able to release sugars from respiratory mucins. Some also produce a hyaluronidase, chondroitin sulfatase, protease, and urease and can, therefore, degrade many of the constituents of respiratory secretions to produce sugars, amino acids, and ammonium ions. The lactate produced as a metabolic end product can serve as a carbon and/or energy source for other members of the respiratory microbiota. Many species are able to inhibit other members of the respiratory microbiota, particularly *Strep. pyogenes*, *Strep. pneumoniae*, *Moraxella catarrhalis*, and *Haemophilus influenzae*, although the mechanisms involved remain unclear. As viridans streptococci are also among the predominant members of the oral microbiota, further information on them is provided in Chapter 8.

Because viridans streptococci are not regarded as major human pathogens (although they do cause serious infections such as bacterial endocarditis), relatively little is known regarding which species are present, or their prevalence, at different sites within the respiratory tract. The situation is complicated by the fact that a number of species are usually present at a particular site and it is difficult to differentiate between them – they are often reported simply as "viridans streptococci". The nasopharynx and oropharynx are major sites of colonization, and the organisms are also frequently present in the nasal cavity and vestibule. Species encountered most frequently in the nasopharynx and oropharynx include *Streptococcus mitis* biovar 1, *Strep. mitis* biovar 2, *Streptococcus oralis*, *Streptococcus salivarius*, *Streptococcus vestibularis*, *Streptococcus gordonii*, *Streptococcus sanguinis*, *Streptococcus anginosus*, and *Strep. mitior*.

4.4.1.2 *Neisseria* spp.

Neisseria spp. are Gram-negative cocci that frequently occur in pairs, and their main characteristics are as follows.

• Gram-negative cocci;
• Aerobes;
• Growth stimulated by CO_2;
• Anaerobic growth in the presence of nitrite as a terminal electron acceptor;
• Nonsporing;
• Nonmotile;
• G+C content of their DNA is 47–52 mol%;
• Growth over the temperature range 22–40°C;
• Optimal growth at 35–38°C;

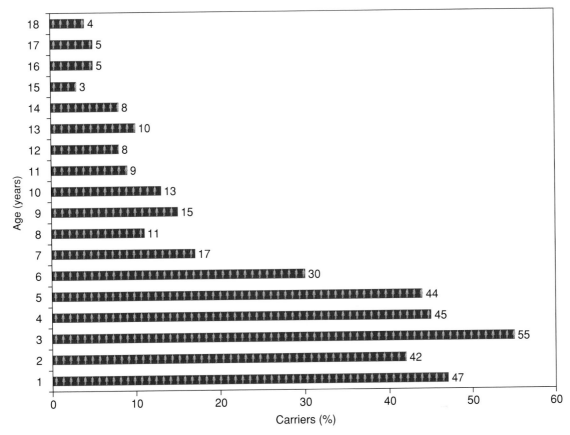

Fig. 4.17 Results of a study of nasopharyngeal carriage of *Strep. pneumoniae* in 3198 healthy children in The Netherlands. The peak incidence occurred during the first 3 years of life, followed by a gradual decline until a stable colonization rate was reached after the age of 10 years (34).

- Catalase-positive;
- Oxidase-positive;
- Oxidize carbohydrates to produce mainly acetate;
- Growth over the pH range 6.0–8.0;
- Optimum growth at pH 7.4–7.6.

Sixteen species are recognized in the genus and, apart from *Neisseria meningitidis* and *Neisseria gonorrhoeae*, they are not nutritionally fastidious.

4.4.1.2.1 N. meningitidis

N. *meningitidis*, also known as the meningococcus, is a member of the indigenous microbiota of the nasopharynx and is an important human pathogen. Its 2.27-Mbp genome has been sequenced (http://cmr.tigr.org/tigr-scripts/CMR/GenomePage.cgi?org=gnm). Thirteen

serogroups (A, B, C, D, etc.) of the organism are recognized on the basis of the antigenicity of the capsular polysaccharide. Further differentiation into 20 serotypes and ten subtypes is based on the antigenic nature of porin proteins and other outer-membrane proteins, respectively. In addition, 13 immunotypes can be distinguished on the basis of the antigenicity of the lipooligosaccharide – these are prefixed by the letter L. Although serotyping has been widely used for epidemiological studies of the organism, this is now being replaced by molecular approaches such as multilocus enzyme electrophoresis, DNA fingerprinting, and PCR. N. *meningitidis* can utilize only a limited number of substrates as carbon and energy sources – these include glucose and lactate. Acetate is a major end product of

Table 4.5 Adhesins of N. *meningitidis* and their complementary receptors.

Adhesin	Receptor
PilC proteins on type IV pilus	CD46 and other unknown proteins
PilE protein on type I pili	CD46 and other unknown proteins
MspA (meningococcal serine protease A)	Unidentified receptor on human bronchial epithelial cells
App (adhesion and penetration protein)	NDA
NadA (Neisseria adhesin A)	NDA
NhhA (Neisseria hia/hsf homologue)	Laminin, heparan sulfate
Opacity (Opa) proteins	Carcinoembryonic antigen cell adhesion molecules (CEACAM or CD66); heparan sulfate; vitronectin
Opc proteins	Heparan sulfate, vitronectin
Lipooligosacccharide	Asialoglycoproteins

NDA, no data available.

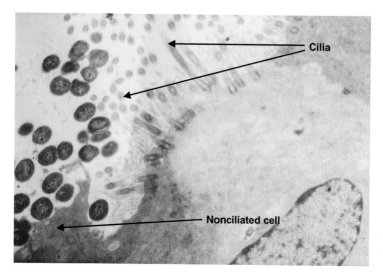

Fig. 4.18 Scanning electron micrograph of human nasopharyngeal explant superinfected with N. *meningitidis* for 24 h. Bacteria can be seen adhering to a nonciliated epithelial cell but not to ciliated cells. Reprinted with the permission of the Society for General Microbiology, Reading, UK, from: Read, R.C., Fox, A., Miller, K., Gray, T., Jones, N., Borrows, R., Jones, D.M. and Finch, R.G. (1995) Experimental infection of human nasal mucosal explants with *Neisseria meningitidis*. J Med Microbiol 42, 353–61.

metabolism. A variety of adhesins have been identified as being important in mediating adhesion of the organism to the respiratory mucosa (Fig. 4.18) and/or macromolecules present in respiratory secretions, and these are listed in Table 4.5. The crystal structure of one important adhesin is shown in Fig. 4.19.

N. *meningitidis* produces an IgA protease and a serine protease which are able to partially hydrolyze proteins present in respiratory secretions. It also produces acetate as a major metabolic end product that can be used as a carbon and energy source by other organisms.

Approximately 10% of the population harbor N. *meningitidis* in their upper respiratory tract at any one time, the main sites colonized being the nasopharynx and oropharynx, although the organism has also been detected in the nasal cavity as well as in the mouth (Fig. 4.20). Carriage rates are highest among teenagers and lowest in infants (Fig. 4.21). Broadly speaking, the carriage rate is usually less than 3% in children younger than 4 years, 24–37% in teenagers and young adults (15–24 years), and less than 10% in older age-groups.

The main diseases caused by the organism are meningitis and septicemia, but it may also be responsible for

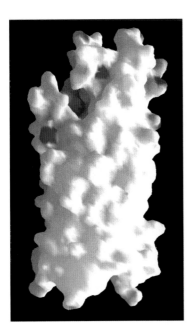

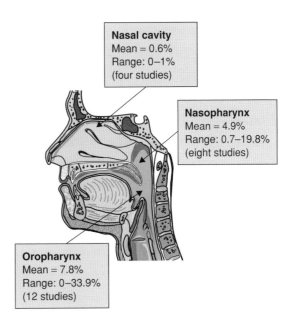

Nasal cavity
Mean = 0.6%
Range: 0–1%
(four studies)

Nasopharynx
Mean = 4.9%
Range: 0.7–19.8%
(eight studies)

Oropharynx
Mean = 7.8%
Range: 0–33.9%
(12 studies)

Fig. 4.19 Surface representations of the *N. meningitidis* adhesin OpcA, colored by electrostatic potential, with blue as positive charge and red as negative charge. The proposed proteoglycan-binding site with its concentration of positive charge is visible at the top left of the molecule. Reproduced with permission from: Prince, S.M., Achtman, M. and Derrick, J.P. (2002) Crystal structure of the OpcA integral membrane adhesin from *Neisseria meningitidis*. *Proc Natl Acad Sci U S A* 99, 3417–21. Copyright 2002 National Academy of Sciences, Washington, DC, USA.

Fig. 4.20 Prevalence of *N. meningitidis* at various sites within the respiratory tract. The figures represent the proportion of healthy individuals (adults and children) harboring the organism at the indicated site and are based on studies carried out in a number of countries (12, 19, 20, 23, 33, 45–55)

septic arthritis, pneumonia, conjunctivitis, pericarditis, otitis, and sinusitis.

4.4.1.2.2 Other Neisseria *spp.*

Colonization of the respiratory tract by *Neisseria lactamica*, an organism closely related to *N. meningitidis*, has attracted considerable interest as it is thought that the presence of this organism protects against colonization by the latter (Fig. 4.21). Such protection is likely to be due to the induction by *N. lactamica* of cross-reactive antibodies. The frequency of isolation of the organism is high during the first 4 years (see Fig. 4.21), but then progressively decreases. Few studies have investigated the occurrence of other *Neisseria* spp. in the respiratory tract. However, the most frequent colonizer of both the oropharynx and nasopharynx appears to be *Neisseria perflava/Neisseria sicca* (Fig. 4.22).

4.4.1.3 *Haemophilus* spp.

Haemophilus spp. are Gram-negative pleomorphic organisms which, depending on the growth conditions, may appear as coccobacilli or filamentous rods. *Haemophilus* spp. are nutritionally exacting, and all species require hemin and/or nicotinamide adenine dinucleotide for growth in vitro. The main characteristics of *Haemophilus* spp. are as follows:

- Gram-negative bacilli;
- Many species are capsulated;
- Aerobes;
- Growth generally enhanced by elevated levels (5–10%) of CO_2;
- Nonmotile;
- Nonsporing;
- G+C content of their DNA is 37–44 mol%;
- Growth over the temperature range 20–40°C;
- Optimum growth at 35–37°C;
- Optimum pH for growth is 7.6;
- Fermentation of carbohydrates to produce succinic, lactic, and acetic acids.

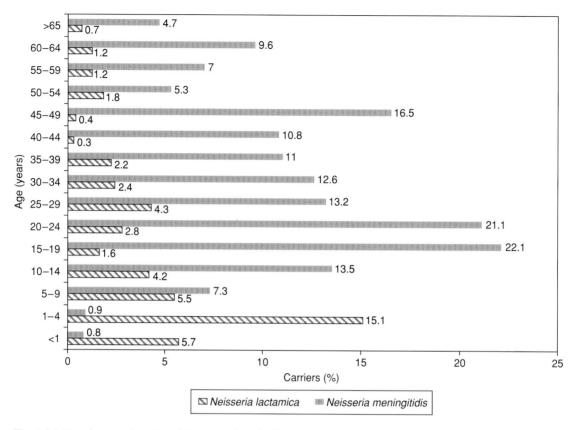

Fig. 4.21 Nasopharyngeal carriage of *N. meningitidis* and *N. lactamica* in 9439 healthy individuals in the UK and the USA. Bars denote the proportion of individuals in each age group who are colonized by the organism (55).

4.4.1.3.1 H. influenzae

Of the haemophili, the species most extensively investigated is *H. influenzae*, and its 1.83-Mbp genome has been sequenced (http://cmr.tigr.org/tigr-scripts/CMR/GenomePage.cgi?org=ghi). On the basis of the structure and antigenicity of their capsules, six serotypes (a to f) of *H. influenzae* are recognized. The most well studied of these is *H. influenzae* type b (Hib), which is responsible for many of the severe, invasive diseases caused by the organism. Strains that do not have a capsule are usually referred to as nontypeable *H. influenzae* (NTHi) although, strictly speaking, they should be termed "nonencapsulated *H. influenzae*".

The main adhesins of the organism are listed in Table 4.6, and adhesion of the organism to human nasopharyngeal cells is shown in Fig. 4.23.

The organism produces an IgA protease which can partially degrade IgA – subsequent hydrolysis by other proteases could release amino acids for use as a carbon, nitrogen, or energy source. It also produces a sialidase. Its metabolic end products (succinic, lactic, and acetic acids) can serve as carbon and energy sources for other respiratory microbes. Most type b strains produce a bacteriocin, hemocin, which is toxic to nearly all non-type b strains of the organism, some other *Haemophilus* spp., and some members of the *Enterobacteriaceae*, e.g. *Escherichia coli*. Hemocin may, therefore, play a role in maintaining type b strains in the respiratory tract.

The main sites within the respiratory tract colonized by *H. influenzae* are the oropharynx and nasopharynx, although the organism may also be present in the nasal cavity and nasal vestibule (Fig. 4.24). It is also found in the mouth.

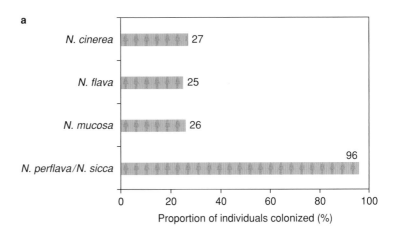

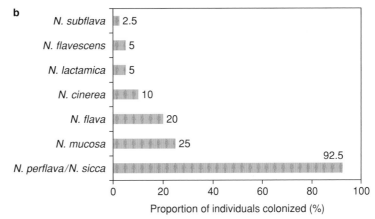

Fig. 4.22 Frequency of isolation of various *Neisseria* spp. from (a) the oropharynx of 175 healthy adults (56) and (b) the nasopharynx of 40 healthy adults (88).

Table 4.6 Adhesins of *H. influenzae* and their complementary receptors.

Adhesin	Receptor(s) and/or target cells
Hap (*Haemophilus* adhesion and penetration protein)	Fibronectin, laminin, and type IV collagen
HMW1 protein (125 kDa)	α2-3 Sialylated glycoproteins
HMW2 protein (120 kDa)	NDA
Hia (*H. influenzae* adhesin) (a 115-kDa protein)	Oropharyngeal and laryngeal cells
Hsf (*Haemophilus* surface fibril) (a 240-kDa protein)	Oropharyngeal and laryngeal cells
HifA protein in pili	Buccal cells, mucus, gangliosides
P2 (outer membrane protein)	Sialic acid-containing moieties
P5 fimbrin	Carcinoembryonic antigen cell adhesion molecule 1 (CEACAM1), intercellular adhesion molecule 1 (ICAM-1; CD54), respiratory mucin
OapA (opacity-associated protein A)	Nasopharyngeal cells

NDA, no data available.

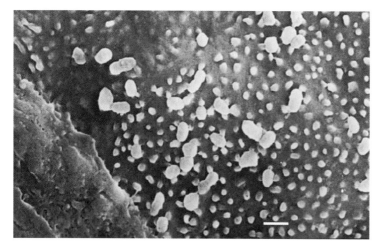

Fig. 4.23 Scanning electron micrograph of *H. influenzae* adhering to the surface elevations (microplicae) of human pharyngeal epithelial cells. Scale bar, 1 μm. Reprinted from: Ndour, C.T., Ahmed, K., Nakagawa, T., Nakano, Y., Ichinose, A., Tarhan, G., Aikawa, M. and Nagatake, T. (2001) Modulating effects of mucoregulating drugs on the attachment of *Haemophilus influenzae*. *Microbial Pathogenesis* 30, 121–7, ©2001, with permission from Elsevier, Amsterdam, The Netherlands.

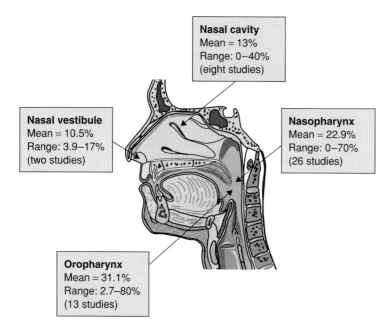

Nasal cavity
Mean = 13%
Range: 0–40%
(eight studies)

Nasal vestibule
Mean = 10.5%
Range: 3.9–17%
(two studies)

Nasopharynx
Mean = 22.9%
Range: 0–70%
(26 studies)

Oropharynx
Mean = 31.1%
Range: 2.7–80%
(13 studies)

Fig. 4.24 Prevalence of *H. influenzae* at various sites within the respiratory tract. The figures represent the proportion of healthy individuals (adults and children) harboring the organism at the indicated site and are based on studies carried out in a number of countries (3, 5–10, 12, 17, 19–23, 25, 27–33, 38, 43, 45, 57–60).

Carriage of Hib has been reduced to very low levels in those countries with comprehensive vaccination programs against the organism. Carriage of other serotypes is generally low in the first 6 months of life, reaches a maximum between the ages of 3 and 5 years, and then gradually declines in adulthood.

H. influenzae is an important human pathogen, and the main diseases for which it is responsible include meningitis, pneumonia, epiglottitis, otitis media, sinusitis, bronchitis, conjunctivitis, and arthritis.

4.4.1.3.2 Other Haemophilus *spp.*
Few studies have been carried out to determine the identity of other *Haemophilus* spp. in the respiratory tract.

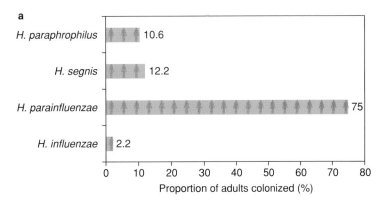

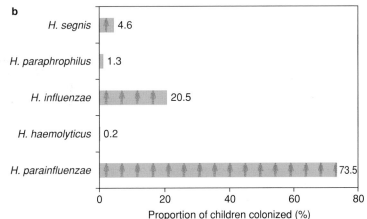

Fig. 4.25 Frequency of isolation of various *Haemophilus* spp. from the oropharynx of (a) ten healthy adults and (b) ten healthy children (2–7 years old) (60).

However, in one such study, *Haemophilus parainfluenzae* was found to be the most frequently detected *Haemophilus* sp. in the oropharynx of adults and children – other *Haemophilus* species present were *H. segnis, H. paraphrophilus,* and *H. haemolyticus* (Fig. 4.25).

4.4.1.4 *Moraxella catarrhalis*

Mor. catarrhalis is an aerobic, nonmotile, Gram-negative coccus, which is usually seen as pairs of cocci. It can grow over the temperature range 20–42°C, but growth is optimal at 37°C. It is catalase- and oxidase-positive. It does not produce acid from glucose or from other carbohydrates. The main adhesins of the organism and their complementary receptors are shown in Table 4.7. As can be seen in Fig. 4.26, the pattern of adhesion of the organism to

human bronchial epithelial cells depends on the particular strain.

Mor. catarrhalis produces a DNAse, phospholipase, and an esterase which, by hydrolyzing macromolecules in respiratory secretions, contributes to the pool of nutrients available to the respiratory microbiota.

The main sites within the respiratory tract colonized by *Mor. catarrhalis* are the oropharynx, the nasopharynx, and the nasal cavity, although the organism may also be found in the nasal vestibule (Fig. 4.27).

Mor. catarrhalis can be detected in most children during the first year of life, whereas few adults (approximately 3%) are colonized by the organism. As is the case with other members of the respiratory microbiota, carriage rates are markedly affected by geographic location and socioeconomic status (see section 4.4.2.3).

Table 4.7 Adhesins of *Mor. catarrhalis* and their complementary receptors.

Adhesin	Receptor and/or target cell
CD protein (an outer membrane protein)	Mucin
UspA1 (ubiquitous surface protein A1)	Epithelial cells
UspA2 (ubiquitous surface protein A2)	Fibronectin, vitronectin
Hag (hemagglutinating) protein	IgD (present in respiratory secretions); epithelial cells
McaP (*Mor. catarrhalis* adherence protein)	Epithelial cells
MID (*Mor. catarrhalis* immunoglobulin D-binding) protein	IgD; epithelial cells
Type IV pili	Gangliosides; cilia
Lipooligosaccharide	Epithelial cells
McmA (*Mor. catarrhalis* metallopeptidase-like adhesin)	Lung and laryngeal cells

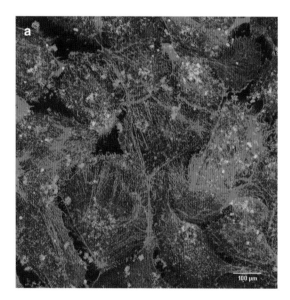 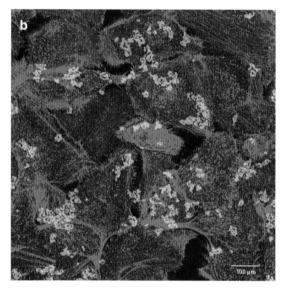

Fig. 4.26 Confocal microscopy of *Mor. catarrhalis* strains ATCC 25238 and O35E adhering to human bronchial epithelial cells. The organisms are stained green, while red F-actin was used for visualization of the epithelial cells. (a) Strain O35E showed mainly a diffuse adherence pattern with only a fair level of aggregation. (b) Strain ATCC 25238 formed multicellular grape-like aggregates on the cell surface. Reproduced from: Slevogt, H., Tiwari, K.N., Schmeck, B., Hocke, A., Opitz, B., Suttorp, N. and Seybold, J. (2006) Adhesion of *Moraxella catarrhalis* to human bronchial epithelium characterized by a novel fluorescence-based assay. *Med Microbiol Immunol* (*Berl*) 195, 73–83 with kind permission of Springer Science and Business Media, Berlin, Germany.

The organism is a frequent cause of sinusitis and otitis media in children, and can also cause lower respiratory tract infections, particularly in adults with chronic obstructive pulmonary disease. It has also been responsible for outbreaks of respiratory tract infections in hospitalized patients.

4.4.1.5 Staphylococci

4.4.1.5.1 Staph. aureus
The main characteristics of the genus *Staphylococcus* have been described in Chapter 2. *Staph. aureus* is distinguished from other members of the genus by

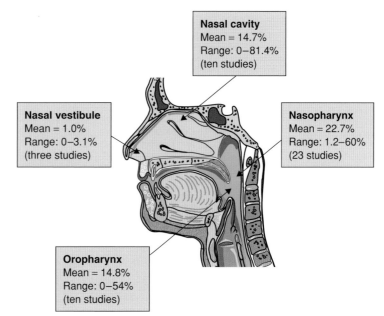

Nasal cavity
Mean = 14.7%
Range: 0–81.4%
(ten studies)

Nasal vestibule
Mean = 1.0%
Range: 0–3.1%
(three studies)

Nasopharynx
Mean = 22.7%
Range: 1.2–60%
(23 studies)

Oropharynx
Mean = 14.8%
Range: 0–54%
(ten studies)

Fig. 4.27 Prevalence of *Mor. catarrhalis* at various sites within the respiratory tract. The figures represent the proportion of healthy individuals (adults and children) harboring the organism at the indicated site and are based on studies carried out in a number of countries (3, 5–10, 12, 17, 20, 23, 27–33, 38, 61–63).

its ability to produce the enzyme coagulase. The genomes of several strains of the organism have been sequenced (http://cmr.tigr.org/tigr-scripts/CMR/GenomePage.cgi?org=ntsa08). The organism produces enzymes (caseinase, gelatinase, elastase, nucleases, and a hyaluronate lyase) able to degrade a number of macromolecules present in respiratory secretions. It also produces several cytotoxins and hemolysins that can kill a variety of host cells, thereby liberating nutrients for use by the respiratory microbiota.

Staph. aureus produces a number of bacteriocins with broad-spectrum activity (Table 4.8).

Staph. aureus has a variety of adhesins (Table 4.9), although which of these are important in maintaining the organism at most of its carriage sites remains to be established. In the nasal vestibule, the precise site of colonization is the moist squamous epithelium on the septum adjacent to the nasal ostium (i.e. the opening), which is devoid of vibrissae. Clumping factor B appears to be the main adhesin involved in maintaining the organism at this site, and its complementary receptor is cytokeratin K10, which is expressed on the surface of nasal epithelial cells and keratinocytes. Studies of its adhesion to human corneocytes have shown that it adheres mainly to the periphery of these cells (Fig. 4.28).

The organism is found mainly in the nasal vestibule, but is also present in the nasal cavity, nasopharynx, and the oropharynx, as well as at sites outside the respiratory tract (Figs 4.29 and 4.30).

Cross-sectional studies have revealed that approximately 27% of healthy adults harbor *Staph. aureus* in the nasal vestibule, but longitudinal investigations have identified two groups of carriers: (1) those in whom the organism can be detected on several occasions (known as persistent carriers) – these individuals are colonized by a particular strain of the organism; and (2) those in whom the organism is identified on one of several sampling occasions (intermittent carriers) – in these individuals, the colonizing strain changes with time. Carriage rates are affected by, amongst other factors, age – with children having higher persistent carriage rates than adults (Fig. 4.31). Recent studies have shown that 1% and 1.5% of the population of the USA and UK, respectively, harbor methicillin-resistant strains of the organism in the nasal vestibule.

The organism is responsible for a wide range of infections, ranging from minor ailments (e.g. boils and skin abscesses) to life-threatening conditions such as pneumonia, osteomyelitis, endocarditis, wound infections, and septicemia. Furthermore, it is responsible for a number of toxinoses, including food poisoning, toxic epidermal necrolysis, and toxic shock syndrome.

Table 4.8 Some of the bacteriocins produced by *Staph. aureus*.

Bacteriocin	Antimicrobial spectrum
Staphylococcin 188	*Micrococcus luteus*, *Strep. pneumoniae*, *Streptococcus faecalis*, viridans streptococci, *Corynebacterium diphtheriae*, and several *Staphylococcus* spp.
Staphylococcin BacR1	Some strains of *Staph. aureus*, some CNS, some viridans streptococci, *Corynebacterium* spp., *Haemophilus parasuis*, *Bordetella* spp., *Neisseria* spp., and *Bacillus* spp.
Staphylococcin Bac1829	Some strains of *Staph. aureus*, *Streptococcus suis*, *Corynebacterium* spp., *H. parasuis*, *Bordetella* spp., *Moraxella bovis*, and *Pasteurella multocida*
Aureocin A70	Many Gram-positive species including *Listeria monocytogenes*
Aureocin A53	Many strains of *Staph. aureus*, *Staphylococcus simulans*, *Corynebacterium* spp., *M. luteus*, *Lis. monocytogenes*
Staphylococcin Bac201	*Streptococcus agalactiae*, *Enterococcus faecalis*, *Acinetobacter calcoaceticus*, *N. meningitidis* and several *Staphylococcus* spp.
Staphylococcins C55α and C55β	Synergic activity against many strains of *Staph. aureus*, *Staphylococcus carnosus*, and *M. luteus*

Table 4.9 Adhesins of *Staph. aureus* and their complementary receptors.

Adhesin	Receptor and/or target cells
FnbpA (fibronectin binding protein A)	Fibronectin; fibrinogen; elastin
FnbpB (fibronectin binding protein B)	Fibronectin; fibrinogen; elastin
Cna (collagen adhesin)	Collagen
ClfA (clumping factor A)	Fibrinogen
ClfB (clumping factor B)	Fibrinogen
IsdA (iron-regulated surface determinant A)	Fibronectin; fibrinogen
Bsp (bone sialoprotein-binding protein)	Bone sialoprotein
Eap (extracellular adherence protein)	Bone sialoprotein; fibronectin; fibrinogen; vitronectin; epithelial cells, fibroblasts; ICAM-1
Emp (extracellular matrix protein-binding protein)	Fibronectin; fibrinogen; collagen; vitronectin
EbpS (elastin-binding protein S)	Elastin
60-kDa protein	Human bronchial mucin
150-kDa and 170-kDa Ca^{2+}-binding proteins	Human salivary gland mucin

4.4.1.5.2 CNS

CNS are described in Chapter 2. Staphylococci other than *Staph. aureus* have been detected in all regions of the upper respiratory tract (Fig. 4.32). However, in most cases, the organisms have not been fully identified and have simply been reported as "coagulase-negative staphylococci". When further identification was undertaken, *Staphylococcus epidermidis* appeared to be the most frequently encountered CNS. Other *Staphylococcus* species reported to be present include *Staph. capitis*, *Staph. haemolyticus*, *Staph. warneri*, *Staph. hominis*, *Staph. lugdunensis*, *Staph. cohnii* subsp. *cohnii*, and *Staph. auricularis*.

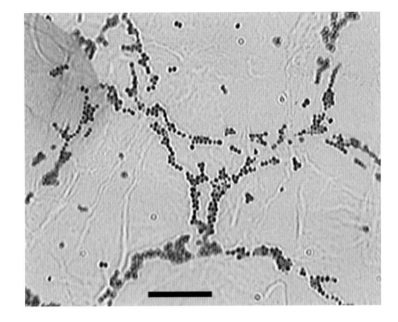

Fig. 4.28 Photomicrograph showing the adhesion of *Staph. aureus* to human corneocytes obtained from a healthy adult. The organism adhered mainly to the peripheries of cells. Scale bar, 10 μm. Reprinted with the permission of Blackwell Publishing Ltd, Oxford, UK, from: Simou, C., Hill, P.B., Forsythe, P.J. and Thoday, K.L. (2005) Species specificity in the adherence of staphylococci to canine and human corneocytes: A preliminary study. *Vet Dermatol* 16, 156–61.

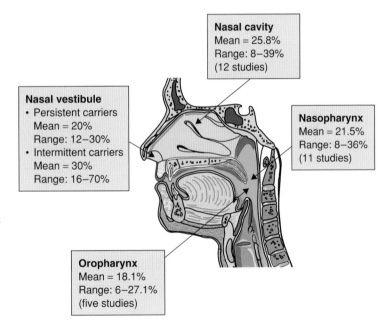

Fig. 4.29 Prevalence of *Staph. aureus* at various sites within the respiratory tract. The figures represent the proportion of healthy individuals (adults and children) harboring the organism at the indicated site and are based on studies carried out in a number of countries (6–8, 18–20, 23, 29, 31, 34, 45, 64–66).

4.4.1.6 *Mollicutes*

The *Mollicutes* are a class of bacteria that do not have a cell wall and are the smallest free-living organisms known, some species having a diameter of only 0.2 μm. They have very small genomes (<600 kbp in some species), which have been sequenced in a number of species (see sections 6.4.1 and 7.4.1). They are slow-growing, nutritionally fastidious, and exacting in their environmental requirements. Three of the eight genera

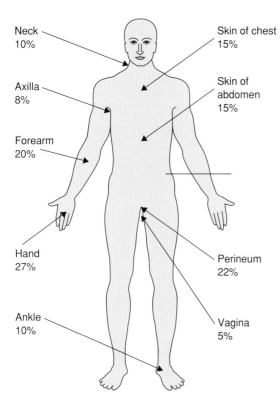

Neck
10%

Skin of chest
15%

Axilla
8%

Skin of
abdomen
15%

Forearm
20%

Hand
27%

Perineum
22%

Ankle
10%

Vagina
5%

Fig. 4.30 Carriage of *Staph. aureus* at sites outside the respiratory tract. The figures represent the proportions of healthy individuals colonized by *Staph. aureus* at the sites indicated (67).

belonging to this class (*Mycoplasma*, *Ureaplasma*, and *Acholeplasma*) have species that are members of the indigenous microbiota of humans. These three genera are facultative anaerobes and are differentiated on the basis of their requirement for sterols and their ability to ferment glucose, to utilize arginine, and to hydrolyze urea. *Ureaplasma* spp. (G+C content = 27–30 mol%) hydrolyze urea, but are unable to metabolize glucose or arginine. On the other hand, *Mycoplasma* spp. (G+C content = 23–41 mol%) cannot hydrolyze urea, but most species can utilize arginine, and many can metabolize glucose. Unlike the other two genera, *Acholeplasma* spp. (G+C content = 27–36 mol%) do not require sterols for growth.

Mollicutes are difficult to grow, and few surveys of the microbiota of any body site have used the specialized media and incubation conditions that are necessary for their detection and quantification. Consequently, little

is known regarding their presence in the various microbial communities inhabiting humans – their main habitats appear to be the mucosal surfaces of the respiratory and genito-urinary tracts.

Mollicutes detected in the respiratory tract of healthy humans include *Acholeplasma laidlawii*, *Mycoplasma buccale*, *Mycoplasma faucium*, *Mycoplasma fermentans*, *Mycoplasma lipophilum*, *Mycoplasma orale*, *Mycoplasma primatum*, and *Mycoplasma salivarium*. Of these, *Myc. fermentans* is able to act as an opportunistic respiratory pathogen.

4.4.1.7 *Kingella kingae*

Kin. kingae is a facultatively anaerobic, non-motile, Gram-negative bacillus with tapered ends that occur in pairs or small chains. Growth is enhanced by CO_2. It is β-hemolytic, oxidase-positive, catalase-negative, and urease-negative. It is nutritionally fastidious and produces acid from glucose and maltose. It produces alkaline- and acid-phosphatases. Its main habitat is the oropharynx, where it can be found in 18% of children in the age group 19–48 months. Peak carriage occurs between the ages of 6 months and 4 years, and it is generally not isolated from children younger than 6 months. In young children, it is an important cause of invasive infections such as septic arthritis, osteomyelitis, spondylodiscitis, bacteremia, and endocarditis.

4.4.2 Community composition at the various sites within the respiratory tract

One of the main determinants of the presence of an organism in the respiratory tract is its ability to adhere to a particular site within this complex system. Apart from the anterior regions of the nasal vestibules, which have an epithelium that differs from that of the rest of the respiratory tract, the problems encountered by an organism attempting to adhere to most regions of the respiratory tract are similar, because most of the mucosal surfaces of the respiratory tract are covered by a layer of mucus. Many members of the respiratory microbiota are able to adhere to respiratory mucus; these include *H. influenzae* (both type b and nontypeable strains), *Strep. pneumoniae*, *Staph. aureus*, *Mor. catarrhalis*, *Strep. pyogenes*, and viridans streptococci. However, other members of the respiratory microbiota, including *N. meningitidis*, do not demonstrate a high affinity for respiratory mucus. The adhesins and receptors

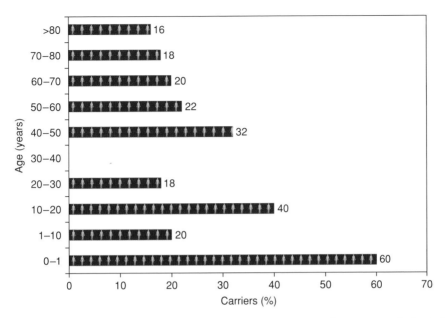

Fig. 4.31 Effect of age on the carriage of *Staph. aureus* in the nasal vestibule. Bars denote the proportion of individuals in each age group who are colonized by the organism. No data were available for the age group of 30–40 years (67).

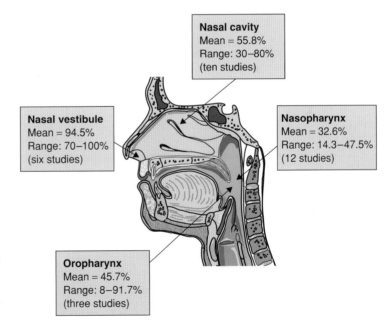

Nasal cavity
Mean = 55.8%
Range: 30–80%
(ten studies)

Nasal vestibule
Mean = 94.5%
Range: 70–100%
(six studies)

Nasopharynx
Mean = 32.6%
Range: 14.3–47.5%
(12 studies)

Oropharynx
Mean = 45.7%
Range: 8–91.7%
(three studies)

Fig. 4.32 Prevalence of CNS at various sites within the respiratory tract. The figures represent the proportion of healthy individuals (adults and children) harboring the organism at the indicated site, and are based on studies carried out in a number of countries (23, 25, 26, 33, 45, 64, 68–79).

involved in the adhesion of members of the respiratory microbiota to mucus are shown in Table 4.10.

Any interference with the efficient functioning of the mucociliary escalator will aid microbial colonization, and a number of organisms are able to do this by damaging ciliary cells, inhibiting ciliary activity, altering mucus viscosity, or affecting the amount of mucus produced. In addition to the mucous layer, other potential

Table 4.10 Adhesins and receptors involved in the colonization of respiratory mucus by important members of the respiratory microbiota.

Organism	Adhesin	Receptor and/or target mucin
Strep. pyogenes	M protein	α2-6-Linked sialic acid residues in mucin
	Surface-bound pullulanase	Salivary mucins
H. influenzae	P2 and P5	α2-3-Linked sialic acid residues; human middle ear mucin; human nasopharyngeal mucin
	HifA (a pilin)	Human tracheobronchial mucin
Strep. pneumoniae	Not known	α2-3-Linked and α2-6-Linked sialic acid residues; GlcNAc(β1-3)Gal
	17.5-kDa protein	Human middle ear mucin; human nasopharyngeal mucin
	20.5-kDa protein	Human middle ear mucin; human nasopharyngeal mucin
	SP1492 protein	Porcine and bovine mucin
Mor. catarrhalis	CD protein	Sialic acid residues; human middle ear mucin; human nasopharyngeal mucin
Staph. aureus	60-kDa protein	Human bronchial mucin
	150-kDa and 170-kDa Ca^{2+}-binding proteins	Human salivary gland mucin
	127-kDa and 138-kDa proteins	Human nasal mucin

sites of colonization include the periciliary layer and the epithelium itself. The periciliary layer is less frequently replaced than the mucous layer, and therefore offers a more permanent site for microbial colonization. The epithelium offers a number of possible sites – ciliated cells (apart from in the oropharynx and alveoli), non-ciliated cells, goblet cells, serous cells, Clara cells, and the mucins and extracellular matrix molecules expressed by many of these cells. However, healthy ciliated cells rarely have attached microbes, although they do become susceptible to microbial colonization once they have been damaged in some way. This is probably due to the uncovering of additional receptors for bacterial adhesins. Furthermore, if damage is extensive, molecules of the underlying extracellular matrix may be exposed, thereby offering additional binding sites. As can be seen in Tables 4.4, 4.5, 4.6, 4.7, 4.9, and 4.10, respiratory organisms display a wide range of adhesins which can mediate binding to a variety of cells and host macromolecules.

Persistence of bacteria in the respiratory tract can also be achieved by invasion of the mucosa. For example, *Strep. pyogenes*, *N. meningitidis*, *Strep. pneumoniae*, and *H. influenzae* can all invade respiratory epithelial cells in vitro (Fig. 4.33).

4.4.2.1 Nasal vestibule

This region provides a number of sites for microbial colonization, including hair follicles, keratinized and nonkeratinized epithelial cells, vibrissae, extracellular matrix molecules, nasal fluid, and other members of the nasal microbiota. The total number of viable bacteria is usually between 10^6 and 10^7 per nostril. The nasal vestibule is the main site of carriage of the important human pathogen *Staph. aureus*, and many investigations have been undertaken to establish carriage rates of the organism at this site. However, remarkably few studies have determined which other organisms are present. *Corynebacterium* spp., CNS, and *Propionibacterium* spp. are the most frequently detected organisms and generally dominate the microbiota (Fig. 4.34). Gram-negative bacteria are not frequently encountered and, when present, usually comprise only a small proportion of the microbiota.

Very few studies have determined the relative proportions of the various microbes that comprise the microbiota of the nasal vestibule. *Corynebacterium* spp., particularly the lipophilic diphtheroids (*Corynebacterium jeikeium*, *Corynebacterium urealyticum*, and

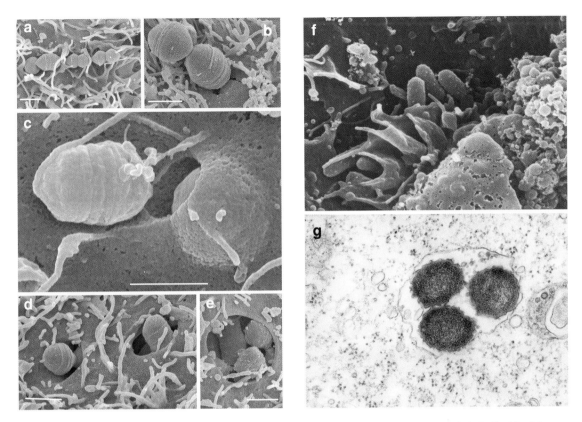

Fig. 4.33 (a–e) Scanning electron micrographs showing the uptake of *Strep. pyogenes* into human epithelial cells. After 1 h of infection, bacteria are (a) attached to eucaryotic cells, or (b–e) are in the process of invading eucaryotic cells. Membrane invaginations are formed at the point of contact between a bacterium and the host cell membrane (visible in b–e). Scale bars, 1 μm. Reproduced with the permission of Blackwell Publishing Ltd, Oxford, UK, from: Molinari, G., Rohde, M., Guzman, C.A. and Chhatwal, G.S. (2000) Two distinct pathways for the invasion of *Streptococcus pyogenes* in nonphagocytic cells. *Cell Microbiol 2*, 145–54. (f) Invasion of epithelial cells of adenoidal tissue in culture by *H. influenzae*. (g) Lamellipodia can be seen engulfing the attached cells which are then internalized within membrane-bound vacuoles. Reprinted with the permission of Blackwell Publishing Ltd, Oxford, UK, from: St Geme, J.W. (2002) Molecular and cellular determinants of nontypeable *Haemophilus influenzae* adherence and invasion, *Cell Microbiol 4*, 191–200.

Corynebacterium afermentans subsp. *lipophilum*) appear to dominate the microbiota (Fig. 4.35).

Most of the CNS present in the nasal vestibule have been found to be strains of *Staph. epidermidis*, although *Staph. hominis* is also frequently detected (Fig. 4.36).

4.4.2.2 Nasal cavity

In most studies of the nasal cavity, the organisms most frequently detected have been CNS, *Corynebacterium* spp., and *Propionibacterium* spp. (Fig. 4.37). However, in one study *Rhodococcus* spp. and *Aureobacterium* spp.

(now included in the genus *Microbacterium*) were found to be frequent colonizers of this region. *Microbacterium* spp. are aerobic, Gram-positive, nonsporing, rod-shaped bacteria which produce acid from glucose. The genus *Rhodococcus* consists of a group of catalase-positive, nonmotile, aerobic, Gram-positive rods. The various species in this genus show wide variation with regard to their morphology, growth patterns, and biochemical characteristics. The G+C content of their DNA is 63–73 mol%.

Staph. epidermidis is the most frequently detected CNS – other *Staphylococcus* species found include *Staph.*

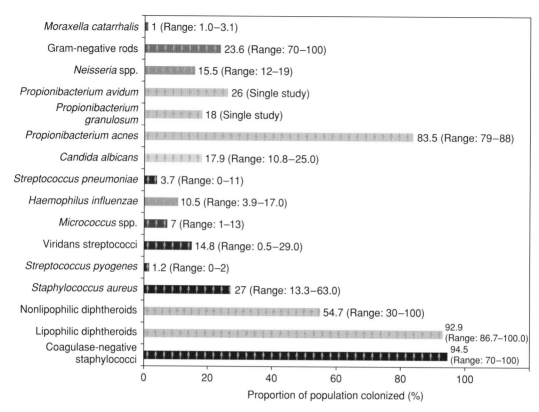

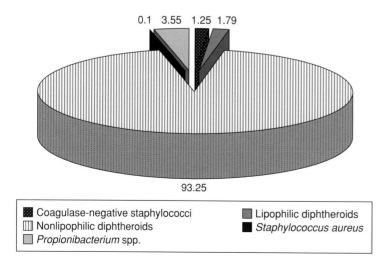

Fig. 4.34 Frequency of detection (mean value and range) of various microbes in the nasal vestibule. Data for *Staph. aureus*, *Strep. pneumoniae*, *N. meningitidis*, *H. influenzae*, *Mor. catarrhalis*, *Staph. aureus*, and CNS are derived from studies used to compile Figs 4.14, 4.16, 4.20, 4.24, 4.27, 4.29, and 4.32. For the other organisms shown, the data are from seven studies carried out on 1078 individuals (including adults and children) in a number of countries (6, 26, 77–81).

Fig. 4.35 Relative proportions (%) of the various microbes that comprise the cultivable microbiota of the nasal vestibule of eight adults (79).

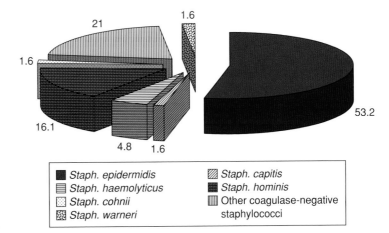

Fig. 4.36 Relative proportions (%) of the various species of CNS found in the nasal vestibule of 16 healthy adults (82).

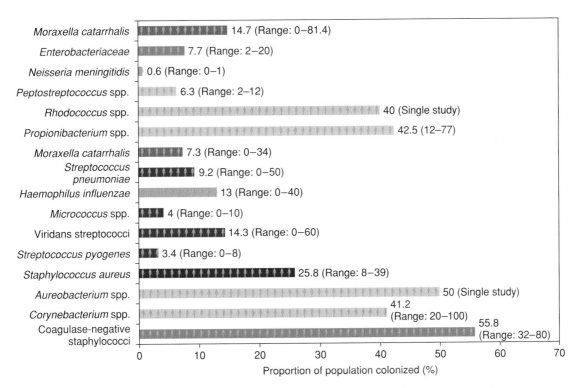

Fig. 4.37 Frequency of detection (mean value and range) of various microbes in the nasal cavity. Data for *Staph. aureus*, *Strep. pneumoniae*, *N. meningitidis*, *H. influenzae*, *Mor. catarrhalis*, *Staph. aureus*, and CNS are derived from studies used to compile Figs 4.14, 4.16, 4.20, 4.24, 4.27, 4.29, and 4.32. For the other organisms shown, the data are from 12 studies carried out on 1210 individuals (including adults and children) in a number of countries (6, 23, 24, 33, 68–75).

capitis, *Staph. haemolyticus*, *Staph. hominis*, and *Staph. lugdunensis*. *Corynebacterium* spp. detected include *C. accolens*, *C. propinquum*, *C. pseudodiphtheriticum*, *C. striatum*, *C. afermentans*, and *C. macginleyi*. Almost all of the *Propionibacterium* spp. detected are strains of *Propionibacterium acnes*. Although *Strep. pneumoniae* and *H. influenzae* are rarely present in the nasal cavity of adults, they are often present in infants.

4.4.2.3 Nasopharynx

The nasopharynx supports a large and varied microbial community – the density of colonization ranges from 3×10^4 to 4×10^8 cfu, with a median value of approximately 6×10^6 cfu. Most of the organisms present are either in the ASL or are attached to nonciliated epithelial cells, each of which may have between ten and 50 adherent bacteria; some cells support microcolonies. It is a major carriage site of several important human pathogens – *Strep. pneumoniae*, *H. influenzae*, *Mor. catarrhalis*, *Strep. pyogenes*, and *N. meningitidis*. Bacteriological studies of the nasopharynx have tended to concentrate on establishing the prevalence of the above-mentioned pathogens in populations rather than trying to determine the overall composition of the complex microbiota inhabiting this site. The presence in the nasopharynx of these potential pathogens is affected by many factors, including age, gender, climate, and social factors (Table 4.11). The proportions

of children and adults harboring these organisms in their nasopharynx are shown in Fig. 4.38. In addition to the above-mentioned organisms, the nasopharynx of adults and children is invariably colonized by viridans streptococci, *Haemophilus* spp., and *Neisseria* spp. *Corynebacterium* spp., CNS, and a variety of obligate anaerobes are also often present (Fig. 4.38 and Table 4.12).

The existence of obligate anaerobes in this predominantly aerobic environment is possible because of oxygen consumption by the aerobic and facultative species present, which results in the formation of anaerobic microhabitats, particularly within the invaginations of epithelial cells and inside the microcolonies that have been observed on epithelial cells (Fig. 4.39). Restricted penetration of oxygen-rich respiratory secretions into these microhabitats, coupled with oxygen consumption by resident microbes, create environments with low oxygen concentrations, thereby enabling the survival of anaerobes.

4.4.2.4 Oropharynx

Nasopharyngeal secretions and saliva, each of which has a dense and varied microbiota, are continually being delivered to the oropharynx. The oropharyngeal microbiota, therefore, is very complex and difficult to define. α-Hemolytic streptococci, nonhemolytic streptococci, *Haemophilus* spp., and *Neisseria* spp. are

Table 4.11 Effect of various factors on the carriage of key members of the indigenous microbiota of the nasopharynx.

Factor	Effect
Age	Prevalence of *Strep. pneumoniae*, *H. influenzae* and *Mor. catarrhalis* decreases in the order infants > children > adults Prevalence of *N. meningitidis* is low in infants and the elderly and is highest in teenagers and young adults
Season	Prevalence of *Strep. pneumoniae* in school children is higher in summer than in winter Prevalence of *Mor. catarrhalis* is higher in winter and autumn than in spring and summer Prevalence of *Staph. aureus* in infants is greater in winter than in summer
Gender	Density of colonization of *Staph. aureus* in infants is greater in males than females
Social factors	Prevalence of *N. meningitidis* is highest in low socioeconomic classes and in institutionalized individuals, e.g. prisoners, military recruits, etc. Prevalence of *Strep. pneumoniae* and *H. influenzae* is highest in low socioeconomic classes Children with siblings have increased carriage of *Strep. pneumoniae*, *H. influenzae*, and *Mor. catarrhalis*

a

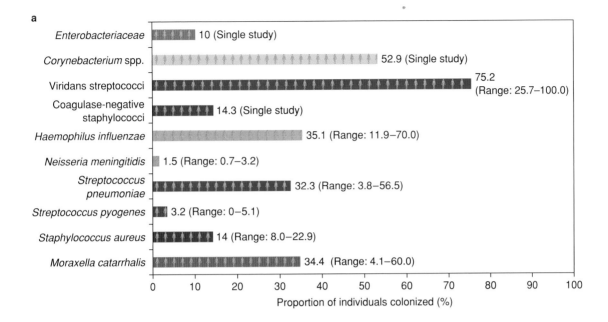

b

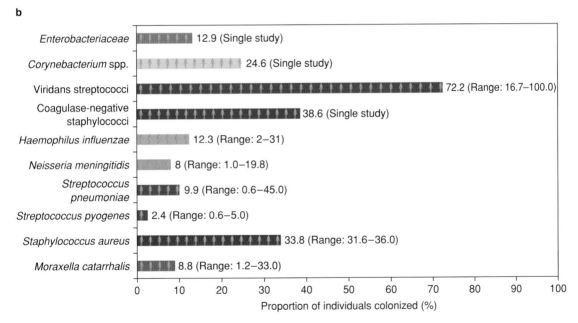

Fig. 4.38 Organisms most frequently detected in the nasopharynx of (a) children (0–7 years old) and (b) adults (>18 years old). Data for *Staph. aureus, Strep. pneumoniae, N. meningitidis, H. influenzae, Mor. catarrhalis, Staph. aureus*, and CNS are derived from studies used to compile Figs 4.14, 4.16, 4.20, 4.24, 4.27, 4.29, and 4.32. For the other organisms shown, the data are from a study involving 380 healthy individuals (8).

Table 4.12 Organisms that may be detected in the nasopharynx of healthy individuals. The list is not exhaustive, but is an indication of the complexity of the nasopharyngeal microbiota.

Aerobes	Facultative anaerobes	Obligate anaerobes
H. influenzae	Strep. pneumoniae	Fusobacterium spp.
H. parainfluenzae	Streptococcus intermedius	Prevotella spp.
Haemophilus parahaemolyticus	Strep. anginosus	Porphyromonas spp.
H. segnis	Streptococcus constellatus	Peptostreptococcus spp.
Haemophilus aphrophilus	Streptococcus sanguis	Propionibacterium spp.
N. meningitidis	Strep. gordonii	Bacteroides spp.
N. cinerea	Strep. mitis	Veillonella spp.
N. sicca	Streptococcus parasanguis	—
N. subflava	Streptococcus crista	—
N. mucosa	Strep. pyogenes	—
N. lactamica	Staph. aureus	—
Neisseria flavescens	Staph. epidermidis	—
Mor. catarrhalis	Staph. cohnii	—
—	Kin. kingae	—

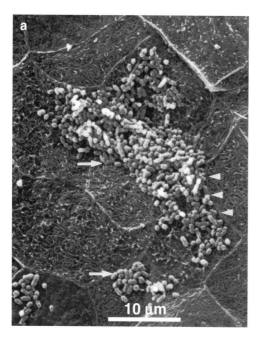 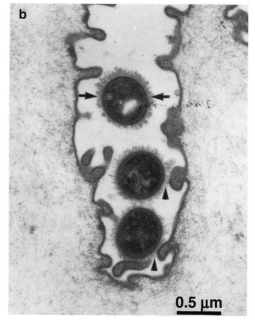

Fig. 4.39 Anaerobic microhabitats in the respiratory tract. (a) Microcolonies of bacteria (arrowed) on the surface of tonsillar epithelial cells. Note the pattern of microridges on the adjacent epithelial cells. Arrowheads denote a junction between adjacent epithelial cells. (b) Transmission electron micrograph showing bacteria within an involution of an epithelial cell. Capsular material (arrows) is visible on the bacterial cells, and this is in contact with projections on the epithelial cell surface (arrowheads). Photomicrographs kindly supplied by Professor Lars-Eric Stenfors, University of Tromso, Tromso, Norway, and reproduced with permission from: Fredriksen, F., Raisanen, S., Myklebust, R. and Stenfors, L.E. (1996) Bacterial adherence to the surface and isolated cell epithelium of the palatine tonsils. *Acta Otolaryngol* 116, 620–6.

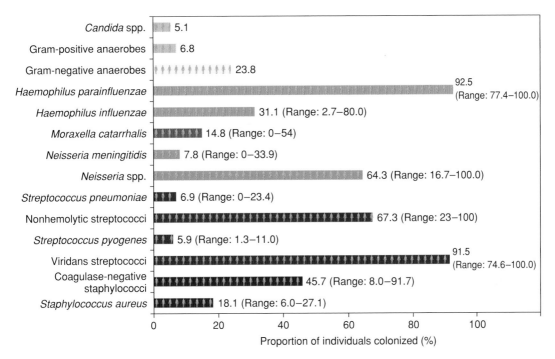

Fig. 4.40 Microbiota of the oropharynx. Data for *Staph. aureus, Strep. pneumoniae, N. meningitidis, H. influenzae, Mor. catarrhalis, Staph. aureus*, and CNS are derived from studies used to compile Figs 4.14, 4.16, 4.20, 4.24, 4.27, 4.29, and 4.32. For the other organisms shown, the data are from four studies involving a total of 201 individuals (adults and children) in a number of countries (45, 60, 64, 83).

invariably present in approximately equal proportions, and together account for as much as 80% of the microbiota (Fig. 4.40).

Like the nasopharynx, the oropharynx frequently harbors the pathogens *Strep. pneumoniae, N. meningitidis, H. influenzae*, and *Mor. catarrhalis* (Figs 4.16, 4.20 4.24, and 4.27). In addition, the oropharynx is the main habitat of the important pathogen *Strep. pyogenes* (Fig. 4.14). Carriage of this organism in healthy adults is usually between 5 and 10%, but is generally higher in schoolchildren, in whom the carriage rate can be as high as 20% during winter and spring.

The species diversity of the oropharyngeal microbiota is considerable, and at least 15 different *Streptococcus* spp. have been isolated, although *Strep. salivarius, Strep. mitis* biovars 1 and 2, and *Strep. anginosus* generally comprise more than 75% of the streptococci present (Fig. 4.41).

H. parainfluenzae is the most frequently detected *Haemophilus* sp. in the oropharynx (Fig. 4.25). Other

Haemophilus species frequently detected include *H. segnis, H. paraphrophilus*, and *H. influenzae*. The most frequently detected *Neisseria* spp. in adults are *N. perflava, N. sicca, N. mucosa, N. flava*, and *N. cinerea* (see Fig. 4.22). Other *Neisseria* spp. present include *N. lactamica* and *N. subflava*. CNS, coryneforms, and anaerobes (particularly Gram-negative species) are also frequently present. *Staph. epidermidis* is the main CNS detected, but *Staph. cohnii* and *Staphylococcus mucilaginosus* have also been isolated from the oropharynx. The only detailed analysis of coryneforms present in the oropharynx has shown that the most frequently detected species are *Corynebacterium durum* and *Rothia dentocariosa* (Fig. 4.42). The most frequently detected anaerobes are *Bacteroides* spp., *Propionibacterium* spp., *Prevotella melaninogenica, Fusobacterium* spp., *Veillonella* spp., and *Peptostreptococcus* spp.

The oropharynx appears to be one of the main habitats of *Mollicutes*, and at least 11 species of *Mollicutes* have been detected at this site, including *Myc. salivarium*,

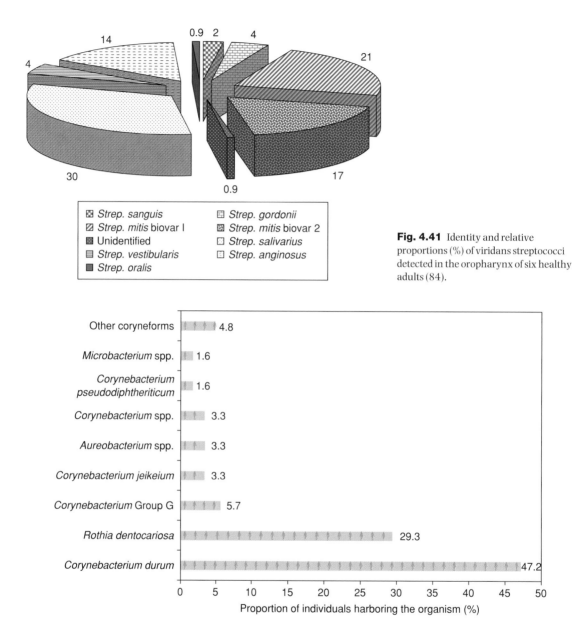

Fig. 4.41 Identity and relative proportions (%) of viridans streptococci detected in the oropharynx of six healthy adults (84).

Fig. 4.42 Identity and prevalence of coryneforms in the oropharynx of 113 healthy adults (85).

Myc. orale, Myc. buccale, Myc. faucium, Myc. lipophilum, and *Acholeplasma laidlawii*. However, little is known regarding their prevalence or population density.

Within the oropharynx are the tonsils, which are important lymphoid tissues. Bacteria are found at three locations on the tonsillar epithelium: in the mucous layer (Fig. 4.43a), attached to epithelial cells (Figs 4.39 and 4.43), and within epithelial cells (Fig. 4.43b).

The microbiota of the surface of the tonsils is similar to that of the rest of the oropharynx, with α-hemolytic streptococci, *Neisseria* spp., *H. influenzae*, *H. parainfluenzae*, and *Staph. aureus* being the most

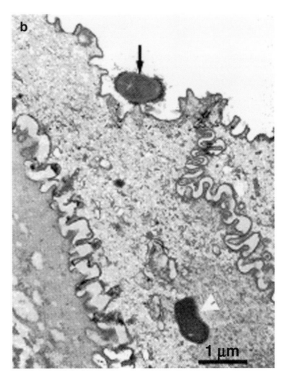

Fig. 4.43 (a) Scanning electron micrograph showing bacteria (arrows) surrounded by mucus which has contracted to form fibrils (arrowheads) during sample processing. Scale bar, 5 μm. (b) Transmission electron micrograph showing bacteria attached to (black arrow) and within (white arrowhead) the tonsillar epithelium. Scale bar, 1 μm. Photomicrographs kindly supplied by Professor Lars-Eric Stenfors, University of Tromso, Tromso, Norway, and reproduced with permission from: Fredriksen, F., Raisanen, S., Myklebust, R. and Stenfors, L.E. (1996) Bacterial adherence to the surface and isolated cell epithelium of the palatine tonsils. *Acta Otolaryngol* 116, 620–6.

frequently isolated organisms. The epithelial cells have a distinctive pattern of microridges (see Fig. 4.39), and bacteria often appear to be "gripped" by these ridges (Fig. 4.44).

Recently, a culture-independent analysis has been undertaken of the indigenous microbiota on the surface of tonsils from five healthy adults (Table 4.13). The microbiota was found to be very diverse – the mean number of species detected per individual being 19. The most frequently detected species were *Strep. mitis biovar* 1, *Strep. mitis biovar* 2, *Gemella sanguinis*, *Gemella haemolysans*, and *Granulicatella adiacens*.

The tonsils have large numbers of deep, convoluted, branching invaginations known as crypts and, because of restricted access to oxygen, these regions support dense populations of anaerobes, which constitute approximately half of the cultivable indigenous microbiota. In a

study of the microbiota of the tonsillar crypts, *Prevotella* spp. were the most frequently isolated organisms and constituted 26% of the total isolates, the most frequently detected species being *Prev. melaninogenica*, *Prev. loescheii*, *Prev. corporis*, and *Prev. oralis* (Fig. 4.45).

4.4.2.5 Lower respiratory tract

The larynx, trachea, bronchi, bronchioles, and alveoli of healthy individuals are not usually colonized by microbes. However, small numbers of bacteria can often be isolated from these regions, particularly the larynx and trachea, as a result of the aspiration of fluids from the upper respiratory tract. Fluid aspiration is a frequent occurrence, particularly during sleep when approximately 50% of healthy individuals aspirate nasopharyngeal secretions into the lungs. Although

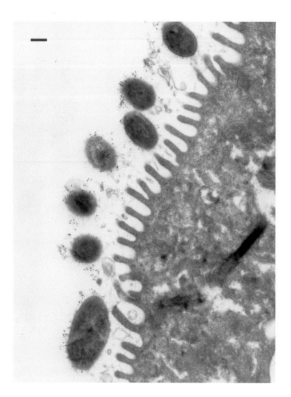

Fig. 4.44 Transmission electron micrograph showing *Strep. pyogenes* attached to the crests of cellular projections (microridges) of the tonsillar epithelium. The bacteria are coated with rabbit antiserum to *Strep. pyogenes* labeled with gold particles. Scale bar, 500 nm. Photomicrograph kindly supplied by Professor Lars-Eric Stenfors, University of Tromso, Tromso, Norway.

the quantity of fluid involved is very small (0.01 to 0.2 ml per night), the high concentration of bacteria (up to 10^8 cfu/ml) in these secretions means that the number of bacteria entering the lower respiratory tract can be significant.

4.4.3 Interactions among members of the respiratory microbiota

The provision of nutrients by the degradation of macromolecules or by the excretion of metabolic end products is the main beneficial interaction occurring among members of the respiratory microbiota. Macromolecules in respiratory secretions known to be degraded by microbes present in the respiratory tract include mucins,

proteins, hyaluronic acid, chondroitin sulfate, and lipids (see Fig. 4.13). Metabolic end products, such as lactate produced by streptococci, staphylococci, and *Haemophilus* spp., can be utilized as carbon and energy sources by *Neisseria* spp. and *Veillonella* spp. Oxygen consumption by *Neisseria* spp., *Haemophilus* spp., and *Mor. catarrhalis* will also contribute to the establishment of microhabitats with low oxygen concentrations, so enabling the growth of microaerophiles and anaerobes.

A number of in vitro studies have shown that many species inhabiting the respiratory tract are able to inhibit or kill other members of the respiratory microbiota. However, the identities of the compounds responsible are generally not known. Antagonistic substances produced by members of the respiratory microbiota include bacteriocins, fatty acids, and hydrogen peroxide. In addition, acid production and oxygen utilization by respiratory microbes can create micro-environments with a low pH and/or low oxygen content, respectively, and this will restrict or prevent the growth of some species. Viridans streptococci have been shown to inhibit many strains of *Strep. pneumoniae*, NTHi, *Mor. catarrhalis*, and *Strep. pyogenes*. Evidence that such interactions occur in vivo comes from studies that have reported an inverse relationship between the presence of α-hemolytic streptococci and a number of respiratory pathogens in the nasopharynx. Such observations have prompted clinical trials of the use of viridans streptococci for preventing and treating infections due to *Strep. pyogenes*. *Prevotella* spp. and *Peptostreptococcus* spp. isolated from the nasopharynx are able to inhibit the growth of *Strep. pneumoniae*, NTHi, *Mor. catarrhalis*, and *Strep. pyogenes*. *Strep. pneumoniae* displays antagonism towards *H. influenzae*, *Mor. catarrhalis*, *Staph. aureus*, and *N. meningitidis*, and this may be due to the production of hydrogen peroxide by *Strep. pneumoniae*.

4.5 OVERVIEW OF THE RESPIRATORY MICROBIOTA

Only the upper regions of the respiratory tract (the nose and pharynx) have resident microbial communities; the lower regions are largely devoid of microbes. In order to avoid expulsion from the respiratory tract, the ability to adhere to a substratum in the tract is an essential prerequisite of potential colonizing microbes. Although the dominant cultivable organisms in the microbial communities found at each site within the tract are known, the exact composition of each community

Table 4.13 Culture-independent analysis of the microbiota on the surfaces of the tonsils of five healthy adults. DNA was extracted from the samples, the bacterial 16S rRNA genes present were amplified by PCR, and the resulting products were cloned and sequenced. The occurrence and levels of bacterial species/phylotypes are shown as either not detected, less than 15% of the total number of clones assayed, or more than 15% of the total number of clones assayed. Novel phylotypes are indicated in bold (86).

Organism	Subject				
	No. 1	**No. 2**	**No. 3**	**No. 4**	**No. 5**
Strep. mitis	ND	<	>	>	>
Streptococcus infantis	<	ND	ND	<	ND
Strep. mitis biovar 2	ND	<	ND	>	<
Streptococcus strain H6	ND	ND	ND	ND	<
Streptococcus parasanguinis	<	<	ND	ND	ND
Streptococcus clone FN051	<	>	ND	ND	ND
Streptococcus australis	ND	<	ND	ND	<
Streptococcus clone AA007	<	ND	ND	ND	ND
Streptococcus isolate T4-E3	<	ND	ND	ND	<
Streptococcus clone FP015	<	>	ND	ND	ND
Strep. oralis	ND	ND	<	<	ND
Streptococcus peroris	<	ND	ND	ND	ND
Strep. intermedius	ND	ND	ND	<	ND
Streptococcus mutans	ND	ND	ND	<	ND
Gran. adiacens	<	>	<	<	<
Granulicatella elegans	ND	ND	<	ND	<
Gemella haemolysans	<	ND	<	>	>
Gem. sanguinis	ND	<	<	ND	<
Gemella morbillorum	ND	ND	<	ND	ND
Leptotrichia buccalis	ND	ND	ND	ND	<
Fusobacterium clone BS011	ND	ND	ND	<	<
Fusobacterium clone R002	ND	ND	ND	<	<
Fusobacterium clone CZ006	ND	ND	ND	<	ND
Peptostreptococcus clone CK035	ND	ND	ND	<	ND
Filifactor alocis	<	ND	<	ND	ND
Eubacterium brachy	ND	ND	<	ND	ND
Selenomonas sputigena	ND	ND	<	ND	ND
Selenomonas clone EY047	<	ND	ND	ND	ND
Dialister pneumosintes	ND	ND	<	ND	ND
Megasphaera micronuciformis	<	ND	ND	ND	ND
Veillonella atypica	<	ND	ND	ND	ND
Veillonella clone AA050	ND	<	ND	ND	ND
Veillonella parvula or *dispar*	<	ND	ND	ND	ND
Atopobium parvulum	<	ND	ND	ND	ND
Eubacterium isolate FTB41	ND	<	ND	ND	ND
Eubacterium clone F058	<	ND	ND	ND	ND
Eubacterium clone BU014	<	ND	ND	ND	ND
N. subflava	ND	ND	<	ND	<
Kingella clone ID059	ND	ND	ND	ND	<
Campylobacter showae	ND	ND	ND	ND	<
Campylobacter gracilis	ND	ND	<	ND	ND
Bacteroidetes clone 28B11	ND	ND	ND	ND	<
Capnocytophaga clone BM058	ND	ND	<	ND	ND
Capnocytophaga clone ID062	ND	ND	ND	ND	<
Porphyromonas clone CW034	ND	ND	<	ND	<

Table 4.13 *(Cont'd)*

Organism	Subject				
	No. 1	**No. 2**	**No. 3**	**No. 4**	**No. 5**
Prevotella **clone HF050**	ND	ND	ND	ND	<
Prevotella **clone ID019**	ND	ND	ND	ND	<
Prevotella clone FW035	ND	ND	ND	ND	<
Prevotella nigrescens	ND	ND	<	ND	ND
Prevotella pallens	<	ND	ND	ND	ND
Prevotella clone DO003	<	ND	ND	ND	ND
Prev. oris	<	ND	<	ND	ND
Prevotella clone BE073	<	ND	ND	ND	ND
Prevotella clone FM005	<	ND	ND	ND	ND
Prevotella **clone GI059**	<	ND	ND	ND	ND
Prev. melaninogenica	<	ND	ND	<	ND

ND, not detected; <, <15% of the total number of clones assayed; >, >15% of the total number of clones assayed.

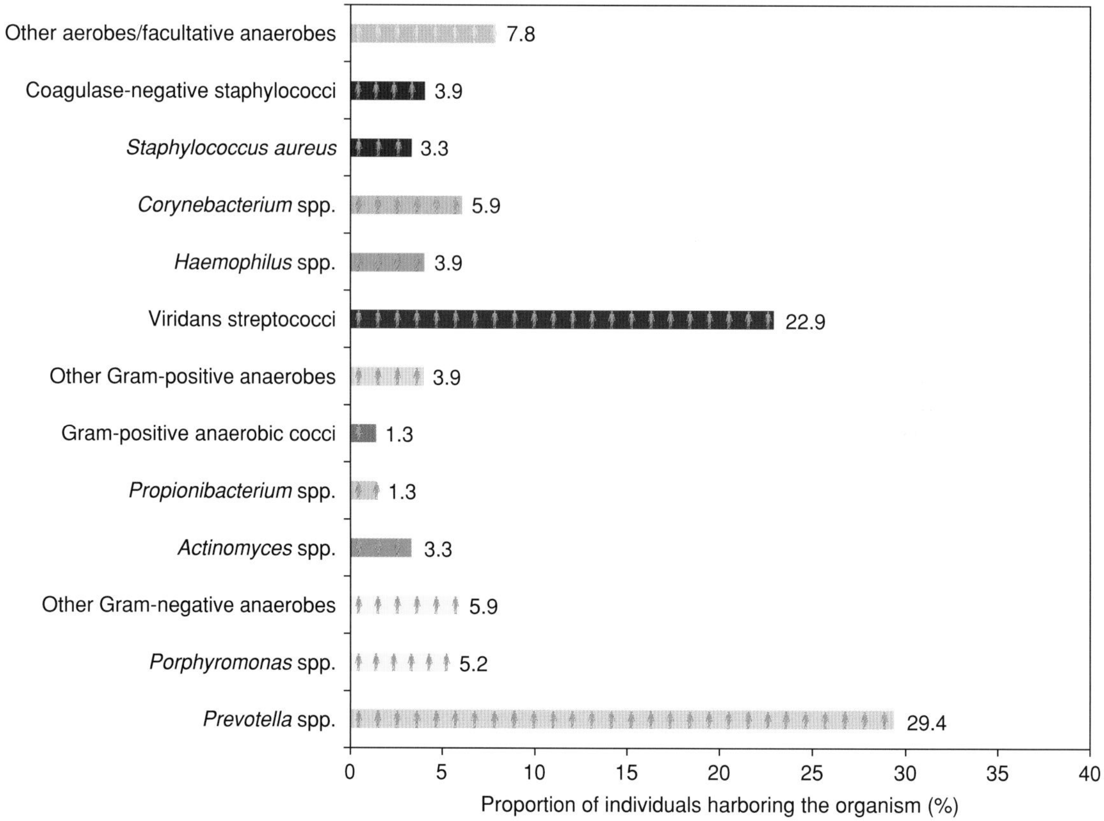

Fig. 4.45 Frequency of isolation of various organisms from the tonsillar crypts of 17 healthy adults (87).

is complex and poorly defined. This is not only because of the complexity of these communities, but is also attributable to the fact that the various regions of the respiratory tract are carriage sites of a number of very important human pathogens (*Strep. pyogenes*, *N. meningitidis*, *Strep. pneumoniae*, *H. influenzae*, *Mor. catarrhalis*, and *Staph. aureus*) and, consequently, most bacteriological studies have focused on the detection of these organisms. Other, more numerous, members of the microbial communities have received little attention. In the oropharynx and nasopharynx, the dominant organisms include species belonging to the genera *Streptococcus* (mainly viridans streptococci), *Haemophilus*, *Neisseria*, *Staphylococcus* (mainly CNS), *Corynebacterium*, *Prevotella*, *Propionibacterium*, *Bacteroides*, *Porphyromonas*, and *Veillonella*. *Mollicutes* are frequently present, but little is known regarding their identity or their exact prevalence. The microbiotas of the nasal vestibule and cavity differ from those of the pharyngeal regions and are dominated by *Corynebacterium* spp., CNS, and *Propionibacterium* spp.

Disappointingly, very few culture-independent studies have been undertaken of the communities found in the respiratory tract.

4.6 SOURCES OF DATA USED TO COMPILE FIGURES

1 Herruzo, R., Chamorro, L., Garcia, M.E., Gonzalez, M.C., Lopez, A.M., Mancenido, N. and Yebenes, L. (2002) *Int J Pediatr Otorhinolaryngol* 65, 117–23.

2 Putnam, S.D., Gray, G.C., Biedenbach, D.J. and Jones, R.N. (2000) *Clin Microbiol Infect* 6, 2–8.

3 Brook, I. and Gober, A.E. (2005) *Chest* 127, 2072–5.

4 Durmaz, R., Durmaz, B., Bayraktar, M., Ozerol, I.H., Kalcioglu, M.T., Aktas, E. and Cizmeci, Z. (2003) *J Clin Microbiol* 41, 5285–7.

5 Gunnarsson, R.K., Holm, S.E. and Soderstrom, M. (1998) *Scand J Prim Health Care* 16, 13–17.

6 Glück, U. and Gebbers, J.O. (2000) *Laryngoscope* 110, 426–8.

7 Principi, N., Marchisio, P., Schito, G.C. and Mannelli, S. (1999) *Pediatr Infect Dis J* 18, 517–23.

8 Konno, M., Baba, S., Mikawa, H., et al. (2006) *J Infect Chemother* 12, 83–96.

9 Christenson, B., Sylvan, S.P. and Noreen, B. (1997) *Scand J Infect Dis* 29, 555–8.

10 Ingvarson, L., Lundgren, K. and Jursing, K. (1986) *Acta Oto-Laryn Supplement* 982, 94–6.

11 Pichichero, M.E., Marsocci, S.M., Murphy, M.L., Hoeger, W., Green, J.L. and Sorrento, A. (1999) *Arch Pediatr Adolesc Med* 153, 624–8.

12 Liassine, N., Gervaix, A., Hegi, R., Strautmann, G., Suter, S. and Auckenthaler, R. (1999) *Eur J Clin Micro Infect Dis* 18, 217–20.

13 Peters, J.E. and Gackstetter, G.D. (1998) *Mil Med* 163, 667–71.

14 Braito, A., Galgani, I., Mohammed, M.R., Iozzi, C., Ame, S.M., Haji, H.S. and Zanchi, A. (2004) *East Afr Med J* 81, 307–12.

15 Fazeli, M.R., Ghaemi, E., Tabarraei, A., Kaplan, E.L., Johnson, D.R., Vakili, M.A. and Khodabakhshi, B. (2003) *Eur J Clin Microbiol Infect Dis* 22, 475–8.

16 Levy, R.M., Huang, E.Y., Roling, D., Leyden, J.J. and Margolis, D.J. (2003) *Arch Dermatol* 139, 467–71.

17 Gazi, H., Kurutepe, S., Surucuoglu, S., Teker, A. and Ozbakkaloglu, B. (2004) *Indian J Med Res* 120, 489–94.

18 Levy, R.M., Leyden, J.J. and Margolis, D.J. (2005) *Clin Microbiol Infect* 11, 153–5.

19 Berkovitch, M., Bulkowstein, M., Zhovtis, D., Greenberg, R., Nitzan, Y., Barzilay, B. and Boldur, I. (2002) *Int J Pediatr Otorhinolaryngol* 63, 19–24.

20 Van Staaij, B.K., Van den Akker, E.H., De Haas Van Dorsser, E.H., Fleer, A., Hoes, A.W. and Schilder, A.G. (2003) *Acta Otolaryngol* 123, 873–8.

21 van der Veen, E.L., Sanders, E.A.M., Videler, W.J.M., van Staaij, B.K., van Benthem, P.P.G. and Schilder, A.G.M. (2006) *Eur Arch Otorhinolaryngol* 263, 750–3.

22 Stjernquist-Desatnik, A., Prellner, K. and Schalen, C. (1991) *J Laryngol Otol* 105, 439–41.

23 Jousimies-Somer, H.R., Savolainen, S. and Ylikoski, J.S. (1989) *J Clin Microbiol* 27, 2736–43.

24 Gluck, U. and Gebbers, J.O. (2003) *Am J Clin Nutr* 77, 517–20.

25 Hoeksma, A. and Winkler, K.C. (1963) *Acta Leiden* 32, 123–33.

26 Lina, G., Boutite, F., Tristan, A., Bes, M., Etienne, J. and Vandenesch, F. (2003) *Appl Environ Microbiol* 69, 18–23.

27 Hjaltested, E.K.R., Bernatoniene, J., Erlendsdottir, H., Kaltenis, P., Bernatoniene, G., Gudnason, T., Haraldsson, A. and Kristinsson, K.G. (2003) *Scand J Infect Dis* 35, 21–6.

28 Marchisio, P., Gironi, S., Esposito, S., Schito, G.C., Mannelli, S. and Principi, N. (2001) *J Med Microbiol* 50, 1095–9.

29 Wolf, B., Gama, A., Rey, L., Fonseca, W., Roord, J., Fleer, A. and Verhoef, J. (1999) *Ann Trop Paediatr* 19, 287–92.

30 Sulikowska, A., Grzesiowski, P., Sadowy, E., Fiett, J. and Hryniewicz, W. (2004) *J Clin Microbiol* 42, 3942–9.

31 Zemlickova, H., Urbaskova, P., Adamkova, V., Motlova, J., Lebedova, V. and Prochazka, B. (2006) *Epidemiol Infect* 134, 1179–87.

32 Chi, D.H., Hendley, J.O., French, P., et al. (2003) *Am J Rhinol* 17, 209–14.

33 Ylikoski, J., Savolainen, S. and Jousimies-Somer, H. (1989) *ORL J Otorhinolaryngol Relat Spec* 51, 50–5.

34 Bogaert, D., van Belkum, A., Sluijter, M., Luijendijk, A., de Groot, R., Rumke, H.C., Verbrugh, H.A. and Hermans, P.W. (2004) *Lancet* 363, 1871–2.

35 Kellner, J.D., McGeer, A., Cetron, M.S., Low, D. E., Butler, J.C., Matlow, A., Talbot, J. and Ford-Jones, E.L. (1998) *Pediatr Infect Dis J* 17, 279–86.

36 Perez, J.L., Linares, J., Bosch, J., Lopez de Goicoechea, M.J. and Martin, R. (1987) *J Antimicrob Chemother* 19, 278–80.

37 Laval, C.B., Andrade, A.L., Pimenta, F.C., de Andrade, J.G., Oliveira, R.M., Silva, S.A., Lima, E.C., Fabio, J.L., Casagrande, S.T. and Brandileone, M.C.C. (2006) *Clin Microbiol Infect* 12, 50–5.

38 Lieberman, D., Shleyfer, E., Castel, H., Terry, A., et al. (2006) *J Clin Microbiol* 44, 525–8.

39 de Lencastre, H. and Tomasz, A. (2002) *J Antimicrob Chemother* 50, 75–81.

40 Faden, H., Heimerl, M., Goodman, G., Winkelstein, P. and Varma, C. (2002) *J Clin Microbiol* 40, 4748–9.

41 Kononen, E., Jousimies-Somer, H., Bryk, A., Kilp, T. and Kilian, M. (2002) *J Med Microbiol* 51, 723–30.

42 Marchisio, P., Esposito, S., Schito, G.C., Marchese, A., Cavagna, R. and Principi, N. (2002) *Emerg Infect Dis* 8, 479–84.

43 Zeta-Capeding, R., Nohynek, H., Sombrero, L., Esparar, G., Mondoy, M., Pascual, L., Esko, E., Leinonen, M. and Ruutu, P. (1995) *J Clin Microbiol* 33, 3077–9.

44 García-Rodríguez, J.A. and Fresnadillo Martínez, M.J. (2002) *J Antimicrob Chemother* 50 (Suppl S2), 59–73.

45 Chapalain, J.C., Dusseau, J.Y., Perrier-Gros-Claude, J.D., Rouby, Y. and Bartoli, M. (1994) *J Antimicrob Chemother* 33, 151–5.

46 Mueller, J.E., Yaro, S., Traoré, Y., Sangaré, L., Tarnagda, Z., Njanpop-Lafourcade, B.M., Borrow, R. and Gessner, B.D. (2006) *J Infect Dis* 193, 812–20.

47 Gazi, H., Surucuoglu, S., Ozbakkaloglu, B., Akcali, S., Ozkutuk, N., Degerli, K. and Kurutepe, S. (2004) *Ann Acad Med Singapore* 33, 758–62.

48 Dominguez, A., Cardenosa, N., Izquierdo, C., Sanchez, F., Margall, N., Vazquez, J.A. and Salleras, L. (2001) *Epidemiol Infect* 127, 425–33.

49 Bakir, M., Yagci, A., Ulger, N., Akbenlioglu, C., Ilki, A. and Soyletir, G. *Eur J Epidemiol* (2001) 17, 1015–18.

50 Gold, R., Goldschneider, I., Lepow, M.L., Draper, T.F. and Randolph, M. (1978) *J Infect Dis* 137, 112–21.

51 Amadou Hamidou, A., Djibo, S., Elhaj Mahamane, A., Moussa, A., Findlow, H., Sidikou, F., Cisse, R., Garba, A., Borrow, R, Chanteau, S. and Boisier, P. (2006) *Microbes Infect* 8, 2098–104.

52 Caugant, D.A., Fogg, C., Bajunirwe, F., Piola, P., Twesigye, R., Mutebi, F., Frøholm, L.O., Rosenqvist, E., Batwala, V., Aaberge, I.S., Rottingen, J.A. and Guerin, P.J. (2006) *Trans R Soc Trop Med Hyg* 100, 1159–63.

53 MacLennan, J., Kafatos, G., Neal, K., Andrews, N., Cameron, J.C., Roberts, R., Evans, M.R., Cann, K., Baxter, D.N., Maiden, M.C. and Stuart, J.M. (2006) *Emerg Infect Dis* 12, 950–7.

54 Bogaert, D., Hermans, P.W.M., Boelens, H., Sluijter, M., Luijendijk, A., Rümke, H.C., Koppen, S., van Belkum, A.,

de Groot, R. and Verbrugh, H.A. (2005) *Clin Infect Dis* 40, 899–902.

55 Trotter, C.L., Gay, N.J. and Edmunds, W. (2006) *Epidemiol Infect* 134, 556–66.

56 Knappi, J.S. and Hook, E.W. (1988) *J Clin Microbiol* 26, 896–900.

57 Kuroki, H., Ishikawa, N., Uehara, S., Himi, K., Sonobe, T. and Niimi, H. (1997) *Acta Paediatr Jpn* 39, 541–5.

58 Bricks, L.F., Mendes, C.M., Lucarevschi, B.R., Oplustil, C.P., Zanella, R.C., Bori, A. and Bertoli, C.J. (2004) *Rev Hosp Clin Fac Med Sao Paulo* 59, 236–43.

59 Barbosa-Cesnik, C., Farjo, R.S., Patel, M., Gilsdorf, J., McCoy, S.I., Pettigrew, M.M., Marrs, C. and Foxman, B. (2006) *Pediatr Infect Dis J* 25, 219–23.

60 Kuklinska, D. and Kilian, M. (1984) *Eur J Clin Microbiol* 3, 249–52.

61 Ejlertsen, T., Thisted, E., Ebbesen, F., Olesen, B. and Renneberg, J. (1994) *J Infect* 29, 23–31.

62 Sehgal, S.C. and al Shaimy, I. (1994) *Infection* 22, 193–6.

63 Quinones, D., Llanes, R., Torano, G. and Perez, M. (2005) *Arch Med Res* 36, 80–2.

64 Golin, V., Mimica, I.M. and Mimica, L.M.J. (1998) *Sao Paulo Med J* 116, 1727–33.

65 Creech, C.B. II., Kernodle, D.S., Alsentzer, A., Wilson, C. and Edwards, K.M. (2005) *Pediatr Infect Dis J* 24, 617–21.

66 Mainous, A.G. 3rd, Hueston, W.J., Everett, C.J. and Diaz, V.A. (2006) *Ann Fam Med* 4, 132–7.

67 Wertheim, H.F., Melles, D.C., Vos, M.C., van Leeuwen, W., van Belkum, A., Verbrugh, H.A. and Nouwen, J.L. (2005) *Lancet Infect Dis* 5, 751–62.

68 Gordts, F., Halewyck, S., Pierard, D., Kaufman, L. and Clement, P.A. (2000) *J Laryngol Otol* 114, 184–8.

69 Klossek, J.M., Dubreuil, L., Richet, H., Richet, B., Sedallian, A. and Beutter, P. (1996) *J Laryngol Otol* 110, 847–9.

70 Nadel, D.M., Lanza, D.C. and Kennedy, D.W. (1999) *Am J Rhinol* 13, 87–90.

71 Rasmussen, T.T., Kirkeby, L.P., Poulsen, K., Reinholdt, J. and Kilian, M. (2000) *APMIS* 108, 663–75.

72 Araujo, E., Palombini, B.C., Cantarelli, V., Pereira, A. and Mariante, A. (2003) *Am J Rhinol* 17, 9–15.

73 Savolainen, S., Ylikoski, J. and Jousimies-Somer, H. (1986) *Rhinology* 24, 249–55.

74 Douglas, M., Nadel, D.M., Lanza, D.C. and Kennedy, D.W. (1999) *Am J Rhinol* 13, 87–90.

75 Gordts, F., Abu Nasser, I., Clement, P.A.R., Pierard, D. and Kaufman, L. (1999) *Int J Pediatr Otorhinolaryngol* 48, 163–7.

76 Kostamo, K., Richardson, M., Virolainen-Julkunen, A., Leivo, I., Malmberg, H., Ylikoski, J. and Toskala, E. (2004) *Rhinology* 42, 213–18.

77 Uehara, Y., Nakama, H., Agematsu, K., Uchida, M., Kawakami, Y., Abdul Fattah, A.S. and Maruchi, N. (2000) *J Hosp Infect* 44, 127–33.

78 Larson, E.L., McGinley, K.J., Foglia, A.R., Talbot, G.H. and Leyden, J.J. (1986) *J Clin Microbiol* 23, 604–8.

79 Heczko, P.B., Hoffler, U., Kasprowicz, A. and Pulverer, G. (1981) *J Med Microbiol* 14, 233–41.

80 Winkler, K.C. and Hoeksma, A. (1963) *Acta Leiden* 32, 123–33.

81 Leyden, J.J. and McGinley, K.J. (1992) Coryneform bacteria. In: Noble W.C (ed.).*The Skin Microflora and Microbial Skin Disease*. Cambridge University Press, Cambridge, UK, pp. 102–17.

82 Marples, R. (1982) *Curr Med Res Opin* 7 (Suppl 2), 67–70.

83 Hokama, T. and Imamura, T. (1998) *J Trop Pediatr* 44, 84–6.

84 Frandsen, E.V., Pedrazzoli, V. and Kilian, M. (1991) *Oral Microbiol Immunol* 6, 129–33.

85 von Graevenitz, A., Punter-Streit, V., Riegel, P. and Funke, G. (1998) *J Clin Microbiol* 36, 2087–8.

86 Aas, J.A., Paster, B.J., Stokes, L.N., Olsen, I. and Dewhirst, F.E. (2005) *J Clin Microbiol* 43, 5721–32.

87 Stjernquist-Desatnik, A. and Holst, E. (1999) *Acta Otolaryngol (Stockh)* 119, 102–6.

88 Saez Nieto, J.A., Marcos, C. and Vindel, A. (1998) *Int Microbiol* 1, 59–63.

4.7 FURTHER READING

Ball, S.L., Siou, G.P., Wilson, J.A., Howard, A., Hirst, B.H. and Hall, J. (2007) Expression and immunolocalization of antimicrobial peptides within human palatine tonsils. *J Laryngol Otol* 26, 1–6

Bals, R. and Hiemstra, P.S. (2004) Innate immunity in the lung: How epithelial cells fight against respiratory pathogens. *Eur Respir J* 23, 327–33.

Bernardini, G., Braconi, D. and Santucci, A. (2007) The analysis of *Neisseria meningitidis* proteomes: Reference maps and their applications. *Proteomics* 7, 2933–46.

Boucher, R.C. (1999) Molecular insights into the physiology of the thin film of airway surface liquid. *J Physiol* 516, 631–8.

Brook, I. and Gober, A.E. (2005) Recovery of potential pathogens and interfering bacteria in the nasopharynx of smokers and nonsmokers. *Chest* 127, 2072–5.

Bullard, B., Lipski, S. and Lafontaine, E.R. (2007) Regions important for the adhesin activity of *Moraxella catarrhalis* Hag. *BMC Microbiol* 7, 65.

Cantin, A.M. (2001) Biology of respiratory epithelial cells: Role in defense against infections. *Pediatric Pulmonol* 26 (Suppl 23), 167–9.

Caugant, D.A., Tzanakaki, G. and Kriz, P. (2007) Lessons from meningococcal carriage studies. *FEMS Microbiol Rev* 31, 52–63.

Choi, C.S., Yin, C.S., Bakar, A.A., Sakewi, Z., Naing, N.N., Jamal, F. and Othman, N. (2006) Nasal carriage of *Staphylococcus aureus* among healthy adults. *J Microbiol Immunol Infect* 39, 458–64.

Clarke, S.R. and Foster, S.J. (2006) Surface adhesins of *Staphylococcus aureus*. *Adv Microb Physiol* 51, 187–224.

Cole, A.M., Dewan, P. and Ganz, T. (1999) Innate antimicrobial activity of nasal secretions. *Infect Immun* 67, 3267–75.

Cole, A.M., Tahk, S., Oren, A., Yoshioka, D., Kim, Y.H., Park, A. and Ganz, T. (2001) Determinants of *Staphylococcus aureus* nasal carriage. *Clin Diagn Lab Immunol* 8, 1064–9.

Courtney, H.S., Hasty, D.L. and Dale, J.B. (2002) Molecular mechanisms of adhesion, colonization, and invasion of group A streptococci *Ann Med* 34, 77–87.

Davidsen, T. and Tonjum, T. (2006) Meningococcal genome dynamics. *Nat Rev Microbiol* 4, 11–22.

De Lencastre, H. and Tomasz, A. (2002) From ecological reservoir to disease: The nasopharynx, day-care centres and drug-resistant clones of *Streptococcus pneumoniae*. *J Antimicrob Chemother* 50 (Suppl S2), 75–81.

Devine, D.A. (2003) Antimicrobial peptides in defence of the oral and respiratory tracts. *Mol Immunol* 40, 431–43.

Diamond, G., Legarda, D. and Ryan, L.K. (2000) The innate immune response of the respiratory epithelium. *Immunol Rev* 173, 27–38.

Diggle, M.A. and Clarke, S.C. (2006) Molecular methods for the detection and characterization of *Neisseria meningitidis*. *Expert Rev Mol Diagn* 6, 79–87.

Erwin, A.L. and Smith, A.L. (2007) Nontypeable *Haemophilus influenzae*: Understanding virulence and commensal behavior. *Trends Microbiol* 15, 355–62.

Ganz, T. (2002) Antimicrobial polypeptides in host defense of the respiratory tract. *J Clin Invest* 109, 693–7.

Garcia-Rodriguez, J.A. and Fresnadillo Martinez, M.J. (2002) Dynamics of nasopharyngeal colonization by potential respiratory pathogens. *J Antimicrob Chemother* 50 (Suppl C), 59–73.

Grubor, B., Meyerholz, D.K. and Ackermann, M.R. (2006) Collectins and cationic antimicrobial peptides of the respiratory epithelia. *Vet Pathol* 43, 595–612.

Gunnarsson, R.K., Holm, S.E. and Soderstrom, M. (1998) The prevalence of potential pathogenic bacteria in nasopharyngeal samples from healthy children and adults. *Scand J Prim Health Care* 16, 13–17.

Hood, D.W. (2003) The genome sequence of *Haemophilus influenzae*. *Methods Mol Med* 71, 147–59.

Karalusa, R. and Campagnaria, A. (2000) *Moraxella catarrhalis*: A review of an important human mucosal pathogen. *Microbes Infect* 2, 547–59.

Konno, M., Baba, S., Mikawa, H., et al. (2006) Study of upper respiratory tract bacterial flora: first report. Variations in upper respiratory tract bacterial flora in patients with acute upper respiratory tract infection and healthy subjects and variations by subject age. *J Infect Chemother* 12, 83–96.

Kononen, E., Jousimies-Somer, H., Bryk, A., Kilp, T. and Kilian, M. (2002) Establishment of streptococci in the upper respiratory tract: Longitudinal changes in the mouth and nasopharynx up to 2 years of age. *J Med Microbiol* 51, 723–30.

Kozlova, I., Vanthanouvong, V., Johannesson, M. and Roomans, G.M. (2006) Composition of airway surface liquid determined by X-ray microanalysis. *Ups J Med Sci* 111, 137–53.

Laube, D.M., Yim, S., Ryan, L.K., Kisich, K.O. and Diamond, G. (2006) Antimicrobial peptides in the airway. *Curr Top Microbiol Immunol* 306, 153–82.

Leiberma, A., Dagan, R., Leibovitz, E., Yagupsky, P. and Fliss, D.M. (1999) The bacteriology of the nasopharynx in childhood. *Int J Pediatr Otorhinolaryngol* 49 (Suppl 1), S151–3.

LeVine, A.M. and Whitsett, J.A. (2001) Pulmonary collectins and innate host defense of the lung. *Microbes Infect* 3, 161–6.

Lopez, R. (2006) Pneumococcus: The sugar-coated bacteria. *Int Microbiol* 9, 179–90.

Lux, T., Nuhn, M., Hakenbeck, R. and Reichmann, P. (2007) Diversity of bacteriocins and activity spectrum in *Streptococcus pneumoniae*. *J Bacteriol* Aug 17; (Epub ahead of print).

Manetti, A.G., Zingaretti, C., Falugi, F., Capo, S., Bombaci, M., Bagnoli, F., Gambellini, G., Bensi, G., Mora, M., Edwards, A.M., Musser, J.M., Graviss, E.A., Telford, J.L., Grandi, G. and Margarit, I. (2007) *Streptococcus pyogenes* pili promote pharyngeal cell adhesion and biofilm formation. *Mol Microbiol* 64, 968–83.

Marrs, C.F., Krasan, G.P., McCrea, K.W., Clemans, D.L. and Gilsdorf, J.R. (2001) *Haemophilus influenzae* – human specific bacteria. *Front Biosci* 6, 41–60.

Mitchell, T.J. (2006) *Streptococcus pneumoniae*: Infection, inflammation and disease. *Adv Exp Med Biol* 582, 111–24.

Mukundan, D., Ecevit, Z., Patel, M., Marrs, C.F. and Gilsdorf, J.R. (2007) Pharyngeal colonization dynamics of *Haemophilus influenzae* and *Haemophilus haemolyticus* in healthy adult carriers. *J Clin Microbiol* Aug 8; (Epub ahead of print).

Nascimento, J. dos S., Giambiagi-deMarval, M., de Oliveira, S.S., Ceotto, H., dos Santos, K.R. and Bastos, Mdo C. (2005) Genomic fingerprinting of bacteriocin-producer strains of *Staphylococcus aureus*. *Res Microbiol* 156, 837–42.

Nassif, X. (2002) Genomics of *Neisseria meningitidis*. *Int J Med Microbiol* 291, 419–23.

Nilsson, P. and Ripa, T. (2006) *Staphylococcus aureus* throat colonization is more frequent than colonization in the anterior nares. *J Clin Microbiol* 44, 3334–9.

Nochi, T. and Kiyono, H. (2006) Innate immunity in the mucosal immune system. *Curr Pharm Des* 12, 4203–13.

Obaro, S. and Adegbola, R. (2002) The pneumococcus: Carriage, disease and conjugate vaccines. *J Med Microbiol* 51, 98–104.

Palmer, L.B., Albulak, K., Fields, S., Filkin, A.M., Simon, S. and Smaldone, G.C. (2001) Oral clearance and pathogenic oropharyngeal colonization in the elderly. *Am J Respir Crit Care Med* 164, 464–8.

Paterson, G.K. and Mitchell, T.J. (2006) Innate immunity and the pneumococcus. *Microbiology* 152, 285–93.

Peacock, S.J., de Silva, I. and Lowy, F.D. (2001) What determines nasal carriage of *Staphylococcus aureus*? *Trends Microbiol* 9, 605–10.

Pericone, C.D., Overweg, K., Hermans, P.W. and Weiser, J.N. (2000) Inhibitory and bactericidal effects of hydrogen peroxide production by *Streptococcus pneumoniae* on other inhabitants of the upper respiratory tract. *Infect Immun* 68, 3990–7.

Randell, S.H. and Boucher, R.C. University of North Carolina Virtual Lung Group. (2006) Effective mucus clearance is essential for respiratory health. *Am J Respir Cell Mol Biol* 35, 20–8.

Rasmussen, T.T., Kirkeby, L.P., Poulsen, K., Reinholdt, J. and Kilian, M. (2000) Resident aerobic microbiota of the adult human nasal cavity. *APMIS* 108, 663–75.

Rogan, M.P., Geraghty, P., Greene, C.M., O'Neill, S.J., Taggart, C.C. and McElvaney, N.G. (2006) Antimicrobial proteins and polypeptides in pulmonary innate defence. *Respir Res* 7, 29.

Rose, M.C. and Voynow, J.A. (2006) Respiratory tract mucin genes and mucin glycoproteins in health and disease. *Physiol Rev* 86, 245–78.

Shopsin, B. and Kreiswirth, B.N. (2001) Molecular epidemiology of methicillin-resistant *Staphylococcus aureus*. *Emerg Infect Dis* 7, 323–6.

Snyder, L.A., Davies, J.K., Ryan, C.S. and Saunders, N.J. (2005) Comparative overview of the genomic and genetic differences between the pathogenic *Neisseria* strains and species. *Plasmid* 54, 191–218.

Stannard, W. and O'Callaghan, C. (2006) Ciliary function and the role of cilia in clearance. *J Aerosol Med* 19, 110–15.

Stjernquist-Desatnik, A. and Holst, E. (1999) Tonsillar microbial flora: Comparison of recurrent tonsillitis and normal tonsils. *Acta Oto-Laryngol* (*Stockholm*) 119, 102–6.

Travis, S.M., Singh, P.K. and Welsh, M.J. (2001) Antimicrobial peptides and proteins in the innate defense of the airway surface. *Curr Opin Immunol* 13, 89–95.

Trotter, C.L., Gay, N.J. and Edmunds, W.J. (2006) The natural history of meningococcal carriage and disease. *Epidemiol Infect* 134, 556–66.

van Belkum, A. (2006) Staphylococcal colonization and infection: Homeostasis versus disbalance of human (innate) immunity and bacterial virulence. *Curr Opin Infect Dis* 19, 339–44.

Verduin, C.M., Hol, C., Fleer, A., van Dijk, H. and van Belkum, A. (2002) *Moraxella catarrhalis*: From emerging to established pathogen. *Clin Microbiol Rev* 15, 125–44.

Yazdankhah, S.P. and Caugant, D.A. (2004) *Neisseria meningitidis*: An overview of the carriage state. *J Med Microbiol* 53, 821–32.

Chapter 5

THE INDIGENOUS MICROBIOTA OF THE URINARY SYSTEM OF FEMALES

The functions of the urinary system are to dispose of the waste products of metabolism and to regulate the chemical composition and volume of body fluids. The system consists of several organs – two kidneys, two ureters, a bladder, and a urethra (Fig. 5.1). The kidneys filter the blood, remove urea and other wastes, and excrete these waste materials as urine. The ureters transport the urine from the kidneys to the bladder where it is stored temporarily. Urine is periodically expelled by the bladder into the urethra, from which it is discharged into the environment.

For convenience, the male and female urinary systems are considered separately because: (1) the anatomy of the system in males and females differs significantly; (2) the terminal portion of the urinary system (the urethra) in males is also part of the reproductive system, and this results in important functional differences from the urethra of females; and (3) the urethral opening in females is closer to the anus than in males, and is also close to the vaginal introitus, and these densely colonized sites are sources of potential microbial colonizers of the urethra. These three factors result in significant differences in the indigenous microbiota of the urinary system of males and females.

The only region of the female urinary system with an indigenous microbiota is the urethra – the kidneys, ureters, and bladder are normally sterile in healthy individuals.

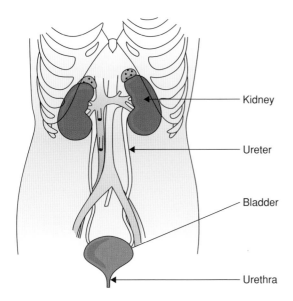

Fig. 5.1 The main components of the urinary system. (From http://www.training.seer.cancer.gov/module_anatomy/anatomy_physiology_home.html; funded by the US National Cancer Institute's Surveillance, Epidemiology and End Results (SEER) Program with Emory University, Atlanta, Georgia, USA).

Kidney

Ureter

Bladder

Urethra

5.1 ANATOMY AND PHYSIOLOGY OF THE URINARY SYSTEM OF FEMALES

In females, the urethra is a short tube (approximately 3.8 cm in length) that leads from the floor of the bladder to the external environment via the external urethral orifice – this lies in front of the vaginal opening (Fig. 5.2). The urethra has a complex structure that includes many glands (paraurethral and mucous) and

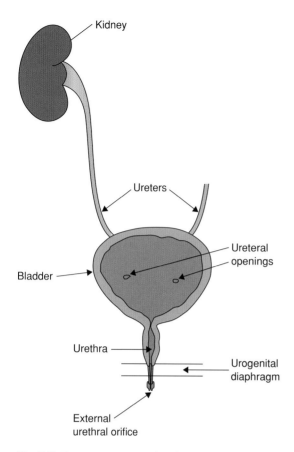

Fig. 5.2 The urinary system in females.

pit-like recesses which open into the lumen. It is lined by a nonkeratinized, stratified, squamous epithelium which is similar to that of the vagina (Fig. 5.3). However, the epithelium becomes keratinized near the external orifice.

5.2 ANTIMICROBIAL DEFENSES OF THE FEMALE URINARY SYSTEM

One of the most important antimicrobial defense mechanisms of the urethra is the shedding of the outermost cells with their adherent microbes. In females, however, this is under hormonal control, and the urethra of menstruating women and post-menopausal women on hormone replacement therapy (HRT) has abundant exfoliating cells. In contrast, exfoliation occurs at a slower rate in the urethra of pre-menarcheal girls and post-menopausal women not on HRT. A layer of mucus (secreted by the paraurethral glands) covers the epithelium, and this traps microbes, thereby preventing them from adhering to the underlying epithelium. Microbial colonization is also hindered by the flushing action of urine. A total of between 1 and 2 liters of urine are expelled each day during urination, in quantities of 200–400 ml, at a flow rate of between 40 and 80 ml/h. Furthermore, approximately mid-way along the urethra (where the urethra passes through the urogenital diaphragm), a high-pressure zone exists, and this hinders microbial ascent from the urethral orifice to the bladder. The epithelium is also coated with a layer of Tamm–Horsfall glycoprotein (also known as uromodulin), which

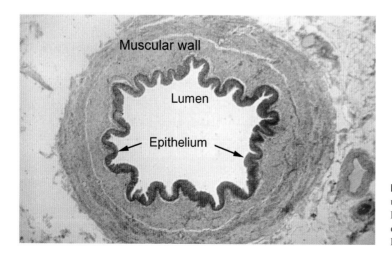

Fig. 5.3 Cross-section through the urethra of a female. Image © 2006, David King, Southern Illinois University School of Medicine, Springfield and Carbondale, Illinois, USA, used with permission.

Table 5.1 Antimicrobial defense mechanisms of the female urinary system.

Defense mechanism	Mode of action
Desquamation	Outermost cells are shed with their attached microbes – but this is under hormonal control
Mucus	Prevents attachment of microbes to the epithelium; aids removal of microbes
Urine flow	Mechanically removes bacteria
High-pressure zone in mid-urethra	Restricts ascent of bacteria into bladder
pH of urine	Low pH can inhibit or kill microbes
Osmolarity of urine	High osmolarity can be microbicidal or microbistatic
Urea	High concentration of urea can be microbicidal or microbistatic
Tamm–Horsfall glycoprotein	Prevents bacterial adhesion to epithelium
Manno-oligosaccharides	Prevent bacterial adhesion to epithelium
GP51 protein	Prevents bacterial adhesion to epithelium
sIgA antibodies	Prevent bacterial adhesion
Hepcidin 20 (10–30 µg/l)	Microbistatic or microbicidal
Cathelicidin (LL-37) (0.2–5.9 µg/l)	Microbistatic or microbicidal
Human β-defensin-1 (10–100 µg/l)	Microbistatic or microbicidal
Human defensin-5 (50 µg/l)	Microbistatic or microbicidal

prevents microbes adhering to its surface. This protein is synthesized in the kidney and is the most abundant protein in urine – between 50 and 100 mg are produced each day. It is continuously deposited from urine onto the epithelium and contains receptors for the adhesins of various enterobacteria, including the type 1 and type S fimbriae of *Escherichia coli* – the organism most frequently responsible for urinary tract infections (UTIs). The protein-bound bacteria are then removed by the flushing action of urine. The protein also binds to polymorphonuclear leukocytes (PMNs) so enhancing phagocytosis of the bacteria–protein complex.

Another protein present in urine is GP51 which is secreted by bladder epithelial cells. This binds to a wide range of organisms including *E. coli*, *Enterobacter cloacae*, *Klebsiella pneumoniae*, *Proteus* spp., *Pseudomonas aeruginosa*, *Serratia marcescens*, *Staphylococcus aureus*, *Staphylococcus epidermidis*, and *Enterococcus faecalis*, thereby preventing their adhesion to epithelial cells. The bacteria form aggregates which are removed by the flushing action of urine. Urine also contains manno-oligosaccharides which bind to the adhesins on the type 1 fimbriae of *E. coli*, thereby preventing the organism from adhering to the epithelium.

The pH of urine in healthy individuals ranges from 4.6 to 7.5, with a mean value of 6.0 (Table 5.1). This low pH prevents the growth of some microbes and can

be microbicidal. The high urea content of urine (200–400 mmoles/l) also exerts a microbicidal or microbistatic effect, as can the high osmolarity (up to 1 300 mOsm/kg). The antibacterial effect of urine is greatest at low pHs.

Urine contains several antimicrobial peptides (Table 5.1). Of these, LL-37/hCAP-18 is constitutively expressed by uroepithelial cells and in renal tubuli, but synthesis and release of the peptide is dramatically up-regulated by the presence of bacteria (Fig. 5.4). Secretory immunoglobulin A (sIgA), together with smaller quantities of IgG and IgM, is present in urine, and is able to block the adhesion of *E. coli* and other bacteria to epithelial cells.

Little information is available on the presence of Toll-like receptors (TLRs) in the urinary tract, although TLRs 2, 3, 4, and 9 are known to be expressed by human uroepithelial cells (Fig. 5.5).

5.3 ENVIRONMENTAL DETERMINANTS WITHIN THE FEMALE URETHRA

The pH of the urethra ranges from 5.8 in the distal region to 6.1 in the proximal regions. However, the intermittent voiding of urine (with a pH of approximately 6) will certainly affect urethral pH. The

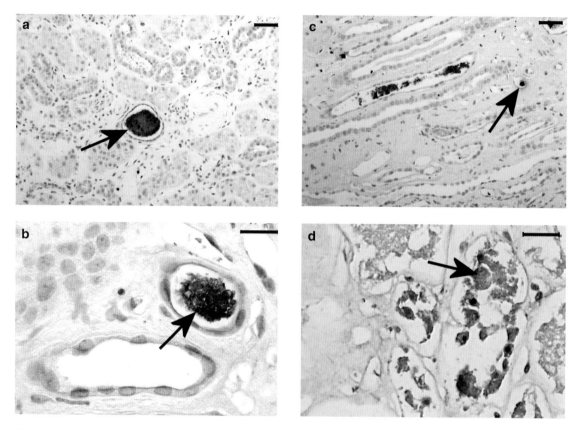

Fig. 5.4 Expression of LL-37 in healthy human renal tissue. The sections were stained using a polyclonal antibody to LL-37. LL-37 is present (brown staining, indicated by arrows) in the hyaline substance in the lumen of renal tubuli (a and b) and in neutrophils (c). After 24 h of in vitro incubation and exposure to bacteria, epithelial cells stained positive for LL-37 (d). Scale bars in (a) and (c), 50 μm; scale bars in (c) and (d), 20 μm. Reprinted with the permission of Macmillan Publishers Ltd, London, UK, from; *Nature Medicine.* Chromek, M., Slamova, Z., Bergman, P., et al. (2006) The antimicrobial peptide cathelicidin protects the urinary tract against invasive bacterial infection. Volume 12, pp. 636–41, © 2006.

urethra is an aerobic region, and the oxygen content of urine is high with a mean partial pressure of approximately 81 mmHg. Nutrient sources for urethral microbes include mucus and other materials secreted by glands lining the urethra, substances present in urine (see Table 5.2) and molecules derived from dead and dying epithelial cells. A number of urethral microbes (including coagulase-negative staphylococci [CNS], lactobacilli, viridans streptococci, Gram-positive anaerobic cocci [GPAC] and *Bacteroides* spp.) can liberate sugars and amino acids from mucins. Many of these organisms are also proteolytic and so can produce amino acids from host proteins for use as a carbon and/or energy source. Urine provides a number of additional nitrogen sources (urea, NH_4^+, and creatinine),

phosphate, and some trace elements (Table 5.2) but no carbohydrates – except in the case of individuals suffering from diabetes.

In order for microbes to survive in the urethra, they must, of course, be able to withstand the numerous antimicrobial mechanisms protecting this region (see Table 5.1).

5.4 THE INDIGENOUS MICROBIOTA OF THE FEMALE URETHRA

The repeated flushing of the urethra during urination is a strong environmental selecting factor making the ability to adhere to the urethral epithelium a key

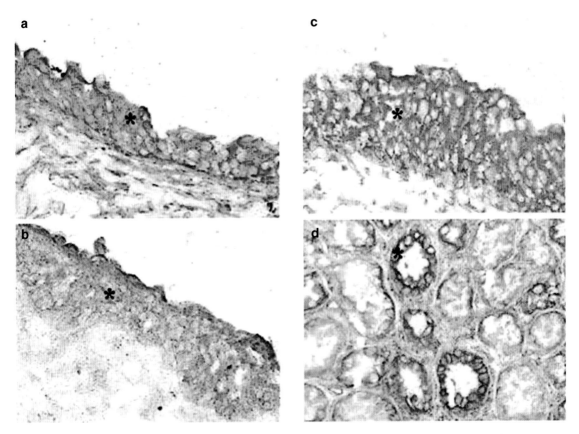

Fig. 5.5 Expression of TLR4 in the human urinary tract epithelium. Biopsies were stained with monoclonal anti-human TLR4 antibody and were visualized with Fast-Red substrate. TLR4 (shown in red) was detected in the epithelium (indicated by asterisks) lining the (a) bladder, (b) ureter, (c) pelvis, and (d) cortex of the kidney. Magnification, ×200. Reprinted with permission from: Samuelsson, P., Hang, L., Wullt, B., Irjala, H. and Svanborg, C. (2004) *Infect Immun* 72, 3179–86.

Table 5.2 Composition of urine.

Solute	Concentration (mmol/l)
Urea	200–400
Sodium	50–130
Chloride	50–130
Potassium	20–70
Ammonium	30–50
Phosphate	25–60
Calcium	10–24
Creatinine	6–20
Uric acid	0.7–8.7
Bicarbonate	0–2
Protein	<10 mg/100 ml
Glucose	Trace

requirement for any successful colonizer of this region. In contrast to males, the urethra of females is usually colonized by microbes along its whole length.

5.4.1 Members of the urethral microbiota

The main organisms colonizing the female urethra are *Lactobacillus* spp., CNS, *Corynebacterium* spp., viridans streptococci, *Bacteroides* spp., GPAC, and *Mollicutes*. *Corynebacterium* spp. and CNS are described in Chapter 2, while viridans streptococci are described in Chapters 4 and 8. *Mollicutes* are described in sections 4.4.1.6 and 6.4.1. *Lactobacillus* spp. and *Bacteroides* spp. are among the dominant organisms of the vagina and colon, respectively – these organisms are described in Chapters 6 and 9.

Table 5.3 Nomenclature of GPAC previously considered to belong to the genus *Peptostreptococcus*.

Previous nomenclature	Current or proposed nomenclature
Peptostreptococcus anaerobius	Unchanged
Peptostreptococcus asaccharolyticus	*Peptoniphilus asaccharolyticus*
Peptostreptococcus indolicus	*Peptoniphilus indolicus*
Peptostreptococcus magnus	*Finegoldia magna*
Peptostreptococcus micros	*Parvimonas micra* (formerly *Micromonas micros*)
Peptostreptococcus prevotii	*Anaerococcus prevotii*
Peptostreptococcus productus	*Ruminococcus productus*
Peptostreptococcus vaginalis	*Anaerococcus vaginalis*
Peptostreptococcus tetradius	*Anaerococcus tetradius*

The GPAC comprise the following genera: *Peptococcus, Peptostreptococcus, Anaerococcus, Peptoniphilus, Gallicola, Finegoldia, Parvimonas, Ruminococcus, Schleiferella, Coprococcus,* and *Sarcina*. Species belonging to these genera are members of the indigenous microbiota of the urinary tract, oral cavity, vagina, skin, and intestinal tract. Until recently, many species of GPAC were assigned to the genera *Peptostreptococcus* or *Peptococcus*. However, the nomenclature of these organisms has undergone dramatic changes and now only one species belonging to the genus *Peptococcus* is recognized – *Peptococcus niger* – while members of the genus *Peptostreptococcus* have been reclassified into five genera. The new nomenclature of species mentioned in this and subsequent chapters is given in Table 5.3. All 11 genera of GPAC are obligately anaerobic, nonsporing cocci which may occur in pairs, tetrads, clusters, or chains. They have complex growth requirements and often need to be supplied with several vitamins and amino acids. Species belonging to the genera *Peptococcus, Peptostreptococcus, Anaerococcus, Peptoniphilus, Gallicola, Schleiferella, Finegoldia,* and *Parvimonas* use peptones as their major energy sources and, although they do not need them for growth, some can ferment carbohydrates. The main end products of metabolism are acetic and butyric acid, although a few species produce mainly isovaleric or caproic acids. *Anaerococcus* spp. are saccharolytic and butyrate-producing, while *Peptoniphilus,* spp. are non-saccharolytic and butyrate-producing. Species belonging to the genera *Ruminococcus, Coprococcus,* and *Sarcina*, however, do require carbohydrates for growth and produce mainly lactate, acetate, formate, and succinate. *Ruminococcus* spp. are predominant members of the intestinal microbiota and are described in more detail in Chapter 9. A range of exoenzymes are

produced by GPAC including proteases, glycosidases, hyaluronidase, DNase, RNase, and hemolysins.

5.4.2 Community composition in the female urethra

The main groups of organisms found in the urethra of females depend very much on the age and sexual maturity of the individual. Nevertheless, they invariably include members of the genera *Lactobacillus, Corynebacterium, Staphylococcus, Bacteroides, Streptococcus, Fusobacterium,* and *Veillonella* as well as GPAC. These organisms are also members of the indigenous microbiota of the skin, vagina, or the intestinal tract and are presumably derived from these regions. Hence, studies have shown that strains of *E. coli* and *Staphylococcus saprophyticus*, both of which are frequent causes of UTIs, are identical to those present in the intestinal tract of the same individual.

The adhesion of organisms to a host structure is an essential prerequisite for the colonization of any region. In the urethra, however, this is particularly important because of the need to withstand the regular flushing action of urine. Few studies have been carried out on the adhesion of urethral microbes to epithelial cells; most studies of this type have involved uropathogenic organisms (e.g. *E. coli* and *Staph. saprophyticus*), and have shown that bacteria are not uniformly distributed over the urethral epithelium and that most cells have no attached bacteria. Cells that are colonized generally have at least ten adherent bacteria which are evenly distributed over the cell surface. The adherent bacteria are usually single or in pairs – microcolonies are only occasionally present.

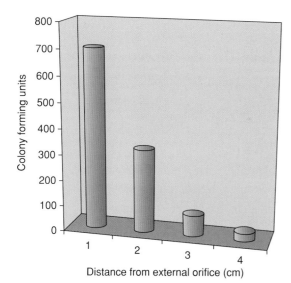

Fig. 5.6 Number of viable bacteria isolated from 1-cm sections of the urethra in 52 adult females. Unfortunately, only aerobic cultivation was carried out in this study (1).

Surprisingly few studies of the urethral microbiota of females have been published, and many of these have not employed anaerobic culture techniques; this is unfortunate, as studies that have done so have demonstrated that anaerobes constitute an appreciable proportion of the microbiota of this region. The total number of viable bacteria colonizing the urethra is approximately 10^5 colony-forming units (cfu). Unlike the urethra of males, in which only the distal portion is colonized by microbes (Chapter 7), bacteria have been detected along the entire length of the female urethra. However, the density of colonization decreases with increasing distance from the external urethral orifice (Fig. 5.6). Furthermore, while all women harbor bacteria in the first 1-cm segment (i.e. nearest the urethral orifice), only approximately 50% of women have bacteria present in the final 1-cm segment.

On the basis of the very limited data that are available, the most frequently detected organisms in the urethra of healthy females appear to be *Corynebacterium* spp., GPAC, *Bacteroides* spp., CNS, and lactobacilli (Fig. 5.7).

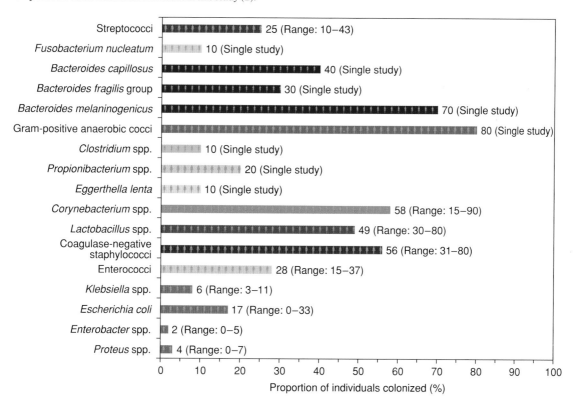

Fig. 5.7 Frequency of isolation of bacteria from the urethra of healthy females. The data shown are the means (and ranges) based on the results of six studies involving 219 pre-menopausal females (2–7).

Table 5.4 Organisms that have been detected in the urethral microbiota of females. This is not a complete list but is indicative of the diversity of the urethral microbiota.

Obligate anaerobes	Mollicutes	Aerobes/facultative anaerobes
Bacteroides capillosus	Mycoplasma hominis	Staph. epidermidis
Bacteroides fragilis group	Mycoplasma genitalium	Staph. aureus
Bacteroides amylophilus	Mycoplasma fermentans	Micrococcus spp.
Bacteroides distasonis	Ureaplasma urealyticum	Corynebacterium spp.
Bacteroides vulgatus	—	Viridans streptococci
Bacteroides ovatus	—	Streptococcus agalactiae
An. prevotii	—	E. coli
Veillonella spp.	—	Proteus spp.
Propionibacterium spp.	—	Enterobacter spp.
Finegoldia magna	—	Citrobacter spp.
Peptoniphilus asaccharolyticus	—	Enterococcus spp.
Fusobacterium nucleatum	—	Acinetobacter spp.
Pep. anaerobius	—	Klebsiella spp.
Rum. productus	—	Lactobacillus spp.
Eggerthella lenta	—	—
Clostridium spp.	—	—
Prev. intermedia	—	—

A wide range of organisms has been detected in the urethral microbiota, although many are present only in small numbers – these are listed in Table 5.4.

The number of cultivable species present in the female urethra, the frequency of detection of the constituent organisms, and their relative proportions are all markedly affected by the sexual maturity of the individual. In general, the number of species detected decreases in the following order: post-menopausal women (not receiving HRT) > pre-menopausal women > pre-menarcheal girls. *Corynebacterium* spp., lactobacilli, and CNS dominate the cultivable urethral microbiota of pre-menarcheal and pre-menopausal individuals, but the cultivable urethral microbiota of post-menopausal females is dominated by *Bacteroides* spp. (Fig. 5.8). *Corynebacterium* spp. and CNS are the most frequently isolated organisms in all groups, while lactobacilli are frequently recovered only from pre-menopausal women (Fig. 5.9). The main changes in the urethral microbiota of females associated with sexual maturity are as follows:

• The diversity of the community and the number of organisms present is greater in older age groups.
• Facultative anaerobes dominate the urethral microbiota of pre-menarcheal girls and pre-menopausal

women, whereas obligate anaerobes are dominant in post-menopausal women.
• Facultative Gram-negative rods are generally absent from the urethras of pre-menarcheal girls and pre-menopausal women, but may be present in post-menopausal women.
• There is an increased frequency of isolation of facultative Gram-negative bacilli in post-menopausal women.
• There is a decreased isolation frequency of lactobacilli in pre-menarcheal girls and post-menopausal women.
• The proportion of CNS and *Corynebacterium* spp. in post-menopausal women decreases.
• The proportion of black-pigmented Gram-negative anaerobic bacilli in post-menopausal women increases. Many of these changes are a consequence of age-related, hormone-induced changes in the urethra similar to those which occur in the vagina (section 6.3.1). Hence, the microbiotas of pre-menopausal women and pre-menarcheal girls have high proportions of lactobacilli, and these organisms display wide-ranging anti-bacterial and anti-adhesive effects (see section 6.4.1.1) and therefore exert a profound effect on the composition of the urethral microbiota.

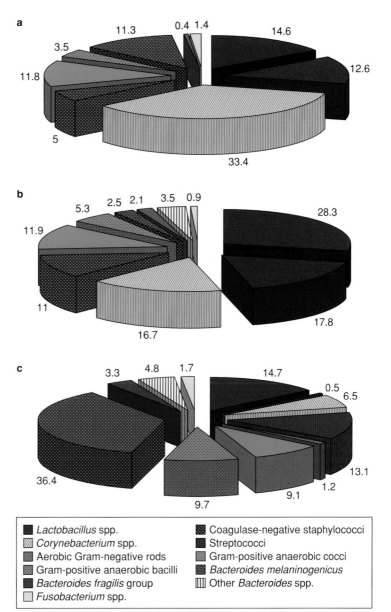

Fig. 5.8 Relative proportions of groups of organisms comprising the cultivable urethral microbiota of (a) nine pre-menarcheal girls, (b) ten pre-menopausal women, and (c) ten post-menopausal women (4).

Legend:

- ■ *Lactobacillus* spp.
- ▨ *Corynebacterium* spp.
- ▨ Aerobic Gram-negative rods
- ▨ Gram-positive anaerobic bacilli
- ■ *Bacteroides fragilis* group
- ▨ *Fusobacterium* spp.
- ▨ Coagulase-negative staphylococci
- ■ Streptococci
- ▨ Gram-positive anaerobic cocci
- ▨ *Bacteroides melaninogenicus*
- ▥ Other *Bacteroides* spp.

5.5 OVERVIEW OF THE MICROBIOTA OF THE URINARY TRACT OF FEMALES

Microbial colonization of the urinary tract of females is hindered by the regular flushing action of urine and by the production of a number of molecules that prevent adhesion of microbes to the epithelium, as well as by the usual array of defense mechanisms found at mucosal surfaces. Only the urethra of the female urinary tract has a resident microbiota and, as this is relatively short, microbes can be detected along its entire length. Remarkably few studies have been directed at

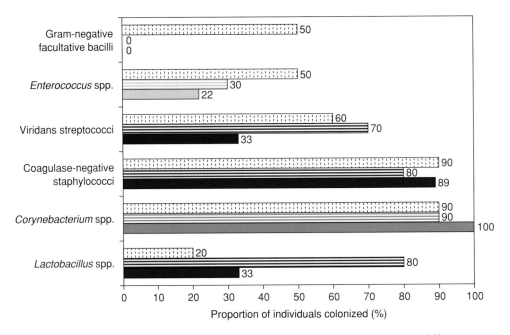

Fig. 5.9 Frequency of isolation of bacteria from the urethras of nine pre-menarcheal girls (denoted by solid bars), ten pre-menopausal women (indicated by lined bars), and ten post-menopausal women (denoted by dashed bars) (4).

ascertaining the composition of the urethral microbiota, and all the studies with this goal were culture-dependent. The organisms most frequently isolated include *Corynebacterium* spp., GPAC, *Bacteroides* spp., CNS, and lactobacilli. However, the sexual maturity of the individual has a profound effect on microbial community composition. *Corynebacterium* spp., CNS, streptococci, and lactobacilli dominate the microbiota of pre-menarcheal girls and pre-menopausal women, whereas Gram-negative anaerobic bacilli and lactobacilli dominate that of post-menopausal individuals.

5.6 SOURCES OF DATA USED TO COMPILE FIGURES

1 Cox, C.E. (1966) *South Med J* 59, 621–6.
2 Levendoglu, F., Ugurlu, H., Ozerbil, O.M., Tuncer, I. and Ural, O. (2004) *Spinal Cord* 42, 106–9.
3 Montgomerie, J.Z., McCary, A., Bennett, C.J., Young, M., Matias, B., Diaz, F., Adkins, R. and Anderson, J. (1997) *Spinal Cord* 35, 282–5.
4 Marrie, T.J., Swantee, C.A. and Hartlen, M. (1980) *J Clin Microbiol* 11, 654–9.
5 Pfau, A. and Sacks, T. (1981) *J Urol* 126, 630–4.
6 Elkins, I.B. and Cox, C.E. (1974) *J Urol* 111, 88–92.
7 Fair, W.R., Timothy, M.M., Millar, M.A. and Stamey, T.A. (1970) *J Urol* 104, 426–31.

5.7 FURTHER READING

Apodaca, G. (2004) The uroepithelium: Not just a passive barrier. *Traffic* 5, 1–12.
Chowdhury, P., Sacks, S.H., Sheerin, N.S. (2006) Toll-like receptors TLR2 and TLR4 initiate the innate immune response of the renal tubular epithelium to bacterial products. *Clin Exp Immunol* 145, 346–56.
Chromek, M., Slamová, Z., Bergman, P., Kovács, L., Podracká, L., Ehrén, I., Hökfelt, T., Gudmundsson, G.H., Gallo, R.L., Agerberth, B. and Brauner, A. (2006) The antimicrobial peptide cathelicidin protects the urinary tract against invasive bacterial infection. *Nat Med* 12, 636–41.
Ganz, T. Defensins in the urinary tract and other tissues. (2001) *J Infect Dis* 183, S41–2.
Kunin, C.M., Evans, C., Bartholomew, D. and Bates, D.G. (2002) The antimicrobial defense mechanism of the female urethra: A reassessment. *J Urol* 168, 413–19.
Levendoglu, F., Ugurlu, H., Ozerbil, O.M., Tuncer, I. and Ural, O. (2004) Urethral cultures in patients with spinal cord injury. *Spinal Cord* 42, 106–9.

Lewis, S.A. (2000) Everything you wanted to know about the bladder epithelium but were afraid to ask. *Am J Physiol: Renal Physiol* 278, F867–74.

Marrie, T.J., Harding, G.K.M. and Ronald, A.R. (1978) Anaerobic and aerobic urethral flora in healthy females. *J Clin Microbiol* 8, 67–72.

Marrie, T.J., Swantee, C.A. and Hartlen, M. (1980) Aerobic and anaerobic urethral flora of healthy females in various physiological age groups and of females with urinary tract infections. *J Clin Microbiol* 11, 654–9.

Park, C.H., Valore, E.V., Waring, A.J. and Ganz, T. (2001) Hepcidin, a urinary antimicrobial peptide synthesized in the liver. *J Biol Chem* 276, 7806–10.

Saemann, M.D., Weichhart, T., Horl, W.H., Zlabinger, G.J. (2005) Tamm-Horsfall protein: A multilayered defence molecule against urinary tract infection. *Eur J Clin Invest* 35, 227–35.

Samuelsson, P., Hang, L., Wullt, B., Irjala, H. and Svanborg, C. (2004) Toll-like receptor 4 expression and cytokine responses in the human urinary tract mucosa. *Infect Immun* 72, 3179–86.

Serafini-Cessi, F., Monti, A. and Cavallone, D. (2005) *N*-Glycans carried by Tamm-Horsfall glycoprotein have a crucial role in the defense against urinary tract diseases. *Glycoconj J* 22, 383–94.

Volgmann, T., Ohlinger, R. and Panzig, B. (2005) *Ureaplasma urealyticum* – Harmless commensal or underestimated enemy of human reproduction? A review. *Arch Gynecol Obstet* 273, 133–9.

Zhang, D., Zhang, G., Hayden, M.S., Greenblatt, M.B., Bussey, C., Flavell, R.A. and Ghosh, S. (2004) A Toll-like receptor that prevents infection by uropathogenic bacteria. *Science* 303, 1522–6.

Chapter 6

THE INDIGENOUS MICROBIOTA OF THE REPRODUCTIVE SYSTEM OF FEMALES

6.1 ANATOMY AND PHYSIOLOGY OF THE FEMALE REPRODUCTIVE SYSTEM

The female reproductive system consists of the ovaries, fallopian tubes, uterus, cervix, vagina, and the vulva (Fig. 6.1). Of these organs, only the vagina, vulva, and the cervix of the uterus are usually colonized by microbes. The indigenous microbiota of the vagina and, to a lesser extent, the cervix, has been studied extensively. In contrast, little is known about the microbial communities residing on the vulva.

The vulva consists of a number of anatomical regions – the mons pubis, labia majora and labia minora, clitoris, and vestibule (Fig. 6.2). The vestibule is lined with a stratified squamous epithelium and has several openings – the vaginal orifice, the external urethral

orifice, and the openings of the mucus-secreting para-urethral and Bartholin's glands. The mons pubis is covered by skin and pubic hair (Table 6.1). The labia majora are longitudinal folds of skin which are covered in pubic hair on their outer surfaces and have sebaceous and sudoriferous glands. In contrast, the labia minora are covered with a stratified squamous epithelium (which may have a thin keratinized layer on its surface), are free of hair, and have numerous sebaceous glands but few sudoriferous glands. From the medial (i.e. innermost) surfaces of the labia majora to the vagina, there is a gradual transition in the structure of the epithelium as it changes from a keratinized epithelium typical of the outer body surface to the mucosal epithelium found in the vagina. The clitoris is covered with a thin stratified squamous

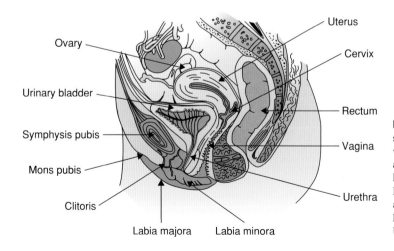

Fig. 6.1 The female reproductive system. (From http://www.training.seer.cancer.gov/module_anatomy/anatomy_physiology_home.html; funded by the US National Cancer Institute's Surveillance, Epidemiology and End Results (SEER) Program with Emory University, Atlanta, Georgia, USA.)

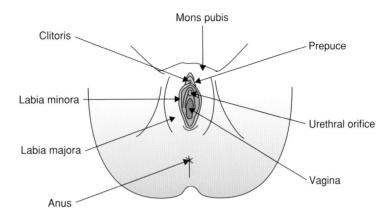

Fig. 6.2 The major regions of the vulva. (From http://www.training.seer.cancer.gov/module_anatomy/anatomy_physiology_home.html; funded by the US National Cancer Institute's Surveillance, Epidemiology and End Results (SEER) Program with Emory University, Atlanta, Georgia, USA.)

Table 6.1 Main anatomical features of the various regions of the vulva.

Structure	Type of epithelium	Hair	Adipose tissue	Sebaceous glands	Sudoriferous glands
Mons pubis	Keratinized	Yes	Yes	Yes	Yes
Labia majora	Keratinized	Yes	Yes	Yes	Yes
Labia minora	Stratified squamous	No	No	Yes	Yes, but few
Clitoris	Stratified squamous	No	No	No	No

epithelium and has a hood of skin known as the prepuce or foreskin.

The vagina is a muscular, tubular, distendable organ approximately 8 cm in length and extends from the cervix to the vulva (Fig. 6.3). It functions as a passageway during childbirth, is an outlet for the menstrual flow, and is a receptacle for the penis during sexual intercourse. It has a folded lining consisting of a stratified, squamous, nonkeratinized epithelium with a surface area of approximately 360 cm^2 (Fig. 6.4).

No glands are present in the vagina, and the surface is lubricated by mucus produced by the cervix. During the menstrual cycle (Fig. 6.5), the epithelium is constantly undergoing re-modeling because of the proliferation, maturation, and desquamation of cells. In general, estrogens (particularly estradiol) stimulate this process, whereas progesterone inhibits maturation of the cells. The low levels of estrogens present in individuals before the menarche (i.e. the onset of menstruation) and after the menopause (i.e. the end of menstruation) result in pre-menarcheal girls and post-menopausal women having a relatively thin vaginal epithelium.

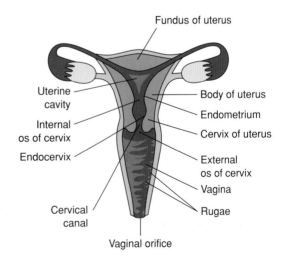

Fig. 6.3 The main structural features of the vagina and cervix.

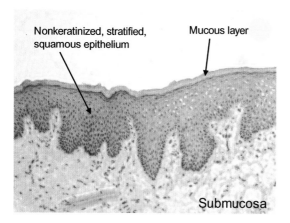

Fig. 6.4 Cross-section through the vaginal mucosa. Reprinted from: Pivarcsi, A., Nagy, I., Koreck, A., Kis, K., Kenderessy-Szabo, A., Szell, M., Dobozy, A., and Kemeny, L. (2005) Microbial compounds induce the expression of pro-inflammatory cytokines, chemokines and human β-defensin-2 in vaginal epithelial cells. *Microbes and Infection 7*, 1117–27, ©2005, with permission from Elsevier, Amsterdam, The Netherlands.

The cervix is the lower, narrow portion of the uterus that opens into the vagina (Fig. 6.3). The upper region of the cervix opens into the uterus via the "internal os". The lumen then widens to form the cervical canal which narrows to produce an opening (the external os) into the vagina. The portion of the cervix that protrudes into the vagina is known as the ectocervix, and this is covered by a stratified, squamous, nonkeratinized epithelium identical to that present in the vagina. This type of epithelium extends into the cervical canal for a distance which varies depending on the age of the individual. It then changes into a simple columnar epithelium which lines the rest of the canal – this mucosal surface is known as the endocervix. Almost all of the cells of the endocervix secrete mucus, while the rest are ciliated. Unlike the mucosa of the uterus, the cervical epithelium is not shed during menstruation. However, it is affected by changes in hormone levels during the menstrual cycle. Hence, the quantity of mucus secreted increases ten-fold (to approximately 700 mg/day) when the level of estrogen peaks at mid-cycle. Following ovulation, the increased level of progesterone causes less mucus to be produced, and it becomes more viscous, so that it forms a plug which seals the canal.

6.2 ANTIMICROBIAL DEFENSE SYSTEMS OF THE FEMALE REPRODUCTIVE SYSTEM

The mons pubis and the labia majora have the innate and acquired defense mechanisms typically present in skin, whereas the labia minora, clitoris, and vestibule have defense mechanisms similar to those found at other mucosal surfaces. Although the vagina and cervix have the usual mucosa-associated defense mechanisms, the innate and acquired immune responses of these regions are subject to hormonal fluctuations, and IgG rather than IgA predominates in secretions.

(a) The menstrual (uterine endometrial) cycle

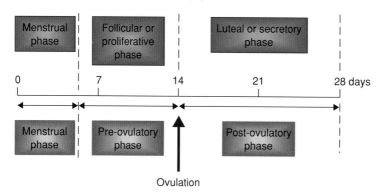

Ovulation

(b) The ovarian cycle

Fig. 6.5 The term female reproductive cycle refers to: (a) the menstrual cycle, a series of changes in the endometrium of the uterus, and (b) the ovarian cycle, a monthly series of events involving the maturation of an oocyte.

Table 6.2 Innate defense mechanisms of the cervix.

Defense mechanism	Functions
Desquamation	Continually removes attached bacteria
Ciliated epithelium covered with mucus	Constitutes a mucociliary escalator, which continually sweeps incoming organisms back into the vagina
Mucus	High viscosity hinders penetration of microbes into underlying epithelium Carbohydrate moieties on mucins, as well as free sugars, bind to bacterial adhesins, thereby blocking attachment to epithelial receptors Contains antimicrobial compounds – lysozyme, lactoferrin, SLPI, calprotectin, hCAP-18/LL-37, HD-5, HD-6, HBD-1, HBD-2, HBD-3, HBD-4, HNP-1, HNP-2, HNP-3, and HNP-4 Contains complement
TLR1 to TLR6	Activation induces expression of pro-inflammatory cytokines, chemokines, and antimicrobial peptides

6.2.1 Innate defense systems

In order to protect any fertilized ovum from infection, it is important that the uterus and upper regions of the genital tract are maintained in a sterile condition. Consequently, the cervix is equipped with a variety of antimicrobial defense mechanisms to prevent ingress of microbes from the heavily colonized lower genital tract (Table 6.2). However, unlike at other body sites, the effectiveness of many of these defense mechanisms varies markedly during the menstrual cycle. Hence, the quantity of mucus secreted increases ten-fold at mid-cycle, while its viscosity reaches a minimum just before ovulation. Likewise, the concentrations of several antimicrobial compounds in cervical mucus vary throughout the cycle (Fig. 6.6). Furthermore, complement levels are at a minimum during mid-cycle, and estrogen stimulates the rate of cell maturation and thereby the rate of desquamation of epithelial cells.

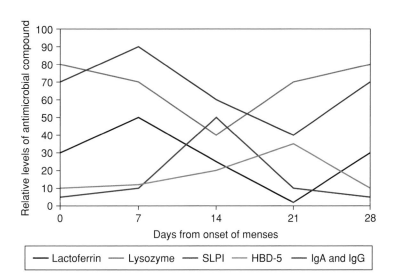

Fig. 6.6 Fluctuations in the concentration of antimicrobial compounds in cervical mucus during the menstrual cycle.

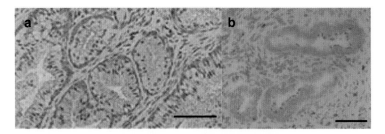

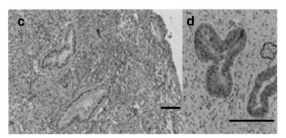

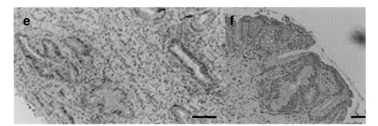

Fig. 6.7 Expression of TLRs (brown-staining regions) in the endocervical epithelium of healthy individuals. Immunostaining was performed using the avidin–biotin–peroxidase technique and polyclonal antibodies specific for each TLR. (a–f) Positive staining for TLR 1 to TLR 6, respectively. Reprinted from: Fazeli, A., Bruce, C., and Anumba, D.O. (2005) Characterization of Toll-like receptors in the female reproductive tract in humans. *Human Reproduction* 20, 1372–78, by permission of the European Society of Human Reproduction and Embryology, Beigem, Belgium.

Toll-like receptors (TLR1 to TLR6) have also been detected in the ectocervix and endocervix of healthy individuals (Fig. 6.7). Recognition of complementary microbial ligands by these TLRs results in the expression of pro-inflammatory cytokines, chemokines, and antimicrobial peptides.

If conception occurs, the cervical mucus thickens and forms a plug that blocks the cervical canal. This plug acts as a physical barrier to prevent ingress of microbes, and also has broad-spectrum antimicrobial activity resulting from the presence of a range of antimicrobial compounds (Fig. 6.8).

The antimicrobial defense mechanisms operating in the vagina are similar to those in the cervix, although no mucociliary escalator is present (Table 6.3 and Fig. 6.9).

As can be seen from Table 6.3, a large variety of antimicrobial peptides and proteins are present in the vagina, and the concentrations of some of these are listed in Table 6.4. However, an important additional defense mechanism is the low pH (approximately 4)

although, again, this is affected by the menstrual cycle and by the sexual maturity of the individual. Between the menarche and the menopause, the vaginal mucosa contains large quantities of glycogen which, during anaerobic respiration, is converted to organic acids – mainly lactic and acetic acids. This produces a very acidic environment, which inhibits the growth of many bacterial species – the acids can also exert a direct antimicrobial effect.

6.2.2 Acquired immune defense systems

The antibodies present in genital tract secretions are not only produced locally but are also derived from serum by transudation. Hence, unlike other mucosal secretions, the predominant class of antibody in genital fluids is IgG rather than IgA. These antibodies are able to prevent microbial adhesion to, and invasion of,

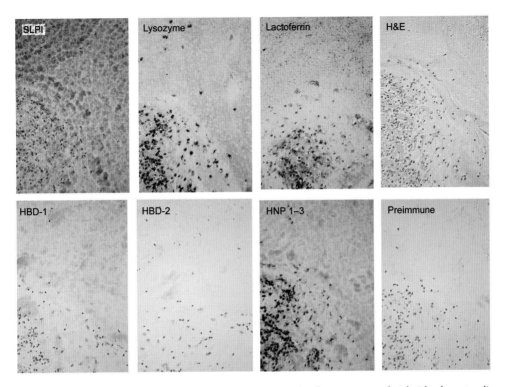

Fig. 6.8 Distribution of antimicrobial peptides in cervical mucous plugs. The plugs were stained with either hematoxylin and eosin (H&E), or with the antibodies against the indicated antimicrobial polypeptides. The immunoreactive areas are stained brown, and the counterstain is blue. Reprinted from: Hein, M., Valore, E.V., Helmig, R.B., Uldbjerg, N. and Ganz, T. (2002) Antimicrobial factors in the cervical mucus plug. *Am J Obstet Gynecol* 187, 137–44, © 2002, with permission from Elsevier, Amsterdam, The Netherlands.

Table 6.3 Innate antimicrobial defense mechanisms operating in the vagina. Details regarding the range of antimicrobial compounds contained in cervical secretions are presented in Table 6.2.

Defense mechanism	Functions
Low pH owing to glycogen metabolism	Inhibits microbial growth; kills some microbes
Desquamation	Continually removes attached bacteria
Cervical secretions	Contains a range of antimicrobial compounds
Production of antimicrobial compounds: lysozyme, lactoferrin, SLPI, calprotectin, calgranulin, histones, hCAP18/LL-37, HD-5, HD-6, HBD-1, HBD-2, HBD-3, HBD-4, HNP-1, HNP-2, HNP-3, and HNP-4, hemocidins (antibacterial fragments of hemoglobin)	Inhibits or kills microbes
TLR1 to TLR6	Activation induces expression of pro-inflammatory cytokines, chemokines, and antimicrobial peptides

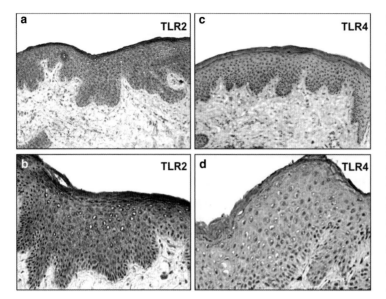

Fig. 6.9 Expression of TLRs in the vaginal mucosa of healthy individuals. Immunostaining was performed using the avidin–biotin–peroxidase technique and polyclonal antibodies specific for each TLR. (a) and (b) TLR2 is abundantly expressed (brown staining) in all cell layers of the epithelium. (c) and (d) Expression of TLR4 is seen mainly in the superficial layer of cells. Magnifications (a and c): 200×. Magnifications (c and d): 400×. Reprinted from: Pivarcsi, A., Nagy, I., Koreck, A., Kis, K., Kenderessy-Szabo, A., Szell, M., Dobozy, A. and Kemeny, L. (2005) Microbial compounds induce the expression of pro-inflammatory cytokines, chemokines and human β-defensin-2 in vaginal epithelial cells. *Microbes and Infection* 7, 1117–27, ©2005, with permission from Elsevier, Amsterdam, The Netherlands.

Table 6.4 Concentrations of some of the antimicrobial peptides/proteins present in vaginal fluid.

Antimicrobial peptide/protein	Concentration in vaginal fluid (μg/ml)
Calprotectin	34 ± 7
Cystatin	32
Lysozyme	13 ± 2
Histone H2	11
Cathepsin	11
Lactoferrin	0.9 ± 0.2
SLPI	0.7 ± 0.1
HBD-2	0.57 ± 0.13
HNP-1, HNP-2, and HNP-3	0.35 ± 0.07
HBD-1	0.04 ± 0.02

6.3 ENVIRONMENTAL DETERMINANTS AT DIFFERENT REGIONS OF THE REPRODUCTIVE SYSTEM

The anatomy and physiology of the vagina and cervix are markedly affected by the concentrations and relative proportions of a number of hormones. Consequently, the environments within these organs differ not only with regard to the sexual maturity of the individual but, in the case of post-menarcheal/pre-menopausal females, also during the monthly reproductive cycle and as a result of pregnancy. Such differences exert a profound effect on the vaginal and cervical microbiotas. The environments within the various regions of the vulva are also likely to be affected by hormonal fluctuations, although less is known about this.

Other factors that can affect the environment of the cervix and vagina include the use of tampons and various contraceptive devices.

6.3.1 Vagina

The vagina is predominantly a microaerophilic environment with a low partial pressure of oxygen (PO_2). However, the PO_2 varies during the menstrual cycle, with low levels of approximately 1.0 mmHg being found in mid-cycle. During menstruation, menstrual blood

the epithelial surface as well as neutralizing toxins and other potentially harmful antigenic materials. They are also involved in antibody-dependent, cell-mediated cytotoxicity and the promotion of phagocytosis by neutrophils. Antibody production is under hormonal control, and the levels of both IgA and IgG are higher in the pre-ovulatory than the post-ovulatory phase.

Table 6.5 Conditions in the vagina at various stages of sexual maturity.

Characteristic	Neonate	Pre-menarche	Post-menarche	Post-menopause
Estrogen level	High	Low	High	Low
pH	Acidic	Neutral	Acidic	Neutral
Glycogen content	High	Low	High	Low
Redox potential	High	Low	High	Low

(with an oxygen content of approximately 42 mmHg) will increase the oxygen content of the vagina, and it has been reported that, in healthy individuals during menstruation, the PO_2 in the vagina ranges between 4 and 35 mmHg – which is 3–22% of that present in air. Other causes of fluctuations in the oxygen content of the vagina include the insertion of a contraceptive diaphragm or tampon – both of these will tend to increase the partial pressure of oxygen, even if only transiently. Such fluctuations may affect the vaginal microbiota by encouraging the growth of aerobes and facultative anaerobes.

The pH of the vagina is affected by the sexual maturity of the individual (Table 6.5) and, in post-menarcheal/pre-menopausal females, by the menstrual cycle (Fig. 6.10). Vaginal pH appears to correlate with the glycogen content of the vaginal mucosa and is lowest in neonates and in post-menarcheal/pre-menopausal individuals.

In Fig. 6.10, it can be seen that, at the onset of menstrual flow, the pH is almost neutral. It then falls during and after menstruation and reaches a minimum at mid-cycle. The pH remains below 4 for approximately 1 week, then steadily increases until the onset of menstruation. Pregnancy does not appear to affect the pH of the vagina, which remains less than 4.0.

The temperature of the vagina is approximately 37°C, although it varies slightly during the menstrual cycle.

The main source of host-derived nutrients for the vaginal microbiota is vaginal fluid, which is produced at a rate of 1–3 g per day by women of reproductive age. This is a complex mixture consisting of cervical mucus, endometrial fluid, a transudate from the vaginal mucosa, secretions of the Bartholin's glands, desquamated vaginal epithelial cells, and leukocytes (Table 6.6 and Fig. 6.11). It is an acidic

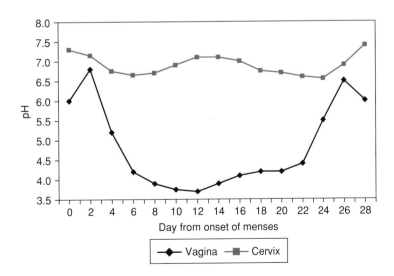

Fig. 6.10 Variations in the pH of the vagina and cervix during the menstrual cycle.

Table 6.6 Principal constituents of vaginal fluid and cervical mucus in post-menarcheal/pre-menopausal women. Approximate concentrations of the constituents are given where these are known.

Constituent	Vaginal fluid	Cervical mucus
Protein/glycoprotein	Mucins, albumin, immunoglobulins (IgG, IgA, IgM), transferrin, lactoferrin, α_2-haptoglobin, α_1-antitrypsin, α_2-macroglobulin	Mucins, albumin, transferrin, lactoferrin, α_2-haptoglobin, elastase, α_2-macroglobulin, β-lipoprotein, α_1-antitrypsin, immunoglobulins
Carbohydrate	Glucose (0.62 g/100 g), glycogen (1.5 g/100 g), oligosaccharides (80–160 mM), mannose, glucosamine, fucose	Mainly glucose and fucose
Lipid	Neutral lipids and phospholipids	Neutral lipids and phospholipids
Low molecular mass organic compounds	Urea (49 mg/100 ml), lactic acid, acetic acid, butanoic acid, propanoic acid	Urea, lactate, acetate
Amino acids	Fourteen amino acids identified including alanine, glycine, histidine, leucine, trytophan	Mainly alanine, glutamate, threonine, taurine
Inorganic ions	Na^+ (23 mmol/l), Cl^- (62 mmol/l), K^+ (61 mmol/l)	Mainly Na^+ and Cl^-

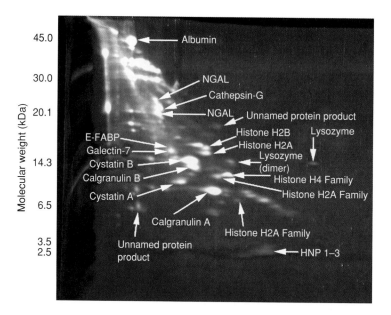

Fig. 6.11 Identification of polypeptides in vaginal fluid. The figure shows a Sypro Ruby-stained two-dimensional gel of human vaginal fluid. In brief, vaginal fluid was extracted using 5% acetic acid, vacuum dried, and resuspended in 0.1% hexa decyl trimethyl ammonium bromide (CETAB)/10% acetic acid/3× acid urea loading dye (9M urea, 5% acetic acid, and methyl green). Vaginal fluid extract (10 μl) was subjected to two-dimensional electrophoresis with acid-urea-polyacrylamide gel electrophoresis (AU-PAGE) as the first dimension, followed by tricine-SDS-PAGE as the second dimension. The protein spots were identified by tandem mass spectrometry (MALDI-TOF/TOF). NGAL, neutrophil gelatinase-associated lipocalin; E-FABP, fatty acid-binding protein 5; HNP 1–3, human neutrophil peptide 1, 2, and 3. Image kindly supplied by Alexander M. Cole and Nitya Venkataraman, Department of Molecular Biology and Microbiology, Burnett College of Biomedical Science, University of Central Florida, Florida, USA.

fluid with a pH of approximately 4.5, and its exact composition varies throughout the menstrual cycle. Other occasional sources of additional nutrients in the vagina would be provided by menstrual fluid and seminal fluid.

In post-menarcheal/pre-menopausal women, approximately 80 ml of menstrual fluid is produced over a period of 3–5 days. The fluid consists of 30–50% whole blood, with the remainder being an endometrial transudate. It therefore provides a range of additional nutrients including carbohydrates, amino acids, proteins, urea, lipids, and fatty acids. In addition, its high hemoglobin content is a valuable source of iron for many organisms

6.3.2 Cervix

The physiology and, ultimately, the environment of the cervix, are affected by estrogen and progesterone levels. It is a microaerophilic environment – the PO_2 being at its lowest (12 mmHg) during the proliferative phase. The pH of the cervix is generally higher than that of the vagina, and ranges from 5.4 to 8.2 with a median pH of 7.0. The pH varies through the menstrual cycle, being more alkaline preceding and during menstruation and at ovulation (Fig. 6.10). The pH of the cervix during pregnancy is approximately 6.5. The temperature of the cervix is approximately 37°C.

The main host-derived source of nutrients for bacteria is cervical mucus (Table 6.6). In the case of post-menarcheal/pre-menopausal women, this is supplemented by menstrual fluid. The amount of mucus produced gradually increases until mid-cycle and then decreases. It consists mainly of water (90–95%), and its content of solutes and its viscosity changes during the menstrual cycle; e.g. the IgA and IgG content is highest during the follicular phase (first

half of the cycle) than during the luteal phase (second half of the cycle).

6.3.3 Vulva

Few studies have investigated the environmental conditions in the various regions within the vulva. Some features of the labia majora and the interlabial fold are listed in Table 6.7.

The oxygen content of the labia minora is low – approximately 12% of that found in air. The main sources of nutrients will vary with the exact location within the vulva, but will include vaginal secretions (Table 6.6) and the nutrients typically found on the skin surface (see Chapter 2).

6.3.4 Contribution of the indigenous microbiota to nutrient supply within the reproductive system

As at other sites, end products of microbial metabolism constitute an important source of nutrients for other microbes. As described in sections 6.3.1 and 6.3.2, mucins are major constituents of cervical and vaginal secretions, and many microbes colonizing the vagina and cervix produce enzymes that can contribute to mucin degradation, even if each species is able to achieve only partial hydrolysis (section 1.5.3). Some of the relevant enzymes produced by members of the resident microbiota are listed in Table 6.8.

Enzymes contributing to mucin hydrolysis have been detected in Gram-negative anaerobic rods isolated from the vagina of high proportions of healthy individuals (Table 6.9). Proteins are another valuable source of nutrients and can be degraded by a variety of vaginal microbes (Table 6.8).

Table 6.7 Environmental determinants of some regions of the vulva.

	Labia majora	Interlabial fold
Temperature (°C)	34.0 ± 0.2°C	34.4 ± 0.6°C
Capacitance (arbitrary units; 0 = very dry, 100 = very moist)	74 ± 10 (i.e. high)	129 ± 2 (i.e. very high)
pH	5.2 ± 0.1	5.5 ± 0.1

Table 6.8 Members of the indigenous microbiota of the reproductive system that produce enzymes that are able to contribute to the hydrolysis of mucins.

Organism	Enzymes produced
Lactobacilli	α- and β-galactosidases, α- and β-D-glucosidases, β-glucuronidase
Mobiluncus spp.	Sialidase
Mycoplasma spp.	α- and β-glucosidases, β-galactosidase, β-*N*-acetylglucosaminidase, sialidase
Fusobacterium spp.	Peptidases, proteases
Can. albicans	*N*-Acetylglucosaminidase, aspartyl proteinase, peptidase
Bifidobacterium spp.	Sialidase, α- and β-glycosidases, α- and β-D-glucosidases, α- and β-D-galactosidase, β-D-fucosidase, proteases
Strep. agalactiae	Peptidase and protease
Prevotella spp.	Sialidase, sulfatase, α-fucosidase, β-galactosidase, *N*-acetyl-β-glucosaminidase, α-galactosidase, proteases, peptidases
Bacteroides spp.	Sialidase, α-fucosidase, β-galactosidase, *N*-acetyl-β-glucosaminidase, *N*-acetyl-α-galactosidase, peptidases, proteases
Porphyromonas spp.	Sialidase, glycosidases, proteases, peptidases
G. vaginalis	Sialidase
E. coli	Sialidase, glycosidases,
Ent. faecalis	Sialidase, glycosidases, proteases, peptidases
P. acnes	Sialidase, protease
Actinomyces spp.	Sialidase, protease
Viridans streptococci	Sialidases, proteases, glycosidases
Clostridium spp.	Sialidase, α- and β-galactosidases, α- and β-D-glucosidases, β-glucuronidase, peptidases
Staphylococcus spp.	Proteases

Table 6.9 Frequency of detection in 68 healthy women of vaginal Gram-negative anaerobic rods that produce enzymes involved in the degradation of mucins (1).

Enzyme	Proportion of women with detected activity (%)	Median concentration of bacteria with detected activity (cfu/ml)
N-Acetylglucosaminidase	53	2000
β-Galactosidase	53	2000
Sialidase	50	2000
α-Fucosidase	50	2000
α-Galactosidase	40	2000
Glycine aminopeptidase	35	2000
Arginine aminopeptidase	34	2000

6.4 THE INDIGENOUS MICROBIOTA OF THE FEMALE REPRODUCTIVE SYSTEM

6.4.1 Members of the microbiota

In the population as a whole, a wide range of organisms have been detected in the female reproductive system (Table 6.10), including *Lactobacillus* spp., *Staphylococcus* spp., *Corynebacterium* spp., *Streptococcus* spp., *Enterococcus* spp., *Candida albicans*, *Bifidobacterium* spp., *Gardnerella vaginalis*, *Propionibacterium* spp., Gram-positive anaerobic cocci (GPAC), *Enterobacteriaceae*, *Bacteroides* spp., *Porphyromonas* spp., *Prevotella* spp., *Clostridium* spp., *Fusobacterium* spp., *Veillonella* spp., *Ureaplasma* spp., and *Mycoplasma* spp. Despite the wide range of species that have been detected, both culture-dependent and culture-independent studies have revealed that both the vaginal and cervical microbiota in a disease-free individual consist of only a limited number of species – usually three or four, but invariably fewer than eight.

Staphylococcus spp., *Propionibacterium* spp., and *Corynebacterium* spp. are described in Chapter 2. Streptococci and *Mycoplasma* spp. are addressed in Chapter 4, while GPAC are discussed in Chapter 5. Gram-negative anaerobic rods, *Bifidobacterium* spp., *Enterococcus* spp., *Enterobacteriaceae*, and *Clostridium* spp. are important members of the microbiota of the colon and so are more appropriately described in Chapter 9. *Veillonella* spp., *Porphyromonas* spp., *Prevotella* spp., and *Fusobacterium* spp. are predominant members of the oral microbiota and are described in Chapter 8.

6.4.1.1 *Lactobacillus* spp.

Lactobacilli are Gram-positive bacilli that are usually long and slender. At least 34 species are currently recognized, and their main characteristics are as follows:
- Gram-positive bacilli;
- Microaerophilic or obligate anaerobes;
- Nonmotile;
- Nonsporing;
- G+C content of DNA is 37–53 mol%;
- Catalase-negative;
- Fermentation of sugars, producing a variety of acids, alcohol, and CO_2;
- Aciduric;
- Acidophilic;
- Growth over the pH range 3.5–6.8;

- Optimum pH for growth is approximately 6.0;
- Growth over the temperature range 15–45°C;
- Little proteolytic activity.

Species that produce mainly lactic acid from glucose are termed "homofermentative", while those that produce smaller proportions of lactic acid together with acetic acid, formic acid, and alcohol are known as "heterofermentative". They have complex nutritional requirements and may need to be supplied with amino acids, peptides, fatty acid esters, salts, nucleic acid derivatives, or vitamins. Lactobacilli display a wide range of adhesins, which enable them to adhere to the vaginal epithelium (Fig. 6.12). These include lipoteichoic acids as well as ill-defined proteinaceous and nonproteinaceous adhesins, some of which are located on fimbriae. *L. acidophilus* and *L. gasseri* appear to utilize glycoproteins as adhesins when binding to vaginal epithelial cells, the receptors being glycolipids. In contrast, *L. jensenii* binds to epithelial cells by means of a carbohydrate adhesin.

Lactobacilli are usually (but not always) present in the vagina and cervix of most healthy post-menarcheal/pre-menopausal females and often dominate the microbiotas of these regions. On average, the vagina is colonized by between 10^8 and 10^9 lactobacilli and, in most women, only a single species is present. Owing to failure to speciate the lactobacilli isolated, difficulties with the identification of *Lactobacillus* spp., and revision of their taxonomy, there is confusion with regard to which actual species is present. Many early studies reported that the predominant species in the vagina was *L. acidophilus*. However, genotypic analysis has revealed that this "species" is comprised of six different species, although their differentiation by traditional phenotypic methods is very difficult, if not impossible. In a recent study (Fig. 6.13), the lactobacilli isolated from the vagina of 101 women were identified using whole-DNA genomic probes from known strains, and the predominant species were found to be *L. jensenii* and *L. crispatus* (a member of the *L. acidophilus* complex).

Other strains that have frequently been isolated include *L. gasseri* (a member of the *L. acidophilus* complex), *L. cellobiosus*, *L. fermentum*, and *L. iners*. Less frequently isolated species include: *L. plantarum*, *L. rhamnosus*, *L. brevis*, *L. casei*, *L. vaginalis*, *L. delbrueckii*, and *L. salivarius*. Another complication regarding the identity of vagina-colonizing lactobacilli arises from the fact that some species are difficult to grow. Hence, a number of culture-independent analyses of the vaginal

Table 6.10 Organisms that have been detected in one or more regions of the female reproductive system. This is not an exhaustive list.

Actinomyces europaeus	Enterococcus durans	Mycoplasma hominis
Actinomyces israelii	Enterococcus faecalis	Neisseria sp.
Actinomyces meyeri	Enterococcus faecium	Peptococcus niger
Actinomyces naeslundii	Escherichia coli	Peptoniphilus asaccharolyticus
Actinomyces urogenitalis	Eubacterium contortum	Peptostreptococcus anaerobius
Acinetobacter sp.	Eubacterium limosum	Peptostreptococcus hareif
Aerococcus sp.	Eubacterium spp.	Peptostreptococcus micros
Anaerococcus prevotii	Eubacterium tenue	Peptostreptococcus vaginalis
Anaerococcus tetradius	Finegoldia magna	Porphyromonas asaccharolytica
Arcanobacterium bernardiae	Fusobacterium necrophorum	Porphyromonas endodontalis
Arthrobacter albus	Fusobacterium nucleatum	Porphyromonas levii
Atopobium parvulum	Fusobacterium spp.	Prevotella bivia
Atopobium vaginae	Gardnerella vaginalis	Prevotella buccalis
Bacteroides capillosus	Gemella morbillorum	Prevotella corporis
Bacteroides fragilis	Group D streptococci	Prevotella disiens
Bacteroides ovatus	Haemophilus ducreyi	Prevotella intermedia
Bacteroides uniformis	Haemophilus influenzae	Prevotella oralis
Bacteroides ureolyticus	Klebsiella aerogenes	Prevotella oulorum
Bifidobacterium biavatii	Klebsiella ozaenae	Prevotella ruminicola
Bifidobacterium bifidum	Klebsiella pneumoniae	Propionibacterium acnes
Bifidobacterium breve	Lactobacillus acidophilus	Propionibacterium avidium
Bifidobacterium dentium	Lactobacillus casei	Proteus mirabilis
Bifidobacterium longum	Lactobacillus cellobiosis	Proteus morganii
Butyrivibrio fibrisolvens	Lactobacillus coleohominis	Proteus vulgaris
Campylobacter gracilis	Lactobacillus crispatus	Pseudomonas aeruginosa
Candida albicans	Lactobacillus delbrueckii	Pseudoramibacter alactolyticus
Candida dubliensis	Lactobacillus fermentum	Serratia sp.
Caulobacter sp.	Lactobacillus gasseri	Staphylococcus aureus
Chlamydia psittaci	Lactobacillus iners	Staphylococcus capitis
Citrobacter sp.	Lactobacillus jensenii	Staphylococcus epidermidis
Clostridium bifermentans	Lactobacillus kalixensis	Staphylococcus haemolyticus
Clostridium colicanis	Lactobacillus mucosae	Staphylococcus hominis
Clostridium difficile	Lactobacillus nagelii	Streptococcus agalactiae
Clostridium haemolyticum	Lactobacillus oris	Streptococcus constellatus
Clostridium perfringens	Lactobacillus plantarum	Streptococcus intermedius
Clostridium ramosum	Lactobacillus pontis	Streptococcus mitis
Corynebacterium amycolatum	Lactobacillus reuteri	Streptococcus morbillorum
Corynebacterium coyleiae	Lactobacillus rhamnosus	Streptococcus pyogenes
Corynebacterium minutissimum	Lactobacillus ruminis	Streptococcus salivarius
Corynebacterium pseudogenitalium	Lactobacillus salivarius	Treponema sp.
Corynebacterium singulare	Lactobacillus vaginalis	Veillonella alkalescens
Dialister sp.	Leptotrichia sp.	Veillonella atypica
Eggerthella lenta	Megasphaera sp.	Veillonella parvula
Enterobacter cloacae	Micrococcus sp.	
Enterobacter spp.	Mobiluncus curtisii	

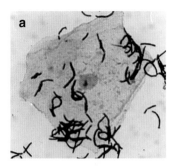

Fig. 6.12 (a) Vaginal epithelial cell from a healthy pre-menopausal female with adherent *Lactobacillus paracasei* subsp. *paracasei*. Reproduced with the permission of Blackwell Publishing Ltd, Oxford, UK, from: Zárate, G. and Nader-Macias, M.E. (2006) Influence of probiotic vaginal lactobacilli on in vitro adhesion of urogenital pathogens to vaginal epithelial cells. *Letters in Applied Microbiology* 43, 174–80. (b) Scanning electron micrograph of *Lactobacillus helveticus* adhering to human epithelial cells. Magnification ×4300. Image kindly supplied by Kathene Johnson-Henry, Research Institute at the Hospital for Sick Children, University of Toronto, Toronto, Ontario, Canada.

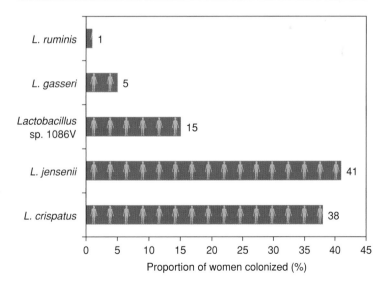

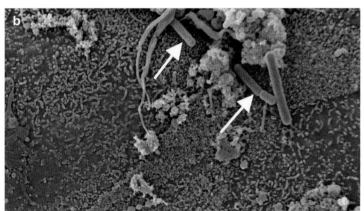

Fig. 6.13 Identity of lactobacilli isolated from the vaginas of a group of 101 healthy post-menarcheal/pre-menopausal females. Following their isolation, the species were identified on the basis of whole-chromosomal DNA homology (2).

microbiota have shown that *L. iners* is a frequent member of the vaginal microbiota (Fig. 6.14). This species does not grow on the usual selective media employed to isolate lactobacilli from complex microbial communities such as those found in the vagina.

On the basis of the results obtained from culture-dependent and culture-independent studies, it would appear that the lactobacilli most frequently present in the vagina are *L. jensenii*, *L. iners*, *L. crispatus*, and *L. gasseri*.

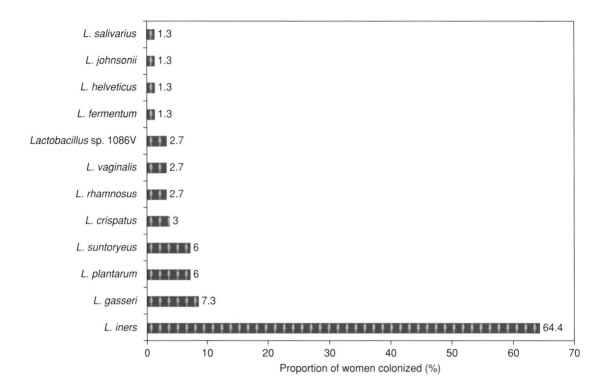

Fig. 6.14 Results of a culture-independent analysis of the lactobacilli colonizing the vaginas of 241 pre-menopausal females (3).

A number of attributes of lactobacilli contribute to their ability to dominate the microbiota of both the vagina and cervix (Table 6.11). The capacity of lactobacilli to reduce the attachment of *Staphylococcus aureus* to vaginal epithelial cells can be seen in Fig. 6.15. Lactobacilli very rarely cause infections in humans.

6.4.1.2 *Gardnerella vaginalis*

G. vaginalis is a Gram-positive bacillus that appears as slender Gram-variable rods or coccobacilli on Gram staining. It is the only member of the genus *Gardnerella*, and its main characteristics are as follows:
- Gram-positive bacillus;
- Nonmotile;
- Nonsporing;
- Facultative anaerobe;
- Optimum growth in CO_2-enriched air;
- G+C content of DNA is 42–44 mol%;
- Piliated;
- Production of an exopolysaccharide;
- Growth over the pH range 6–7;
- Optimum pH for growth is 6.0–6.5;

- Growth over the temperature range 25–42°C;
- Optimum growth at 35–37°C;
- Catalase-negative;
- Urease-negative.

It is nutritionally fastidious and ferments glucose, maltose, and sucrose to produce mainly acetate and lactate. It produces a sialidase, implying that it can partially degrade mucins to obtain sugars that can be used as a carbon and energy source. The pili and surface polysaccharide are involved in adhesion of *G. vaginalis* to epithelial and red blood cells. It can directly bind a number of iron-containing compounds including heme, hemoglobin, catalase, and lactoferrin – all of which can be used as an iron source.

While the organism is often found in the reproductive system of healthy females, *G. vaginalis*, together with several other organisms, is associated with bacterial vaginosis.

6.4.1.3 *Candida albicans*

Can. albicans is a yeast with oval-shaped cells with a diameter of 3–6 µm. It reproduces by forming buds

Table 6.11 Characteristics that contribute to the dominance of lactobacilli within the female reproductive system.

Characteristic	Role in vaginal colonization
Acidophilic	Enables survival at the low pH found in the vagina and, to a lesser extent, in the cervix
Acidogenic	Contributes to the low pH found in the vagina; inhibits competing organisms
Production of lactate and acetate	Exerts a microbicidal effect
Microaerophiles or obligate anaerobes	Enable survival at the low oxygen concentrations present in the vagina and cervix
Hydrogen peroxide production (e.g. by *L. crispatus*, *L. jensenii*, *L. gasseri*, *L. delbrueckii*, *L. casei*, *L. plantarum*, *L. vaginalis*, *L. pentosus*)	Kills competing organisms directly; also oxidizes chloride ions to hypochloric acid and/or chlorine – both of which are potent microbicides
Variety of adhesins	Enable adhesion to mucosa
Bacteriocin production (e.g. by *L. gasseri*, *L. plantarum*, *L. fermentum*, *L. casei*, *L. delbrueckii*, *L. reuteri*, *L. salivarius*)	Inhibits growth of competing organisms; effective against many organisms, including other lactobacilli, *G. vaginalis*, *N. gonorrhoeae*, *Ent. faecalis*, *Staph. aureus*
Production of bacteriolytic enzymes (e.g. *L. fermentum*)	Active against *Strep. pyogenes*, *Strep. agalactiae*, *Staph. aureus*, and some lactobacilli
Biosurfactant production (e.g. by *L. rhamnosus*, *L. fermentum*, *L. acidophilus*)	Prevents attachment of other species to epithelium; induces detachment of bacteria adhering to epithelial cells
Competitive exclusion	Generates competition for receptors on epithelial cells

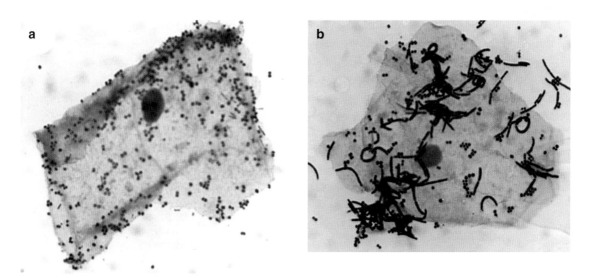

Fig. 6.15 Competitive exclusion of *Staph. aureus* by lactobacilli. The photomicrographs show: (a) large numbers of *Staph. aureus* adhering to a vaginal epithelial cell from a healthy adult and (b) a reduction in the number of *Staph aureus* adhering to cells in the presence of *Lactobacillus paracasei* subsp. *paracasei*. Reproduced with the permission of Blackwell Publishing Ltd, Oxford, UK, from: Zárate, G. and Nader-Macias, M.E. (2006) Influence of probiotic vaginal lactobacilli on in vitro adhesion of urogenital pathogens to vaginal epithelial cells. *Lett Appl Microbiol* 43, 174–80.

(blastoconidia), but these often fail to detach, resulting in a chain of cells known as a pseudohypha. It is a dimorphic fungus and can also produce true septate hyphae. Hyphal production is also induced by incubating the organism in serum at 37°C; after 3 h of incubation, short hyphae, known as germ tubes, are produced, and this is an important, rapid test for the presumptive identification of the organism. The characteristics of *Can. albicans* are as follows:

• Yeast;
• Reproduces by budding;
• Dimorphic fungus;
• Produces refractile chlamydospores;
• Facultative anaerobe;
• Optimum growth under aerobic conditions;
• Growth over the temperature range 20–40°C;
• Growth over the pH range 2–8;
• Optimum pH for growth is 5.1–6.9;
• Fermentation of a wide range of sugars.

Can. albicans can utilize a wide range of sugars as a carbon and energy source, including glucose, maltose, sucrose, galactose, and xylose. It produces an N-acetyl-glucosaminidase and a number of aspartyl proteinases, which implies that it may be able to partially degrade mucins as well as oligosaccharides and proteins present in the vagina, and thereby obtain sugars and amino acids for use as carbon, nitrogen, and energy sources. It also secretes phospholipases and other lipases which enable it to hydrolyze phospholipids and lipids.

A large number of adhesins have been identified in *Can. albicans*, including mannoproteins and integrin analogs which mediate adhesion to epithelial cells and to extracellular matrix molecules such as collagen, fibronectin, and laminin.

Many strains of the organism produce a gliotoxin which has antibacterial and immunosuppressive activities. Hence, it is able to kill *Micrococcus* spp. and can inhibit chemotaxis and phagocytosis by polymorphonuclear leukocytes (PMNs). The organism is associated with a number of infections including candidiasis, balanitis, and denture stomatitis.

6.4.1.4 *Streptococcus agalactiae* (Group B streptococcus)

Strep. agalactiae is a β-hemolytic streptococcus which contains the Lancefield's group B antigen. The organism produces a hyaluronidase and proteases, which means that it can obtain sugars and amino acids from the hyaluronic acid and proteins present in vaginal and cervical secretions. Lipoteichoic acid and unidentified proteins are the main adhesins enabling its adhesion to vaginal and cervical epithelial cells in vitro. It can also bind to extracellular matrix components such as fibronectin and fibrinogen. A proteinaceous adhesin, Lmb, has been detected on the surface of the organism, which mediates binding to human placental laminin.

The organism is able to inhibit the growth of many species found in the vagina, including lactobacilli, *G. vaginalis*, coryneforms, viridans streptococci, enterococci, and peptostreptococci. It is unable to inhibit the growth of staphylococci, but is itself inhibited by coagulase-negative staphylococci (CNS).

While the organism is frequently present in the reproductive tract of healthy females, *Strep. agalactiae* is responsible for a number of infections, including chorioamnionitis, endometritis, and induction of pre-term birth, as well as neonatal infections such as sepsis, pneumonia, and meningitis. A number of serotypes can be distinguished, and one of these, serotype III, is the most frequently associated with neonatal invasive diseases.

6.4.1.5 *Mycoplasma hominis*

The general characteristics of *Mollicutes* are described in section 4.4.1.6. Unlike some mycoplasmas, *Myc. hominis* cells do not have an elongated flask-shaped morphology, but are small (0.2–0.3 μm diameter) and round. The organism does not ferment glucose but can metabolize arginine. It has been found in the vagina and cervix of healthy females, with the frequency of detection in the vagina of healthy pre-menopausal females ranging from 4 to 25%. The organism has also been detected in the cervix of 18% of healthy pregnant females. Although it has been associated with a number of genito-urinary infections (bacterial vaginosis, pelvic inflammatory disease, vaginitis, pre-term labor), its exact role is uncertain, and there is doubt as to whether it is involved in the pathogenesis of these diseases. Interestingly, it can survive within, and form a symbiosis with, *Trichomonas vaginalis*, which is a well-established protozoal pathogen of the vagina.

6.4.1.6 *Ureaplasma urealyticum*

The genus *Ureaplasma* belongs to the *Mollicutes* (section 4.4.1.6) and consists of eight species, two of which have been isolated from humans. *U. urealyticum* is a small (0.2–0.3 μm in diameter), pleomorphic, facultative anaerobe, which is nutritionally fastidious and requires sterols and an enriched medium for growth. The G+C content of its DNA is 27–28 mol%. It can grow over a broad pH range of pH 5–9. The organism is unable to metabolize glucose or arginine, but produces a urease which is involved in energy generation. It secretes a number of enzymes including elastase, several phospholipases, and an IgA protease. It exhibits high resistance to dryness and to temperature change.

The organism was originally classified into 14 distinct serovars, but these have now been grouped into two distinct species – *U. urealyticum* (serovars 2, 4, 5, 7–13) and *Ureaplasma parvum* (serovars 1, 3, 6, 14). Unfortunately, few epidemiological studies have distinguished between the two species. The genome of *U. urealyticum* has recently been sequenced (http://cmr.tigr.org/tigr-scripts/CMR/GenomePage.cgi?org=ntuu01) and has 613 genes encoding proteins and 39 that encode RNA. Fifty-three percent of the protein-encoding genes have been assigned biological functions, 19% are similar to genes with no known function, and the remaining genes have no significant similarity to genes from other organisms. Genes encoding a large number of transporters are present which, presumably, are involved in the uptake of a range of essential nutrients.

6.4.1.7 *Atopobium vaginae*

Atopobium vaginae is a nonmotile, facultatively anaerobic Gram-positive coccus that occurs singly, in pairs or short chains. The G+C content of its DNA is 44 mol%. Its main metabolic end product is lactate. The organism is present in the vagina of between 7 and 20% of healthy post-menarcheal/pre-menopausal individuals. *At. vaginae* is one of several organisms considered to play a role in the pathogenesis of bacterial vaginosis.

6.4.1.8 *Mobiluncus* spp.

Mobiluncus species are motile, anaerobic, nonbranching, nonsporing, curved bacilli that stain Gram-variable or Gram-negative. The genus consists of two species – *Mob. curtisii* and *Mob. mulieris*. The G+C content of their DNA is 49–55.2 mol%. Optimum growth occurs at 35–37°C. They ferment glucose, producing succinic, acetic, and lactic acids as major end products.

While some studies have failed to detect *Mobiluncus* spp. in the vagina of healthy females, others have reported the presence of both species at frequencies ranging from 0.7 to 38%, with *Mob. mulieris* being detected more often. *Mobiluncus* spp. are members of the microbial consortium responsible for bacterial vaginosis.

6.4.2 Community composition at different sites within the female reproductive system

In Chapter 1, the problems associated with defining the indigenous microbiota of a particular site were described. The situation is even more difficult with regard to the female reproductive system, as a variety of additional complicating factors come into play, all of which affect the environment of the various regions and hence the composition of the resident microbial communities. The factors influencing the microbiotas of the various regions of the female reproductive system are as follows, and are in addition to those factors that also affect the microbiota at other body sites (e.g. age, personal hygiene, etc.).

- The sexual maturity of the individual;
- Whether or not the individual is pregnant;
- The stage of the reproductive cycle in post-menarcheal/pre-menopausal individuals;
- Whether or not the individual is sexually active;
- Whether or not the individual is using contraceptives and, if so, what method is being used;
- The sexual behavior of the individual;
- Whether or not hormone replacement therapy is being used (in the case of post-menopausal individuals).

6.4.2.1 Vagina

6.4.2.1.1 Post-menarcheal/pre-menopausal females
It is generally accepted that the vaginal microbiota of healthy, post-menarcheal/pre-menopausal, sexually active, nonpregnant individuals is dominated by lactobacilli (Fig. 6.16). However, recent research has found that the majority of such individuals do not have a *Lactobacillus*-dominated microbiota throughout their menstrual cycle, even though they show no symptoms of vaginal infection. The composition of the vaginal microbiota appears to fluctuate on a daily basis between one that is regarded as being compatible with health and one that is considered to be "abnormal" – thereby exacerbating the problem of defining the indigenous microbiota of this region.

With regard to the structural organization of the vaginal microbiota, the organisms appear to form microcolonies attached to epithelial cells (Fig. 6.17) and embedded in the viscous material associated with

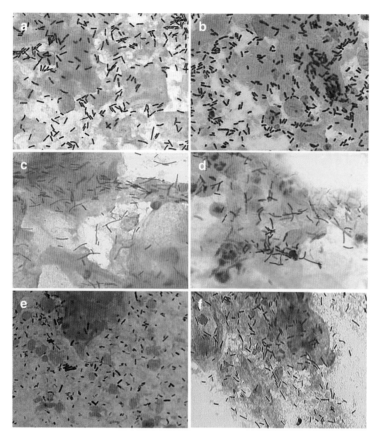

Fig. 6.16 Gram-stained vaginal smears from different individuals, showing epithelial cells with: (a, b) mainly *Lactobacillus crispatus* cell types, i.e. plump homogeneous lactobacilli; (c, d) lactobacilli other than *L. crispatus*; (e, f) mixtures of *L. crispatus* and non-*L. crispatus* cell types. Magnification ×1000. Reproduced from: Verhelst, R., Verstraelen, H., Claeys, G. et al. (2005) Comparison between Gram stain and culture for the characterization of vaginal microflora: Definition of a distinct grade that resembles grade I microflora and revised categorization of grade I microflora. *BMC Microbiol* 5, 61. © 2005 Verhelst et al.; licensee BioMed Central Ltd, London, UK. This is an Open Access article distributed under the terms of the Creative Commons Attribution License. doi:10.1186/1471-2180-5-61.

the epithelial surface. Three important members of the vaginal microbiota, *L. acidophilus*, *L. gasseri*, and *L. jensenii* are capable of self-aggregation, and this is likely to be important in the formation of such microcolonies. Many of the other constituents of the community appear to be more loosely adherent. The structure of the vaginal microbiota has also been investigated using fluorescence in situ hybridization (FISH), and this technique has shown it to consist mainly of an unstructured community within the mucous layer overlying the vaginal epithelium – there was no evidence of biofilm formation in healthy individuals (Fig. 6.18). Whether or not the vaginal microbiota exists as a biofilm is a controversial issue – there is little evidence from microscopy studies that this is the case. However, vaginal biofilms do appear to be formed in individuals with vaginosis.

In most studies of the cultivable vaginal microbiota, the number of organisms isolated from a healthy individual is less than ten and is usually between three and five. In general, more than half of the species detected are obligate anaerobes. The prevalence of the various organisms detected in the vaginal microbiota is shown in Fig. 6.19, from which it can be seen that lactobacilli, CNS, coryneforms, GPAC and Gram-negative anaerobic bacilli (GNAB) are the most frequently encountered groups of cultivable organisms. The anaerobes most frequently detected include species from the following genera: *Bacteroides*, *Prevotella*, *Peptostreptococcus* (including former members of this genus), *Peptococcus*, *Propionibacterium*, *Eubacterium* (including former members of this genus), *Clostridium*, *Bifidobacterium*, *Fusobacterium*, *Porphyromonas*, and *Veillonella* (Table 6.10).

Lactobacilli usually dominate the cultivable microbiota (Fig. 6.20), and invariably only a single species is present – frequently this is *L. jensenii*, *L. crispatus*, *L. gasseri*, or *L. iners*. Other organisms often found in high

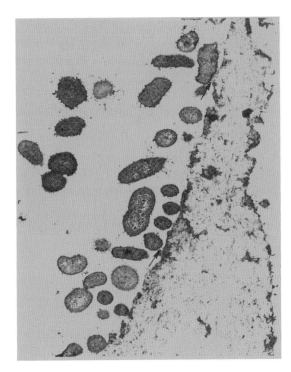

Fig. 6.17 Transmission electron micrograph, showing microcolonies of bacteria adhering to the vaginal epithelium. The sample was subjected to vortex mixing, centrifugation, and sonication, yet still had microcolonies attached to epithelial cells. Arrows indicate the condensed exopolysaccharide glycocalyx on the bacterial surface, which appear to mediate bacterial adhesion to the epithelium. Reproduced with the permission of Taylor & Francis, New York, New York, USA, from: Sadhu, K., Domingue, P.A.G., Chow, A.W. et al. (1989) A morphological study of the in situ tissue-associated autochthonous microbiota of the human vagina. *Microbial Ecol Health Dis* 2, 99–106.

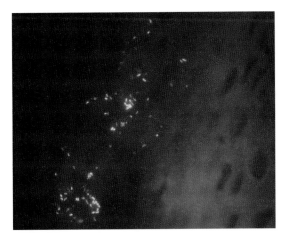

Fig. 6.18 Fluorescence in situ hybridization analysis of a vaginal biopsy from a healthy individual showing organisms associated with the surface of the vaginal mucosa. The bacteria appear to be mainly within the mucous layer overlying the epithelium and include lactobacilli (orange), *G. vaginalis* (red), and other species (green). Magnification ×1000. Reproduced from Swidsinski, A., Mendling, W., Loening-Baucke, V. et al. (2005) Adherent biofilms in bacterial vaginosis. *Obstet Gynecol* 106, 1013–23, with the permission of Lippincott, Williams & Wilkins, Philadelphia, Pennsylvania, USA.

proportions include *G. vaginalis*, streptococci, coryneforms, GPAC, and GNAB.

Culture-independent approaches have confirmed the presence of only a limited number of organisms in the vaginal microbiota of an individual and also that the community is usually dominated by a single *Lactobacillus* sp. Hence, in a study involving 19 premenopausal women, PCR-DGGE (polymerase chain reaction–denaturing gradient gel electrophoresis) analysis (section 1.4.3) of the DNA extracted from the vaginal communities revealed that, in most subjects, the profiles obtained were dominated by between one and three DNA bands. In 79% of the women, at least one of these dominant bands had a sequence corresponding to that of a *Lactobacillus* species, with *L. iners* being the most frequently detected. Similar results were obtained in a study based on 16S rRNA clone libraries prepared from eight healthy adults (Fig. 6.21). The mean number of phylotypes per individual was 3.3 (range: 1–6). Lactobacilli (particularly *L. crispatus* and *L. iners*) were the dominant bacteria detected and constituted between 83 and 100% of the clones per library. Most of the bacterial 16S rRNA sequences closely matched those of known bacteria.

However, the results of one of the most extensive culture-independent studies undertaken to date have shown that the vaginal microbiota of healthy individuals is not always dominated by lactobacilli (Fig. 6.22). No lactobacilli were detected in 40% of this group of 20 disease-free post-menarcheal/pre-menopausal females. In another 40% of the group, although lactobacilli were detected, the microbiota was very complex. Overall, the vaginal microbiota of the 20 individuals was highly diverse, with phylotypes from approximately 70 genera

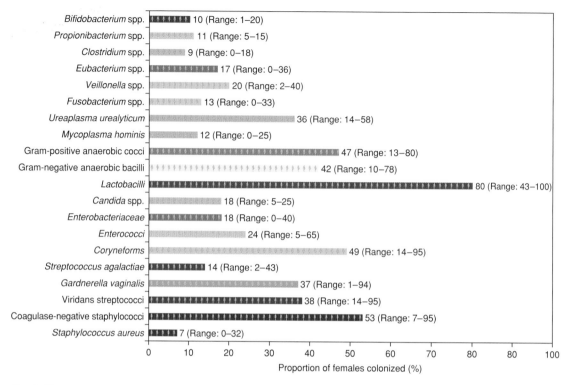

Fig. 6.19 Frequency of detection of various microbes in the vagina of post-menarcheal/pre-menopausal, healthy, nonpregnant females. The data shown are mean values (and ranges) derived from the results of 32 studies involving 1756 individuals (2, 4–34).

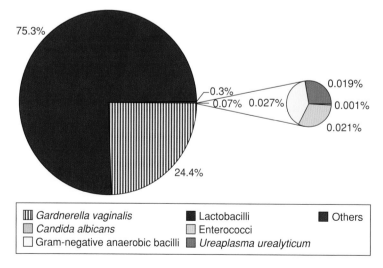

Fig. 6.20 Relative proportions of the predominant organisms constituting the vaginal microbiota of 21 post-menarcheal/pre-menopausal, healthy, nonpregnant females (24).

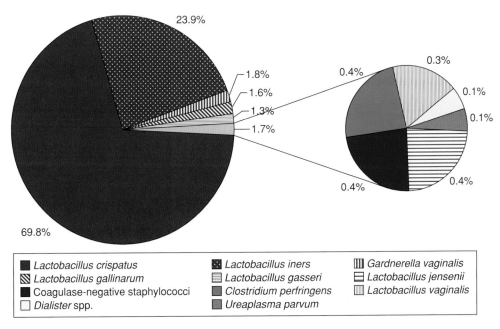

Fig. 6.21 Culture-independent analysis of the vaginal microbiota. DNA was extracted from vaginal fluid obtained from eight healthy post-menarcheal/pre-menopausal individuals, the bacterial 16S rRNA genes present in each sample were amplified by PCR, and the resulting products were cloned and sequenced. One hundred clones were analyzed from each library, and the figures denote the overall proportion (%) of clones of each phylotype detected (35).

being detected. Furthermore, unidentified and/or as-yet-uncultured phylotypes were present in every individual. A surprising finding was the frequent occurrence of *Janthinobacterium* spp. These are motile Gram-negative bacilli that occur singly, in pairs, and in short chains. They are obligate aerobes, catalase-positive and oxidase-positive, and often produce the violet pigment, violacein. They are frequently present in soil and water.

In a study involving 16 healthy, post-menarcheal/pre-menopausal females, libraries were prepared from the PCR amplification products of the *cpn60* gene in DNA extracted from vaginal samples. Sequencing of 6869 clones from the 16 libraries obtained identified 57 different nucleotide sequences, and phylogenetic analysis of these sequences resulted in the identification of 13 distinct sequence clusters (Fig. 6.23). Seventy-eight percent of the clones, representing 32 distinct sequences, corresponded to *Lactobacillus* spp., while 11% corresponded to *G. vaginalis*. The remaining sequences corresponded to *Porphyromonas levii*

(one clone), *Chlamydia psittaci* (10% of clones), and *Megasphaera elsdenii* (one clone).

As described in section 6.3.1, the vaginal environment alters dramatically during the menstrual cycle (e.g. pH, nutrient availability, oxygen content), and this has a profound effect on the indigenous microbiota (Fig. 6.24). The effect of the menstrual cycle on the vaginal microbiota and the main detectable changes are as follows:

• Fewer women have high levels of lactobacilli during menses;

• Recovery of high levels of non-*Lactobacillus* spp. is greatest during menses;

• The proportion of women with *Prevotella* spp. is greatest during menses;

• The proportion of women with *Bacteroides fragilis* is at its lowest during menses.

6.4.2.1.2 Pre-menarcheal girls

Because the vaginal environment of pre-menarcheal girls differs markedly from that of post-menarcheal

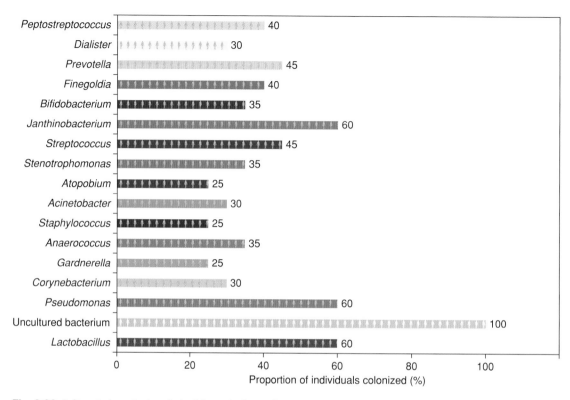

Fig. 6.22 Culture-independent analysis of the vaginal microbiota. DNA was extracted from samples taken from the vaginal mucosa of 20 healthy post-menarcheal/pre-menopausal individuals, the bacterial 16S rRNA genes present in each sample were amplified by PCR, and the resulting products were cloned and sequenced. One thousand clones were analyzed from each library, and the figures denote the frequency of detection of those phylotypes that were present in at least four individuals (36).

women (section 6.3.1), there are marked differences between the vaginal microbiotas of these two groups (compare Figs 6.19 and 6.25). The most notable differences are that in pre-menarcheal girls, the frequency of detection of anaerobes is very high, while coryneforms and CNS are also frequent isolates; lactobacilli, on the other hand, are less common. *Escherichia coli* is also more frequently detected, and G. *vaginalis* is usually absent.

In contrast to the situation in post-menarcheal females, the vaginal microbiota of pre-menarcheal girls is dominated by anaerobes – mainly *Actinomyces* spp., *Peptostreptococcus* spp. (including former members of this genus), and GNAB – rather than by lactobacilli (Fig. 6.26).

6.4.2.1.3 Post-menopausal women
Owing to a number of factors, the onset of the

menopause leads to changes in the vaginal microbiota. Menopause-related changes in the vaginal environment are as follows:
• Decreased glycogen content of the vaginal epithelium due to the low estrogen levels, thereby resulting in an increased pH;
• Thinning of the vaginal epithelium;
• Absence of, or changes in, the receptors available for bacterial adhesins;
• Changes in the innate and/or adaptive immune response;
• Alterations in the production of vaginal fluid, thereby affecting the availability of bacterial nutrients.
The main changes that are detectable in post-menopausal females who are not receiving estrogen replacement therapy are that lactobacilli, G. *vaginalis*, yeasts, and mycoplasmas are recovered less frequently than from pre-menopausal women, whereas GNAB,

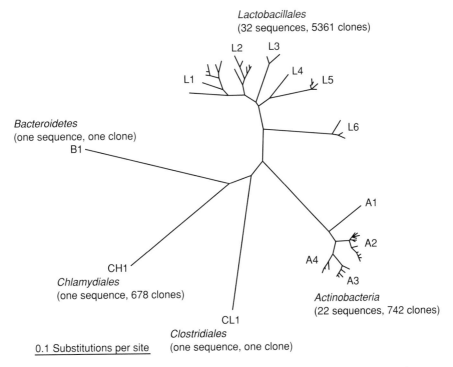

Fig. 6.23 Phylogenetic relationships of *cpn60* sequences obtained from 16 vaginal microbiota libraries. The tree was calculated with the use of a maximum-likelihood distance calculation, followed by neighbor-joining. Phylogenetic clusters L1–L6, A1–A4, B1, CH1, and CL1 are indicated within the major taxonomic groups. Reprinted from: Hill, J.E., Goh, S.H., Money, D.M., Doyle, M., Li, A., Crosby, W.L., Links, M., Leung, A., Chan, D. and Hemmingsen, S.M. (2005) Characterization of vaginal microflora of healthy, nonpregnant women by chaperonin-60 sequence-based methods. *Am J Obstet Gynecol* 193, 682–92, © 2005, with permission from Elsevier, Amsterdam, The Netherlands.

GPAC, and viridans streptococci are recovered more frequently (Fig. 6.27).

There are also differences in the composition of the microbiota – lactobacilli are no longer the dominant organisms as is the case with pre-menopausal individuals (Fig. 6.28). In post-menopausal women who receive estrogen replacement therapy, the vaginal pH is lower than in those not receiving the hormone, and there is a corresponding decrease in the proportions of obligate anaerobes in the vaginal microbiota.

6.4.2.1.4 Vaginal microbiota during pregnancy
During pregnancy, hormonal changes induce an increase in the concentration of both glycogen and lactic acid, and this encourages the proliferation of lactobacilli which are generally detected more frequently in the vaginal microbiota of pregnant women than in nonpregnant individuals. The frequency of detection

of *U. urealyticum* and GPAC also increases in pregnant females, while that of coryneforms and CNS decrease (Fig. 6.29).

As is the case with post-menarcheal/pre-menopausal females, lactobacilli dominate the vaginal microbiota of pregnant females (Fig. 6.30).

6.4.2.2 Cervix

The cervix may be colonized by any of a large variety of microbes (Table 6.10). However, the cultivable microbiota of the cervix consists, on average, of approximately six species, which are present as microcolonies attached to the epithelial surface (Fig. 6.31). The cultivable cervical microbiota is similar to that of the vagina, in that lactobacilli are the most frequently isolated species and coryneforms, staphylococci, streptococci, and GPAC are also usually present. Several

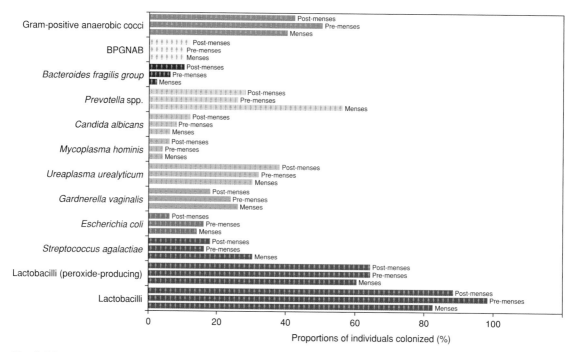

Fig. 6.24 Frequency of detection of various microbes in the vagina at different stages of the menstrual cycle. In this study, the vaginal microbiota of 50 women aged 18–40 years was investigated. Samples were taken as follows: (a) during menses (menses); (b) 7–12 days from onset of menses, i.e. during the pre-ovulatory phase (pre-menses); and (c) 19–24 days from onset of menses, i.e. during the post-ovulatory phase (post-menses). BPGNAB, black-pigmented Gram-negative anaerobic bacilli (37).

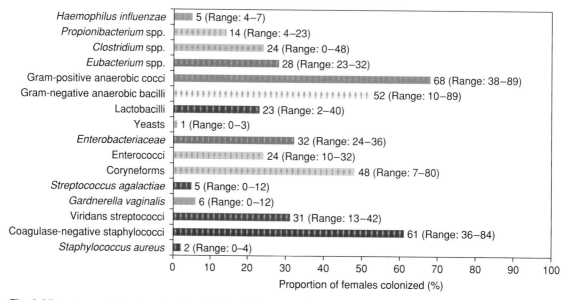

Fig. 6.25 Frequency of detection of various microbes in the vagina of pre-menarcheal, healthy girls. The data shown are mean values (and ranges) derived from the results of four studies involving 125 individuals (38–41).

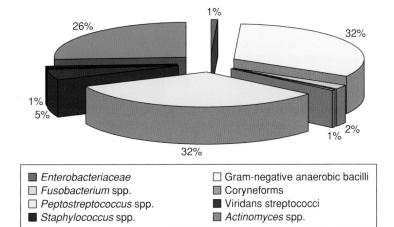

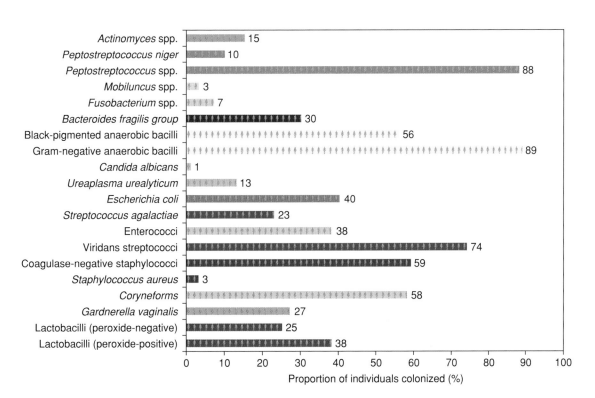

Fig. 6.26 Relative proportions of microbes comprising the vaginal microbiota of 19 pre-menarcheal girls (38).

Legend:
- ■ *Enterobacteriaceae*
- ▫ *Fusobacterium* spp.
- ▫ *Peptostreptococcus* spp.
- ■ *Staphylococcus* spp.
- □ Gram-negative anaerobic bacilli
- ■ Coryneforms
- ■ Viridans streptococci
- ■ *Actinomyces* spp.

Fig. 6.27 Frequency of detection of various microbes in the vagina of post-menopausal, healthy females. The data shown are from a study involving 73 individuals who were not receiving estrogen replacement therapy (42).

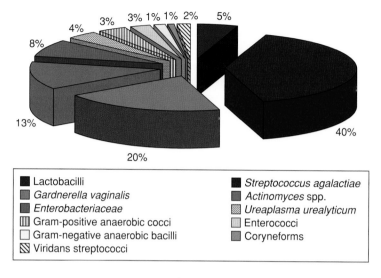

Legend:

- ■ Lactobacilli
- ■ *Gardnerella vaginalis*
- ■ *Enterobacteriaceae*
- ▥ Gram-positive anaerobic cocci
- □ Gram-negative anaerobic bacilli
- ▧ Viridans streptococci
- ■ *Streptococcus agalactiae*
- ■ *Actinomyces* spp.
- ▨ *Ureaplasma urealyticum*
- □ Enterococci
- ■ Coryneforms

Fig. 6.28 Composition of the vaginal microbiota in post-menopausal, healthy females. The data shown are from a study involving 73 individuals who were not receiving estrogen replacement therapy (42).

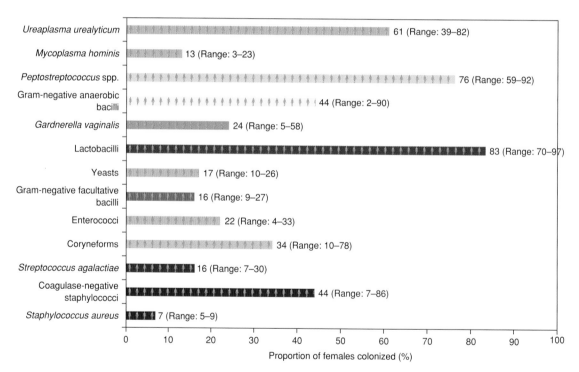

Fig. 6.29 Frequency of detection of various microbes in the vagina of pregnant females. The data shown are mean values (and ranges) derived from the results of eight studies involving 7827 individuals (43–50).

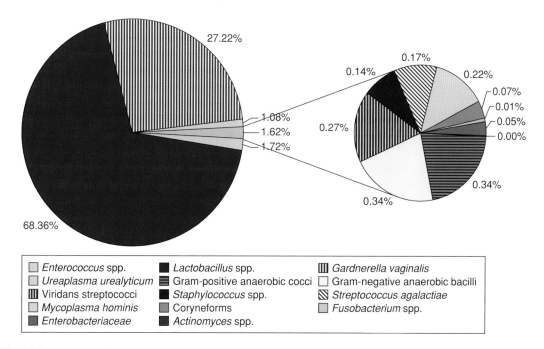

▢ *Enterococcus* spp.	■ *Lactobacillus* spp.	⦀ *Gardnerella vaginalis*
▢ *Ureaplasma urealyticum*	▤ Gram-positive anaerobic cocci	▢ Gram-negative anaerobic bacilli
⦀ Viridans streptococci	■ *Staphylococcus* spp.	▨ *Streptococcus agalactiae*
▢ *Mycoplasma hominis*	▨ Coryneforms	▤ *Fusobacterium* spp.
■ *Enterobacteriaceae*	■ *Actinomyces* spp.	

Fig. 6.30 Composition of the vaginal microbiota in pregnant females. The data shown are from a study involving 132 individuals (47).

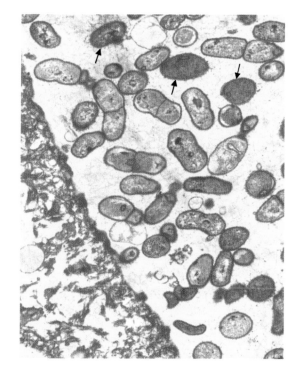

members of the microbiota (notably the obligate anaerobes) are able to cause infections of the genital tract and infections associated with childbirth.

6.4.2.2.1 Post-menarcheal/pre-menopausal females

In culture-based studies, lactobacilli are the most frequently isolated organisms (Fig. 6.32) and comprise the highest proportion of the microbiota (Fig. 6.33).

6.4.2.2.2 Cervical microbiota during pregnancy

Pregnancy has a significant effect on the cervical microbiota (Fig. 6.34). The frequency of isolation of *E. coli*, GPAC, *B. fragilis*, and other *Bacteroides* spp.

Fig. 6.31 (*left*) Transmission electron micrograph of material from the cervix of a healthy volunteer. The bacteria are present as a microcolony on the epithelial surface. Most of the organisms are Gram-positive, but some Gram-negative cells (arrowed) can be seen. Reproduced with the permission of Taylor & Francis, New York, New York, USA, from: Sadhu, K., Domingue, P.A.G., Chow, A.W. et al. (1989) A morphological study of the in situ tissue-associated autochthonous microbiota of the human vagina. *Microbial Ecol Health Dis* 2, 99–106.

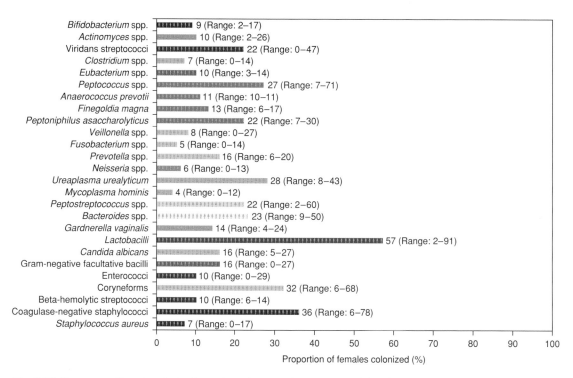

Fig. 6.32 Frequency of detection of microbes in the endocervix of pre-menopausal females. The data represent mean values (and ranges) derived from 11 studies involving a total of 465 individuals (24, 26, 33, 51–58).

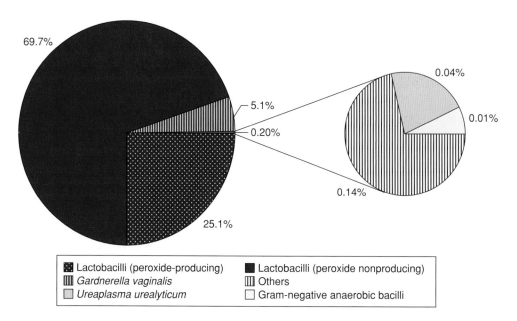

Fig. 6.33 Relative proportions of species comprising the cultivable cervical microbiota of 21 healthy, pre-menopausal females (24).

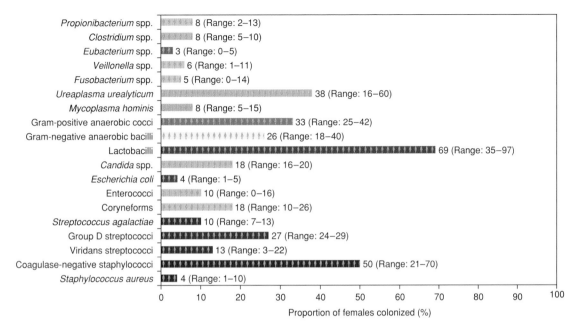

Fig. 6.34 Prevalence of microbes found in the endocervix of pregnant females. The data represent mean values (and ranges) derived from five culture-based studies involving a total of 290 individuals (59–63).

decreases progressively from the first trimester to term. The decreased isolation of anaerobes may be a consequence of the increased vascularity of the genital tract during pregnancy, which results in more oxygen being delivered to the tissues. Consequently, the environment of the cervix would be more aerobic and have a higher redox potential, thereby inhibiting the growth of obligate anaerobes.

6.4.2.3 Vulva

The vulva is an anatomically complex region that provides a variety of habitats, each differing with respect to the nature of the epithelium, moisture content, type and density of glands, degree of occlusion, proximity to other densely populated sites, etc. (sections 6.1 and 6.3.3). Each of these habitats is likely to have a characteristic microbiota. However, because of their close proximity to the anus, as well as the presence of the external openings of the vagina and urethra, the various regions of the vulva are likely to be contaminated with members of the microbiotas of these sites. The extent to which this occurs will depend on the particular anatomy of the individual, the type of clothing worn, and personal hygiene standards, etc.

Unfortunately, few studies of the different regions within the vulva have been carried out. Nevertheless, the data available indicate that the resident microbial communities consist of organisms that are characteristic of both the cutaneous and vaginal microbiotas. This is a consequence of the high moisture content of the vulva and its proximity to the vagina, from which it will be continually supplied with densely populated secretions. Of the various regions within the vulva, the microbiota of the labium majus has been investigated to the greatest extent. This is, basically, a specialized skin region, and therefore CNS and coryneforms are invariably present (Fig. 6.35). However, unlike most cutaneous regions, lactobacilli, viridans streptococci, and *Strep. agalactiae* are also frequently isolated. Of interest is the high carriage rate (approximately 15%) of *Staph. aureus* – this suggests that the labia majora, as well as the anterior nares and perineum, should also be sampled when screening for carriers of the organism.

The cultivable microbiota of the labia majora is dominated by CNS, although lactobacilli and viridans streptococci are also usually present in appreciable proportions (Fig. 6.36).

The results of a culture-independent study of the microbiotas of the labia majora and labia minora in

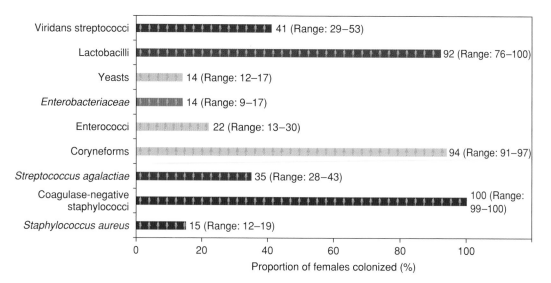

Fig. 6.35 Prevalence of microbes colonizing the labia majora of post-menarcheal/pre-menopausal females. The data represent mean values (and ranges) derived from three culture-based studies involving a total of 316 individuals (11, 64, 65).

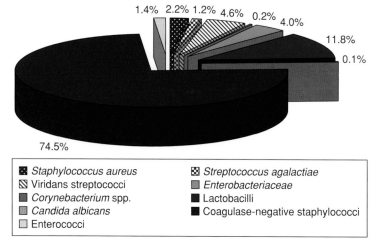

Fig. 6.36 Relative proportions of the various microbes that comprise the cultivable microbiota of the labia majora of post-menarcheal/pre-menopausal females. Data represent the mean values obtained in a study involving 102 individuals (64).

four healthy adults have shown that the species diversity of the former was greater than that of the latter (Fig. 6.37). Even within this small group, considerable diversity was found in the microbiotas from different individuals – no single phylotype was common to all four subjects. Furthermore, the number of phylotypes detected at each site varied widely and ranged from 2 to 27. Overall, the results bore some similarity to those derived from culture-dependent studies, with cutaneous microbes being detected in the microbiota of the labia majora. The microbiota of the labia minora was similar to that of the vagina, with lactobacilli being dominant.

6.4.3 Interactions between organisms colonizing the female reproductive system

The ability of lactobacilli to dominate the microbial communities of the female reproductive tract has been

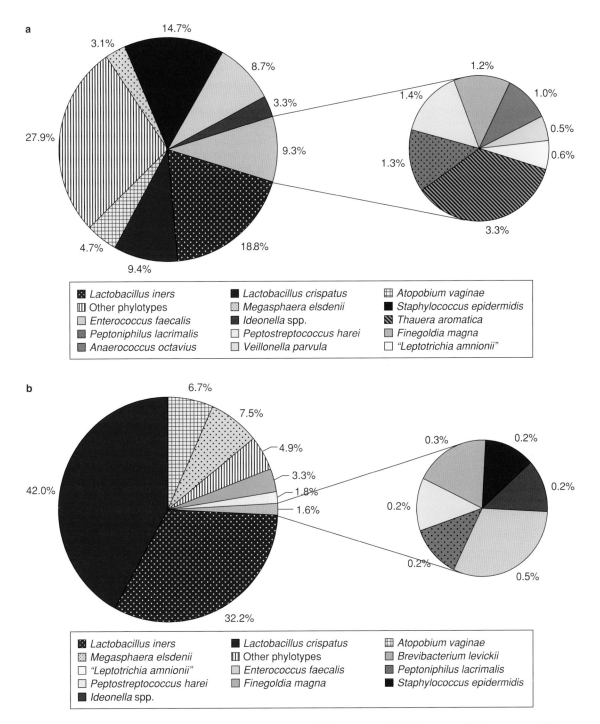

Fig. 6.37 Culture-independent analysis of the microbiota of the (a) labia majora and (b) labia minora. DNA was extracted from samples obtained from four healthy post-menarcheal/pre-menopausal individuals, the bacterial 16S rRNA genes present in each sample were amplified by PCR, and the resulting products were cloned and sequenced. Approximately 200 clones were analyzed from each library, and the figures denote the proportion (%) of clones of each phylotype detected (66).

Table 6.12 Potential beneficial interactions that may occur between organisms inhabiting the female reproductive system.

Activity	Potential benefits
Utilization of oxygen by aerobes and facultative anaerobes (e.g. staphylococci, coryneforms)	Provides an anaerobic environment for obligate anaerobes, e.g. GPAC, GNAB
Production of proteases and peptidases by proteolytic organisms (e.g. *Bifidobacterium* spp., *Bacteroides* spp., *Prevotella* spp., *Porphyromonas* spp., viridans streptococci, enterococci)	Liberates amino acids from proteins in vaginal and cervical secretions for use as carbon, nitrogen, and energy sources by other microbes
Degradation of mucins by bacterial consortia, producing sialidases, glycosidases and proteases	Releases sugars and amino acids from mucins in vaginal and cervical secretions for use as carbon, energy, and nitrogen sources
Production of metabolic end products (e.g. lactic acid by lactobacilli, streptococci, and bifidobacteria; succinic acid is produced by *Bacteroides* spp.)	Can be used as carbon and/or energy sources by other microbes (e.g. *Veillonella* spp. and *Propionibacterium* spp.)
Production of an acidic environment (e.g. lactobacilli, streptococci, bifidobacteria, enterococci)	Creates conditions suitable for the growth of acidophiles (e.g. lactobacilli, streptococci)

the subject of extensive research, and the mechanisms involved are summarized in Table 6.11. Less attention has been directed towards the antagonistic activities of other species. *Strep. agalactiae* has been reported to inhibit the growth of lactobacilli, *G. vaginalis*, *Streptococcus pyogenes*, and many coryneforms, while CNS are inhibitory to *Strep. agalactiae*.

Few studies have reported on the possible positive interactions (e.g. cross-feeding) that may occur in this complex community. Some likely possibilities are listed in Table 6.12.

6.5 OVERVIEW OF THE MICROBIOTA OF THE FEMALE REPRODUCTIVE SYSTEM

Regions of the female reproductive system that are colonized by microbes are the vulva, vagina, and cervix. In addition to being affected by the usual inter-individual variations (e.g. age, socioeconomic factors, etc.), the composition of the microbial communities at these sites is also profoundly influenced by the sexual maturity of the individual and, in females of reproductive age, the menstrual cycle. Unfortunately, few culture-independent investigations of the microbial communities present in the female reproductive system have been undertaken. Our knowledge of the composition of these communities,

therefore, is based mainly on data obtained from culture-dependent studies. Although a wide variety of species have been detected within each of these communities in the population as a whole, in an individual female, each microbial community is generally dominated by a limited number of species. The species most frequently isolated from the vagina and cervix in females of reproductive age include lactobacilli, CNS, GPAC, GNAB, coryneforms, and *Mollicutes*. Lactobacilli are generally the numerically dominant organisms in both the vaginal and cervical microbiotas – but this is not the case in pre-menarcheal girls and postmenopausal women who are not on hormone replacement therapy. The microbiota of the labia minora is similar to that of the vagina, whereas that of the labia majora consists of both vaginal and cutaneous species.

6.6 SOURCES OF DATA USED TO COMPILE FIGURES

1 Olmsted, S.S., Meyn, L.A., Rohan, L.C. and Hillier, S.L. (2003) *Sex Transm Dis* 30, 257–61.
2 Vallor, A.C., Antonio, M.A., Hawes, S.E. and Hillier, S.L. (2001) *J Infect Dis* 184, 1431–6
3 Anukam, K.C., Osazuwa, E.O., Ahonkhai, I. and Reid, G. (2006) *Sex Transm Dis* 33, 59–62.

4 Sullivan, A., Fianu-Jonasson, A., Landgren, B.M. and Nord, C.E. (2005) *Antimicrob Agents Chemother* 49, 170–5.

5 Burton, J.P., Dixon, J.L. and Reid, G. (2003) *Int J Gynecol Obstet* 81, 61–3.

6 Klebanoff, M.A., Andrews, W.W., Yu, K.F., Brotman, R.M., Nansel, T.R., Zhang, J., Cliver, S.P. and Schwebke, J.R. (2006) *Sex Transm Dis* 33, 610–13.

7 Ferris, M.J., Masztal, A., Aldridge, K.E., Fortenberry, J.D., Fidel, Jr. P.L. and Martin, D.H. (2004) *BMC Infect Dis* 4, 5.

8 Verhelst, R., Verstraelen, H., Claeys, G., Verschraegen, G., Delanghe, J., Van Simaey, L., De Ganck, C., Temmerman, M. and Vaneechoutte, M. (2004) *BMC Microbiol* 4, 16.

9 Mijac, V.D., Đukic, S.V., Opavski, N.Z., Đukic, M.K. and Ranin, L.T. (2006) *Eur J Obstet Gynecol Reprod Biol* 129, 69–76.

10 Reid, G., McGroarty, J.A., Tomeczek, L. and Bruce A.W. (1996) *FEMS Immunol Med Microbiol* 15, 23–6.

11 Ronnqvist, P.D., Forsgren-Brusk, U.B. and Grahn-Hakansson, E.E. (2006) *Acta Obstetricia et Gynecologica* 85, 726–35.

12 Schwebke, J.R. and Lawing, L.F. (2001) *Sex Transm Dis* 28, 195–9.

13 Zariffard, M.R., Saifuddin, M., Sha, B.E. and Spear, G.T. (2002) *FEMS Immunol Med Microbiol* 34, 277–81.

14 Arya, O.P., Tong, C.Y.W., Hart, C.A., Pratt, B.C., Hughes, S., Roberts, P., Kirby, P., Howel, J., McCormick, A. and Goddard, A.D. (2001) *Sex Transm Inf* 77, 58–62.

15 Hill, J.E., Goh, S.H., Money, D.M., Doyle, M., Li, A., Crosby, W.L., Links, M., Leung, A., Chan, D. and Hemmingsen, S.M. (2005) *Am J Obstet Gynecol* 193, 682–92.

16 Zhou, X., Bent, S.J., Schneider, M.G., Davis, C.C., Islam, M.R. and Forney, L.J. (2004) *Microbiol* 150, 2565–73.

17 Schreiber, C.A., Meyn, L.A., Creinin, M.D., Barnhart, K.T. and Hillier, S.L. (2006) *Obstet Gynecol* 107, 136–43.

18 Ferris, M.J., Masztal, A. and Martin, D.H. (2004) *J Clin Microbiol* 42, 5892–4.

19 Sobel, J.D. and Chaim, W. (1996) *J Clin Microbiol* 34, 2497–9.

20 Burton, J.P., Cadieux, P.A. and Reid, G. (2003) *Appl Environ Microbiol* 69, 97–101.

21 Clarke, J.G., Peipert, J.F., Hillier, S.L., Heber, W., Boardman, L., Moench, T.R. and Mayer, K. (2002) *Sex Transm Dis* 29, 288–93.

22 Smayevsky, J., Fernández Canigia, L., Lanza, A. and Bianchini, H. (2001) *Infect Dis Obstet Gynecol* 9, 17–22.

23 Eschenbach, D.A., Davick, P.R., Williams, B.L., Klebanoff, S.J., Young-Smith, K., Critchlow, C.M. and Holmes, K.K. (1989) *J Clin Microbiol* 27, 251–6.

24 Mauck, C.K., Creinin, M.D., Rountree, W., Callahan, M.M. and Hillier, S.L. (2005) *Contraception* 72, 53–9.

25 Gupta, K., Hillier, S.L., Hooton, T.M., Roberts, P.L. and Stamm, W.E. (2000) *J Infect Dis* 181, 595–601.

26 Corbishley, C.M. (1977) *J Clin Path* 30, 745–8.

27 Hill, G.B. (1980) Anaerobic flora of the female genital tract. In: Lambe, Jr. D.W., Genco, R.J. and Mayberry-Carson K.J. (eds). *Anaerobic bacteria: Selected topics*. Plenum Publishing Corp., New York, New York, USA, pp. 39–50.

28 Bartlett, J.G., Onderdonk, A.B., Drude, E., Goldstein, C., Anderka, M., Alpert, S. and McCormack, W.M. (1977) *J Infect Dis* 136, 271–7.

29 Tashjian, J.H., Coulam, C.B. and Washington, J.A. (1976) *2nd Mayo Clin Proc* 51, 557–61.

30 Osborne, N.G., Wright, R.C. and Grubin, L. (1979) *Am J Obstet Gynecol* 135, 195–8.

31 Masfari, A.N., Duerden, B.I. and Kinghorn, G.R. (1986) *Genitourin Med* 62, 256–63.

32 Bartlett, J.G. and Polk, B.F. (1984) *Rev Infect Dis* 6 (Suppl. 1), S67–72.

33 Bartlett, J.G., Moon, N.E., Goldstein, P.R., Goren, B., Onderdonk, A.B. and Polk, B.F. (1978) *Am J Obstet Gynecol* 130, 658–61.

34 Shubair, M., Stanek, R., White, S. and Larsen, B. (1992) *Gynecol Obstet Invest* 34, 229–33.

35 Fredricks, D.N., Fiedler, T.L. and Marrazzo, J.M. (2005) *N Engl J Med* 353, 1899–911.

36 Hyman, R.W., Fukushima, M., Diamond, L., Kumm, J., Giudice, L.C. and Davis, R.W. (2005) *Proc Natl Acad Sci U S A* 102, 7952–7.

37 Eschenbach, D.A., Thwin, S.S., Patton, D.L., Hooton, T.M., Stapleton, A.E., Agnew, K., Winter, C., Meier, A. and Stamm, W.E. (2000) *Clin Infect Dis* 30, 901–7.

38 Hill, G.B., St. Claire, K.K. and Gutman, L.T. (1995) *Clin Infect Dis* 20 (Suppl. 2), S269–70.

39 Jaquiery, A., Stylianopoulos, A., Hogg, G. and Grover, S. (1999) *Arch Dis Child* 81, 64–7.

40 Gerstner, G.J., Grunberger, W., Boschitsch, E. and Rotter, M. (1982) *Arch Gynecol* 231, 247–52.

41 Hammerschlag, M.R., Alpert, S., Onderdonk, A.B., Thurston, P., Drude, E., McCormack, W.M. and Bartlett, J.G. (1978) *Am J Obstet Gynecol* 131, 853–6.

42 Hillier, S.L. and Lau, R.J. (1997) *Clin Infect Dis* 25 (Suppl. 2), S123–6.

43 Campbell, J.R., Hillier, S.L., Krohn, M.A., Ferrieri, P., Zaleznik, D.F. and Baker, C.J. (2000) *Obstet Gynecol* 96, 498–503.

44 Votava, M., Tejkalova, M., Drabkova, M., Unzeitig, V. and Braveny, I. (2001) *Eur J Clin Microbiol Infect Dis* 20, 120–2.

45 El-Kersh, T.A., Al-Nuaim, L.A., Kharfy, T.A., Al-Shammary, F.J., Al-Saleh, S.S. and Al-Zamel, F.A. (2002) *Saudi Med J* 23, 56–61.

46 Gil, E.G., Rodriguez, M.C., Bartolome, R., Berjano, B., Cabero, L. and Andreu, A. (1999) *J Clin Microbiol* 37, 2648–51.

47 Hillier, S.L., Krohn, M.A., Rabe, L.K., Klebanoff, S.J. and Eschenbach, D.A. (1993) *Clin Infect Dis* 16 (Suppl. 4), S273–81.

48 Delaney, M.L. and Onderdonk, A.B. (2001) *Obstet Gynecol* 98, 79–84.

49 Bayo, M., Berlanga, M. and Agut, M. (2002) *Int Microbiol* 5, 87–90.

50 Thorsen, P., Jensen, I.P., Jeune, B., Ebbesen, N., Arpi, M., Bremmelgaard, A. and Møller, B.R. (1998) *Am J Obstet Gynecol* 178, 580–7.

51 Thadepalli, H., Savage, Jr. E.W., Salem, F.A., Roy, I. and Davidson, Jr. E.C. (1982) *Gynecol Obstet Invest* 14, 176–83.

52 Keith, L., England, D., Bartizal, F., Brown, E. and Fields, C. (1972) *Br J Vener Dis* 48, 51–6.

53 Haukkamaa, M., Stranden, P., Jousimies-Somer, H. and Siitonen, A. (1987) *Contraception* 36, 527–34.

54 Gorbach, S.L., Menda, K.B., Thadepalli, H. and Keith, L. (1973) *Am J Obstet Gynecol* 117, 1053–5.

55 Haukkamaa, M., Stranden, P., Jousimies-Somer, H. and Siitonen, A. (1986) *Am J Obstet Gynecol* 154, 520–4.

56 Kinghorn, G.R., Duerden, B.I. and Hafiz, S. (1986) *Br J Obstet Gynaecol* 93, 869–80.

57 Osborne, N.G. and Wright, R.C. (1977) *Obstet Gynecol* 50, 148–51.

58 Viberga, I., Odlind, V., Lazdane, G., Kroica, J., Berglund, and Olofsson, S. (2005) *Infect Dis Obstet Gynecol* 13, 183–190.

59 Patterson, R.M., Blanco, J.D., Gibbs, R.S. and St. Clair, P.J. (1986) *Am J Obstet Gynecol* 154, 1111–12.

60 Stiver, H.G., Forward, K.R., Tyrrell, D.L., Krip, G., Livingstone, R.A., Fugere, P., Lemay, M., Verschelden, G., Hunter, J.D. and Carson, G.D. (1984) *Am J Obstet Gynecol* 149, 718–21.

61 Silva, M.G., Peracoli, J.C., Sadatsune, T., Abreu, E.S. and Peracoli, M.T.S. (2003) *Int J Gynecol Obstet* 81, 175–82.

62 Moberg, G., Eneroth, P., Harlin, J., Ljung-Wadstrom, A. and Nord, C.E. (1978) *Med Microbiol Immunol* 165, 139–45.

63 Goplerud, C.P., Ohm, M.J. and Galask, R.P. (1976) *Am J Obstet Gynecol* 126, 858–68.

64 Runeman, B., Rybo, G., Forsgren-Brusk, U., Larko, O., Larsson, P. and Faergemann, J. (2004) *Acta Derm Venereol* 84, 277–84.

65 Runeman, B., Rybo, G., Forsgren-Brusk, U., Larko, O., Larsson, P. and Faergemann, J. (2005) *Acta Derm Venereal* 85, 118–22.

66 Brown, C.J., Wong, M., Davis, C.C., Kanti, A., Zhou, X. and Forney, L.J. (2007) *J Med Microbiol* 56, 271–6.

6.7 FURTHER READING

6.7.1 Books

Berg, R.W. and Davis, C.C. (2006) Microbial ecology of the vulva. In: Farage, M.A. and Maibach, H.I. (eds). *The Vulva: Anatomy, Physiology, and Pathology.* Informa Healthcare, London, UK.

Ledger, W.J. and Witkin, S.S. (2007) Microbiology of the vagina. In: *Vulvovaginal Infections.* ASM Press, Washington, DC, USA.

Ledger, W.J. and Witkin, S.S. (2007) Vaginal immunology. In: *Vulvovaginal Infections.* ASM Press, Washington, DC, USA.

6.7.2 Reviews and papers

Agbakoba, N.R., Adetosoye, A.I. and Adewole, I.F. (2007) Presence of mycoplasma and ureaplasma species in the vagina of women of reproductive age. *West Afr J Med* 26, 28–31.

Anukam, K.C., Osazuwa, E.O., Ahonkhai, I. and Reid, G. (2006) *Lactobacillus* vaginal microbiota of women attending a reproductive health care service in Benin City, Nigeria. *Sex Transm Dis* 33, 59–62.

Aroutcheva, A., Gariti, D., Simon, M., Shott, S., Faro, J., Simoes, J.A., Gurguis, A., and Faro, S. (2001) Defense factors of vaginal lactobacilli. *Am J Obstet Gynecol* 185, 375–9.

Arya, O.P., Tong, C.Y., Hart, C.A., Pratt, B.C., Hughes, S., Roberts, P., Kirby, P., Howel, J., McCormick, A. and Goddard, A.D. (2001) Is *Mycoplasma hominis* a vaginal pathogen? *Sex Transm Infect* 77, 58–62.

Balu, R.B., Savitz, D.A., Ananth, C.V., Hartmann, K.E., Miller, W.C., Thorp, J.M. and Heine, R.P. (2002) Bacterial vaginosis and vaginal fluid defensins during pregnancy. *Am J Obstet Gynecol* 187, 1267–71.

Bayo, M., Berlanga, M. and Agut, M. (2002) Vaginal microbiota in healthy pregnant women and prenatal screening of group B streptococci (GBS). *Int Microbiol* 5, 87–90.

Boris, S.B. (2000) Role played by lactobacilli in controlling the population of vaginal pathogens. *Microbes Infect* 2, 543–6.

Boskey, E.R., Telsch, K.M., Whaley, K.J., Moench, T.R. and Cone, R.A. (1999) Acid production by vaginal flora in vitro is consistent with the rate and extent of vaginal acidification. *Infect Immun* 67, 5170–5.

Brown, C.J., Wong, M., Davis, C.C., Kanti, A., Zhou, X. and Forney, L.J. (2007) Preliminary characterization of the normal microbiota of the human vulva using cultivation-independent methods. *J Med Microbiol* 56, 271–6.

Burton, J.P. and Reid, G. (2002) Evaluation of the bacterial vaginal flora of 20 postmenopausal women by direct (nugent score) and molecular (polymerase chain reaction and denaturing gradient gel electrophoresis) techniques. *J Infect Dis* 186, 1770–80.

Burton, J.P., Cadieux, P.A. and Reid, G. (2003) Improved understanding of the bacterial vaginal microbiota of women before and after probiotic instillation. *Appl Environ Microbiol* 69, 97–101.

Cauci, S., Driussi, S., De Santo, D., Penacchioni, P., Lannicelli, T., Lanzafame, P., De Seta, F., Quadrifoglio, F., de Aloysio, D. and Guaschino, S. (2002) Prevalence of bacterial vaginosis and vaginal flora changes in peri- and postmenopausal women. *J Clin Microbiol* 40, 2147–52.

Chase, D.J., Schenkel, B.P., Fahr, A.M., Eigner, U; Tampon Study Group. (2007) A prospective, randomized, double-blind study of vaginal microflora and epithelium in women using a tampon with an apertured film cover compared with those in women using a commercial tampon with a cover of nonwoven fleece. *J Clin Microbiol* 45, 1219–24.

Cole, A.M. (2006) Innate host defense of human vaginal and cervical mucosae. *Curr Top Microbiol Immunol* 306, 199–230.

Delaney, M.L. and Onderdonk, A.B. (2001) Nugent score related to vaginal culture in pregnant women. *Obstet Gynecol* 98, 79–84.

Donders, G.G. (2007) Definition and classification of abnormal vaginal flora. *Best Pract Res Clin Obstet Gynaecol* Apr 13.

Eggert-Kruse, W., Botz, I., Pohl, S., Rohr, G. and Strowitzki, T. (2000) Antimicrobial activity of human cervical mucus. *Hum Reprod* 15, 778–84.

El-Kersh, T.A., Al-Nuaim, L.A., Kharfy, T.A., Al-Shammary, F.J., Al-Saleh, S.S. and Al-Zamel, F.A. (2002) Detection of genital colonization of group B streptococci during late pregnancy. *Saudi Med J* 23, 56–61.

Eschenbach, D.A., Thwin, S.S., Patton, D.L., Hooton, T.M., Stapleton, A.E., Agnew, K., Winter, C., Meier, A. and Stamm, W.E. (2000) Influence of the normal menstrual cycle on vaginal tissue, discharge, and microflora. *Clin Infect Dis* 30, 901–7.

Ferris, M.J., Masztal, A., Aldridge, K.E., Fortenberry, J.D., Fidel, Jr. P.L. and Martin, D.H. (2004) Association of *Atopobium vaginae*, a recently described metronidazole resistant anaerobe, with bacterial vaginosis. *BMC Infect Dis* 4, 5.

Ferris, M.J., Masztal, A. and Martin, D.H. (2004) Use of species-directed 16S rRNA gene PCR primers for detection of *Atopobium vaginae* in patients with bacterial vaginosis. *J Clin Microbiol* 42, 5892–4.

Fredricks, D.N., Fiedler, T.L. and Marrazzo, J.M. (2005) Molecular identification of bacteria associated with bacterial vaginosis. *N Engl J Med* 353, 1899–911.

Grzesko, J., Elias, M., Manowiec, M. and Gabrys, M.S. (2006) Genital mycoplasmas – morbidity and a potential influence on human fertility. *Med Wieku Rozwoj* 10, 985–92.

Gupta, K., Hillier, S.L., Hooton, T.M., Roberts, P.L. and Stamm, W.E. (2000) Effects of contraceptive method on the vaginal microbial flora: A prospective evaluation. *J Infect Dis* 181, 595–601.

Hein, M., Valore, E.V., Helmig, R.B., Uldbjerg, N. and Ganz, T. (2002) Antimicrobial factors in the cervical mucus plug. *Am J Obstet Gynecol* 187, 137–44.

Hill, D.R., Brunner, M.E., Schmitz, D.C., Davis, C.C., Flood, J.A., Schlievert, P.M., Wang-Weigand, S.Z. and Osborn, T.W. (2005) In vivo assessment of human vaginal oxygen and carbon dioxide levels during and post menses. *J Appl Physiol* 99, 1582–91.

Hill, J.E., Goh, S.H., Money, D.M., Doyle, M., Li, A., Crosby, W.L., Links, M., Leung, A., Chan, D. and Hemmingsen, S.M.

(2005) Characterization of vaginal microflora of healthy, nonpregnant women by chaperonin-60 sequence-based methods. *Am J Obstet Gynecol* 193, 682–92.

Hillier, S. L. and Lau, R.J. (1997) Vaginal microflora in post-menopausal women who have not received estrogen replacement therapy. *Clin Infect Dis* 25 (Suppl. 2), S123–6.

Hyman, R.W., Fukushima, M., Diamond, L., Kumm, J., Giudice, L.C. and Davis, R.W. (2005) Microbes on the human vaginal epithelium. *Proc Natl Acad Sci U S A* 102, 7952–7.

Jaquiery, A., Stylianopoulos, A., Hogg, G. and Grover, S. (1999) Vulvovaginitis: Clinical features, aetiology, and microbiology of the genital tract. *Arch Dis Child* 81, 64–7.

Johansson, M. and Lycke, N. (2003) Immunology of the human genital tract. *Curr Opin Infect Dis* 16, 43–9.

Kiss, H., Kogler, B., Petricevic, L., Sauerzapf, I., Klayraung, S., Domig, K., Viernstein, H. and Kneifel, W. (2007). Vaginal *Lactobacillus* microbiota of healthy women in the late first trimester of pregnancy. *BJOG* Sep 18; (Epub ahead of print).

Kjaergaard. N., Hein. M., Hyttel, L., Helmig, R.B., Schønheyder, H.C., Uldbjerg, N. and Madsen, H. (2001) Antibacterial properties of human amnion and chorion in vitro. *Eur J Obstet Gynecol Reprod Biol* 94, 224–9.

Klebanoff, M.A., Andrews, W.W., Yu, K.F., Brotman, R.M., Nansel, T.R., Zhang, J., Cliver, S.P. and Schwebke, J.R. (2006) A pilot study of vaginal flora changes with randomization to cessation of douching. *Sex Transm Dis* 33, 610–13.

Kubota, T., Nojima, M. and Itoh, S. (2002) Vaginal bacterial flora of pregnant women colonized with group B streptococcus. *J Infect Chemother* 8, 326–30.

Larsen, B. and Monif, G.R. (2001) Understanding the bacterial flora of the female genital tract. *Clin Infect Dis* 32, 69–77.

Mijac, V.D., Đukic, S.V., Opavski, N.Z., Đukic, M.K. and Ranin, L.T. (2006) Hydrogen peroxide producing lactobacilli in women with vaginal infections. *Eur J Obstet Gynecol Reprod Biol* 129, 69–76

Mikamo, H., Sato, Y., Hayasaki, Y., Hua, Y.X. and Tamaya, T. (2000) Vaginal microflora in healthy women with *Gardnerella vaginalis*. *J Infect Chemother* 6, 173–7.

Newton, E.R., Piper, J.M., Shain, R.N., Perdue, S.T. and Peairs, W. (2001) Predictors of the vaginal microflora. *Am J Obstet Gynecol* 184, 845–53.

Olmsted, S.S., Meyn, L.A., Rohan, L.C. and Hillier, S.L. (2003) Glycosidase and proteinase activity of anaerobic gram-negative bacteria isolated from women with bacterial vaginosis. *Sex Transm Dis* 30, 257–61.

Paavonen, J. (1983) Physiology and ecology of the vagina. *Scand J Infect Dis* 40 (Suppl.), 31–5.

Pybus, V. and Onderdonk, A.B. (1999) Microbial interactions in the vaginal ecosystem, with emphasis on the pathogenesis of bacterial vaginosis. *Microbes Infect* 1, 285–92.

Rajan, N., Cao, Q., Anderson, B.E., Pruden, D.L., Sensibar, J., Duncan, J.L. and Schaeffer, A.J. (1999) Roles of glycoproteins and oligosaccharides found in human vaginal fluid in bacterial adherence. *Infect Immun* 67, 5027–32.

Ronnqvist, P.D., Forsgren-Brusk, U.B. and Grahn-Hakansson, E.E. (2006) Lactobacilli in the female genital tract in relation to other genital microbes and vaginal pH. *Acta Obstetricia et Gynecologica* 85, 726–35.

Runeman, B., Rybo, G., Forsgren-Brusk, U., Larko, O., Larsson, P. and Faergemann, J. (2004) The vulvar skin microenvironment: Influence of different panty liners on temperature, pH and microflora. *Acta Derm Venereol* 84, 277–84.

Runeman, B., Rybo, G., Forsgren-Brusk, U., Larko, O., Larsson, P. and Faergemann, J. (2005) The vulvar skin microenvironment: Impact of tight-fitting underwear on microclimate, pH and microflora. *Acta Derm Venereal* 85, 118–22.

Schreiber, C.A., Meyn, L.A., Creinin, M.D., Barnhart, K.T. and Hillier, S.L. (2006) Effects of long-term use of nonoxynol-9 on vaginal flora. *Obstet Gynecol* 107, 136–43.

Schwebke, J.R. and Weiss, H. (2001) Influence of the normal menstrual cycle on vaginal microflora. *Clin Infect Dis* 32, 325.

Schwebke, J.R., Richey, C.M. and Weiss, H.L. (1999) Correlation of behaviors with microbiological changes in vaginal flora. *J Infect Dis* 180, 1632–6.

Sullivan, A., Fianu-Jonasson, A., Landgren, B.M. and Nord, C.E. (2005) Ecological effects of perorally administered pivmecillinam on the normal vaginal microflora. *Antimicrob Agents Chemother* 49, 170–5.

Sundstrom, P. (1999) Adhesins in *Candida albicans*. *Curr Opin Microbiol* 2, 353–7.

Thies, F.L., Konig, W. and Konig, B. (2007) Rapid characterization of the normal and disturbed vaginal microbiota by application of 16S rRNA gene terminal RFLP fingerprinting. *J Med Microbiol* 56, 755–61

Vallor, A.C., Antonio, M.A., Hawes, S.E. and Hillier, S.L. (2001) Factors associated with acquisition of or persistent colonization by, vaginal lactobacilli: Role of hydrogen peroxide production. *J Infect Dis* 184, 1431–6.

Venkataraman, N., Cole, A.L., Svoboda, P., Pohl, J. and Cole, A.M. (2005) Cationic polypeptides are required for anti-HIV-1 activity of human vaginal fluid. *J Immunol* 175, 7560–7.

Verhelst, R., Verstraelen, H., Claeys, G., Verschraegen, G., Delanghe, J., Van Simaey, L., De Ganck, C., Temmerman, M. and Vaneechoutte, M. (2004) Comparison between Gram stain and culture for the characterization of vaginal microflora: Definition of a distinct grade that resembles grade I microflora and revised categorization of grade I microflora. *BMC Microbiol* 4, 16.

Vitali, B., Pugliese, C., Biagi, E., Candela, M., Turroni, S., Bellen, G., Donders, G.G. and Brigidi, P. (2007) Dynamics of vaginal bacterial communities in women developing bacterial vaginosis, candidiasis, or no infection, analyzed by PCR-denaturing gradient gel electrophoresis and real-time PCR. *Appl Environ Microbiol* 73, 5731–41.

Wilson, J.D., Lee, R.A., Balen, A.H. and Rutherford, A.J. (2007) Bacterial vaginal flora in relation to changing oestrogen levels. *Int J STD AIDS* 18, 308–11

Witkin, S.S., Linhares, I.M. and Giraldo, P. (2007) Bacterial flora of the female genital tract: Function and immune regulation. *Best Pract Res Clin Obstet Gynaecol* 21, 347–54.

Zhou, X., Bent, S.J., Schneider, M.G., Davis, C.C., Islam, M.R. and Forney, L.J. (2004) Characterization of vaginal microbial communities in adult healthy women using cultivation independent methods. *Microbiol* 150, 2565–73.

Chapter 7

THE INDIGENOUS MICROBIOTA OF THE URINARY AND REPRODUCTIVE SYSTEMS OF MALES

7.1 ANATOMY AND PHYSIOLOGY

In males, unlike the situation in females, the urethra is also part of the reproductive system (Fig. 7.1). Consequently, it is a passageway for both urine and semen.

The male urethra is approximately 20 cm long and consists of three main regions: (1) the prostatic urethra – this section lies within the prostate and is 3–4 cm long; (2) the membranous urethra – this section is surrounded by the sphincter urethra muscle and is approximately 2 cm long; and (3) the penile (or spongy) urethra – this section lies within the bulb and body of the penis and is approximately 15 cm long. Most of the urethra is lined by a pseudo-stratified or stratified columnar epithelium, apart from the navicular fossa (i.e. the section within the glans penis), which has a stratified, squamous epithelium that becomes keratinized near the external meatus. Numerous glands (urethral glands) are present along the epithelium, which also contains many small collections of mucous cells (glands of Littré). The external orifice of the urethra is located on the glans penis which, in uncircumcised individuals, is covered by a retractable fold of skin – the prepuce or foreskin. On the corona of the glans penis are sebaceous glands (glands of Tyson), which secrete a material known as smegma. The urethra of males is approximately five times longer than that of females, which is one of the reasons why ascending infections (i.e. infections caused by microbes originating from the urethra and peri-urethral region)

of the bladder and kidneys are far less frequent in males than in females.

7.2 ANTIMICROBIAL DEFENSES OF THE MALE URINARY AND REPRODUCTIVE SYSTEMS

Many of the defense mechanisms operating in the male urethra are similar to those in the female urethra and have been described in section 5.2. They include: desquamation of epithelial cells, urinary flow, epithelium-derived antimicrobial peptides, and the presence in urine of anti-adhesive factors (Tamm–Horsfall glycoprotein, secretory immunoglobulin A [sIgA]) and molecules with antimicrobial activity such as urea, antibodies, and antimicrobial peptides (Fig. 7.2). However, the urethra of males also acts as a passageway for semen, which contains additional antimicrobial compounds.

Semen consists of spermatozoa suspended in seminal plasma. The seminal plasma is a mixture of secretions from the male accessory sex glands (i.e. the seminal vesicles, the prostate gland, and the Cowper's glands), Littré glands (which line the urethra), the ampulla (the terminal portion of the vas deferens), and the epididymis. With regard to the antimicrobial properties of seminal plasma, prostatic secretions contain high concentrations of zinc, which can inhibit the growth of *Escherichia coli*, *Chlamydia trachomatis*, *Candida albicans*, and *Trichomonas vaginalis* at the levels found in vivo.

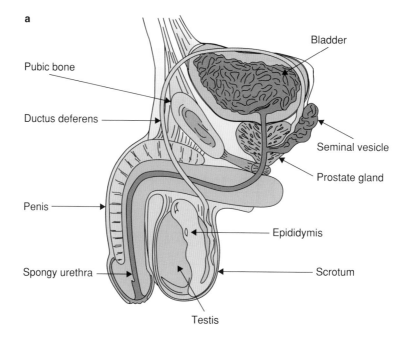

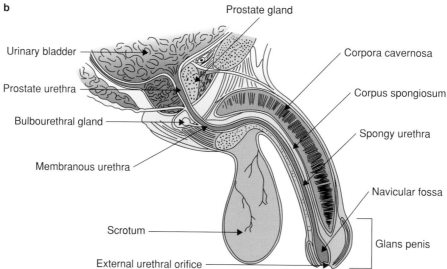

Fig. 7.1 The reproductive system of the male. (a) Main components of the system. (b) Structure of the penis. (From http://www.training.seer.cancer.gov/module_anatomy/anatomy_physiology_home.html; funded by the US National Cancer Institute's Surveillance, Epidemiology and End Results (SEER) Program with Emory University, Atlanta, Georgia, USA.)

Prostatic fluid also contains phospholipase A_2. Other antimicrobial compounds found in seminal plasma are spermidine, spermine (which has activity against a wide range of organisms including *Chlam. trachmomatis*),

lactoferrin, HBD-1 (Fig. 7.3), HD-5, SLPI (400 ng/ml), HBD-4, lysozyme, hCAP-18/LL-37 (900 ng/ml), and SKALP/elafin (400 ng/ml). Recently, it has been shown that peptides produced by cleavage of one of the

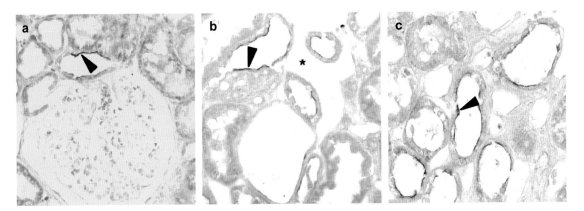

Fig. 7.2 Immunohistochemical localization of hepcidin in the human urinary tract. Hepcidin expression (arrowheads) can be seen in various regions of the kidney, including (a) thick ascending limb, (b) connecting tubule, and (c) distal tubule. Magnification: ×450 approximately. Reproduced with permission from: Kulaksiz, H., Theilig, F., Bachmann, S., Gehrke, S.G., Rost, D., Janetzko, A., Cetin, Y. and Stremmel, W. (2005) *J Endocrinol* 184, 361–70. © Society for Endocrinology, Bristol, UK (2005).

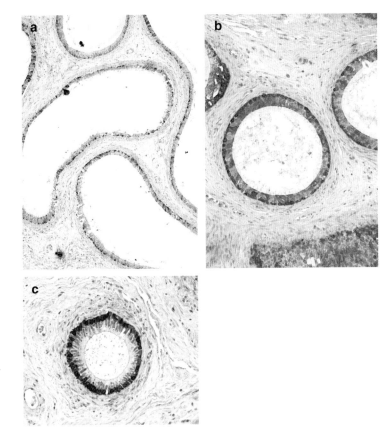

Fig. 7.3 Immunolocalization (dark-brown-staining regions) of human beta-defensin-1 in the (a) head, (b) body, and (c) tail of the epididymis. Images kindly supplied by Charles Pineau, Inserm U625, Université de Rennes, France.

proteins present in semen, semenogelin I, have antibacterial activity.

IgA- and IgM-producing plasma cells are present along the entire length of the penile urethra. IgA, IgG and, to a lesser extent, IgM are also present in seminal fluid. Most of the IgG present is derived from the circulation, whereas the IgA is produced locally. Immunohistochemistry has revealed that the penile urethra is covered in mucus containing IgA, IgG, and IgM, and this constitutes an effective barrier against microbial adhesion and invasion.

7.3 ENVIRONMENTAL DETERMINANTS WITHIN THE MALE URINARY AND REPRODUCTIVE SYSTEMS

Because of the frequent passage of urine with its high oxygen content (approximately 81 mmHg), the urethra is predominantly an aerobic region.

Owing to the intermittent presence of urine (which is acidic) and semen which is slightly alkaline (pH 7.1 to 7.5), marked variations occur in the pH of the urethra. The pH of the glans penis ranges from 4.8 to 7.2 with a median value of 6.02.

A variety of nutrient sources are available to urethral microbes; these include secretions of the various urethral glands, desquamated epithelial cells, urine, and semen. As described in section 5.3, urine provides a number of nitrogen sources (urea, NH_4^+, and creatinine), phosphate, and a variety of trace elements (Table 5.2). Seminal plasma consists mainly of water (92%), but also contains lipids, proteins, carbohydrates, and amino acids (Table 7.1). The passage of urine and semen along the urethra, therefore, will deposit a wide range of nutrients that can be utilized by the urethral microbiota (Table 7.1).

A number of mucins (MUC1, MUC3, MUC4, MUC5AC, MUC13, MUC15, MUC17, and MUC20) are produced by the urethra, and their localization is shown in Fig. 7.4.

The mucins, proteins, and lipids present in the urethra can be degraded by a number of resident microbes (e.g. coagulase-negative staphylococcus [CNS], *Corynebacterium* spp., viridans streptococci, clostridia, bifidobacteria, Gram-positive anaerobic cocci [GPAC], *Prevotella* spp., and *Bacteroides* spp.), thereby providing sugars, amino acids, and fatty acids.

The glans penis is a lipid-rich environment in which an additional source of nutrients for microbes is smegma,

Table 7.1 Substances present in the male urethra that can serve as microbial nutrients.

Nutrient	Main source
Urea	Urine
NH_4^+	Urine
Creatinine	Urine
Uric acid	Urine
Trace elements (Na$^+$, K$^+$, Cl$^-$, Mg^{2+}, SO$_4^{2-}$, PO$_4^{3-}$, Ca^{2+}, Zn^{2+})	Urine, seminal fluid
Citric acid	Prostate
Spermine	Seminal fluid
Fructose	Seminal fluid
Glucose	Seminal fluid
Glycogen	Epididymis
Citric acid	Seminal fluid
Amino acids	Seminal fluid
Proteins (albumin, prealbumin, insulin, α_1-antitrypsin, α_1-glycoprotein, transferrin, IgA, IgG)	Seminal fluid, plasma cells in urethra
Lipids (phosphatidyl choline, phosphatidyl ethanolamine, sphingomyelin)	Epididymis, prostate
Mucins	Mucus produced by several types of glands
Various	Dead/dying epithelial cells

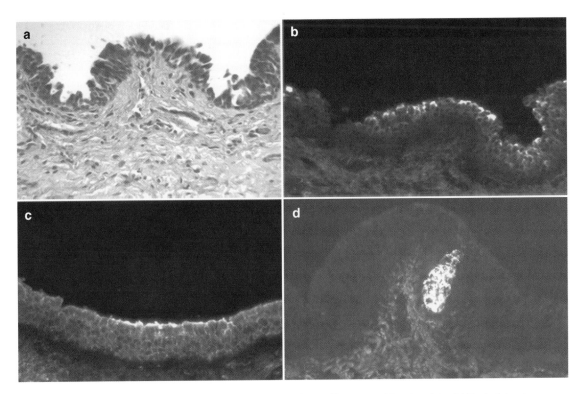

Fig. 7.4 Distribution of various mucins within the male urethra detected by immunohistochemistry. (a) Control showing a cross-section through the male urethra. Prominent apical binding of (b) MUC1 and (c) MUC4 can be seen (white regions) with, in each case, some localization in the subapical portions of the epithelium. (d) MUC5AC can be seen to be present mainly in the glands of Littré, with little or no binding detected on the apical surfaces of the urethral epithelium. Reproduced with permission from: Russo, C.L, Spurr-Michaud, S., Tisdale, A., Pudney, J., Anderson, D. and Gipson, I.K. (2006) Mucin gene expression in human male urogenital tract epithelia. *Hum Reprod* 21, 2783–93. Copyright © 2006 European Society of Human Reproduction and Embryology, Beigem, Belgium.

which contains approximately 27% fat and 13% protein. In uncircumcised individuals, the prepuce reduces moisture loss and desquamation which means that the preputial space (the region between the glans penis and the overlying prepuce) is a warm, moist region with a neutral to slightly alkaline pH and a reduced oxygen content.

7.4 THE INDIGENOUS MICROBIOTA OF THE MALE URINARY AND REPRODUCTIVE SYSTEMS

7.4.1 Members of the microbiota

The main organisms colonizing the male urethra are CNS, *Corynebacterium* spp., viridans streptococci,

GPAC, and *Mollicutes*. CNS and *Corynebacterium* spp. are described in Chapter 2, while viridans streptococci are described in Chapters 4 and 8. *Mollicutes* are discussed in Chapters 4 and 6, while GPAC are described in Chapter 5.

Mycoplasma genitalium is a frequent resident of the male urethra and is the smallest, self-replicating cell known. Its genome was one of the first to be sequenced and is only 580 kbp – the smallest of any free-living organism capable of growing in axenic culture (http://cmr.tigr.org/tigr-scripts/CMR/GenomePage.cgi?org=gmg). Of the 517 genes identified, 482 encode proteins. It can ferment glucose but has very limited biosynthetic capability, is highly fastidious, and requires complex media for growth. It is aerobic, although growth is stimulated by 5% CO_2, and its optimum temperature for growth is $36-38°C$. It has a complex tip structure, known as a "terminal organelle", consisting of the adhesins responsible for mediating adhesion to host cells. Although it does not possess flagella or pili, it can move across solid surfaces by an unknown mechanism.

7.4.2 Microbiota of the male urethra

Many of the organisms most frequently isolated from the male urethra (staphylococci, coryneforms, and streptococci) are members of the cutaneous microbiota and are, presumably, acquired from this source. The urethra of sexually active males has a more diverse microbiota than that of nonsexually active individuals, and many of the organisms present will have been acquired from the vagina and cervix of female sexual partners (see sections 6.4.2.1 and 6.4.2.2).

Only the distal 6 cm of the male urethra is normally colonized by microbes, and the types of organism present are affected by the usual variables (section 1.1) and also by a large number of factors peculiar to this site. The factors influencing the composition of the urethral microbiota are as follows:
• The sexual maturity of the individual;
• Whether the individual is circumcised;
• Whether the individual engages in sexual intercourse;
• The gender of the sexual partner;
• The frequency of sexual intercourse;
• The number of sexual partners;
• The nature of any birth-control measure used (if any) during intercourse;
• Whether the individual engages in oral, anal, or vaginal intercourse.
Studies of the urethral microbiota are usually based on the analysis of single urethral swabs, semen samples, and samples of first-voided urine. Such approaches fail to detect possible variations in the microbial communities along the length of the urethra. The few studies that have sampled different regions of the urethra have demonstrated that most organisms are present within the navicular fossa. The numbers then decrease with increasing distance from the orifice until, after approximately 6 cm, bacteria are generally not detectable. The potential diversity of the urethral microbiota can be appreciated from Table 7.2, which lists some of the organisms that have been detected. However, the number of species isolated from an individual adult is usually five or less.

In a study of the aerobically cultivable urethral microbiota, the most frequently isolated organisms from the external orifice were found to be staphylococci, viridans streptococci, and *Corynebacterium* spp. (Fig. 7.5). The frequency of isolation of all three bacterial groups, as well as the complexity of the community present, decreased with increasing distance from the orifice. Hence, at the orifice, CNS, *Staphylococcus aureus*, viridans streptococci, *Corynebacterium* spp., β-hemolytic streptococci, *Enterobacter* spp., and micrococci were cultivated. In contrast, only CNS, viridans streptococci, *Corynebacterium* spp., and enterococci were detectable within the penile urethra. This is one of the few studies to have carried out an analysis of the relative proportions of the organisms that comprise the urethral microbiota, and viridans streptococci were found to be the dominant species at all three sites. It must be remembered, however, that the study did not take into account the anaerobic members of the microbiota. Nevertheless, when analysis has included anaerobic organisms, CNS, streptococci, and coryneforms still appear to be the dominant organisms (Fig. 7.6).

Figure 7.7 shows the results of a number of studies of the prevalence of a range of microbes, including anaerobic species, in the male urethra. From this, it can be seen that anaerobes, particularly GPAC, Gram-negative anaerobic bacilli (GNAB), and *Veillonella* spp., are frequent constituents of the urethral microbiota of males.

Coryneforms are often present in the urethra, but have rarely been characterized to the species level. In one study where this was carried out, the most frequently detected species was found to be *Corynebacterium seminale* (also known as *Corynebacterium glucuronolyticum*), an organism whose sole habitat appears to be the male urogenital tract (Fig. 7.8).

Engaging in sexual intercourse greatly increases the complexity of the urethral microbiota, the mean number of isolates obtained from a sexually active individual being almost twice that isolated from one who is sexually inactive. Other notable changes include: (1) an increase in the frequency of isolation of *Gardnerella vaginalis*, Gram-negative aerobes and facultative anaerobes, *Streptococcus agalactiae*, Group D streptococci, GPAC, *Veillonella parvula*, GNAB, *Mycoplasma* spp., and *Ureaplasma urealyticum*; and (2) a decrease in the frequency of isolation of α-hemolytic streptococci, *Actinomyces* spp., and *Staph. aureus* (Fig. 7.9). Many of the organisms with a higher prevalence in the urethra of sexually active males are members of the vaginal

Table 7.2 Organisms that have been detected in the urethral microbiota of males. This is not a complete list, but is indicative of the diversity of the urethral microbiota.

Acinetobacter calcoaceticus	*Escherichia coli*	*Propionibacterium anaerobius*
Actinomyces meyeri	*Eubacterium* spp.	*Propionibacterium acnes*
Actinomyces viscosus	*Finegoldia magna*	*Proteus mirabilis*
Anaerococcus prevotii	*Fusobacterium mortiferum*	*Ruminococcus productus*
Anaerococcus tetradius	*Fusobacterium nucleatum*	*Staphylococcus aureus*
Arthrobacter sp.	*Fusobacterium varium*	*Staphylococcus auricularis*
Bacteroides capillosus	*Gardnerella vaginalis*	*Staphylococcus epidermidis*
Bacteroides fragilis	*Haemophilus influenzae*	*Staphylococcus haemolyticus*
Bacteroides ovatus	*Haemophilus parainfluenzae*	*Staphylococcus hominis*
Bacteroides ureolyticus	*Lactobacillus* spp.	*Staphylococcus simulans*
Bacteroides vulgatus	*Mobiluncus curtisii*	*Streptococcus agalactiae*
Bifidobacterium spp.	*Mycoplasma genitalium*	*Streptococcus constellatus*
Brevibacterium sp.	*Mycoplasma hominis*	*Streptococcus intermedius*
Clostridium arciformis	*Peptococcus* spp.	*Streptococcus milleri*
Corynebacterium spp.	*Peptoniphilus asaccharolyticus*	*Streptococcus morbillorum*
Corynebacterium jeikeium	*Porphyromonas gingivalis*	*Streptococcus mutans*
Corynebacterium mucifaciens	*Prevotella bivia*	*Streptococcus pyogenes*
Corynebacterium seminale	*Prevotella bucalis*	*Streptococcus salivarius*
Corynebacterium striatum	*Prevotella corporis*	*Streptococcus sanguinis*
Dermabacter hominis	*Prevotella disiens*	*Turicella otitidis*
Eikenella corrodens	*Prevotella intermedia*	*Ureaplasma urealyticum*
Enterococcus faecalis	*Prevotella melaninogenica*	*Ureaplasma parvum*
Enterobacter cloacae	*Prevotella oralis*	*Veillonella alkalescens*
	Prevotella ruminicola	*Veillonella parvula*

microbiota (e.g. *Strep. agalactiae*, *G. vaginalis*, GNAB, and *U. urealyticum*) and, therefore, are likely to have been acquired during sexual intercourse.

At the time of writing (mid-2007), only one culture-independent analysis of the urethral microbiota appears to have been published. In this study, restriction fragment length polymorphism (RFLP) analysis was performed on the cloned products from the PCR amplification of the 16S rRNA genes present in DNA obtained from the urethral microbiota of five healthy individuals. A total of 71 different RFLP types were found in this group of individuals, and the sequences of these corresponded to a wide variety of organisms as well as to several novel phylotypes.

The following list comprises only some of the 71 RFLP types detected – those that were dominant in the samples (13).

- *Pseudomonas gessardii*;
- *Pseudomonas libanensis*;
- *Pseudomonas synxantha*;
- *Pseudomonas fluorescens*;
- *Pseudomonas veronii*;
- *Pseudomonas pickettii*;
- *Pseudomonas* sp. strain ICO38 771;
- *Pseudomonas* sp. clone NBO.1H 852;
- *Streptococcus* sp. oral strain H6 1043.

Many of the sequences corresponded to organisms not previously detected in the urethra and included a number of *Pseudomonas*, *Ralstonia*, and *Sphingomonas* spp., which are frequently found in water samples and so could represent contaminants (derived from water used to wash the urethra prior to sampling) rather than being true members of the urethral microbiota. Of the organisms cultivated from the male urethra (listed

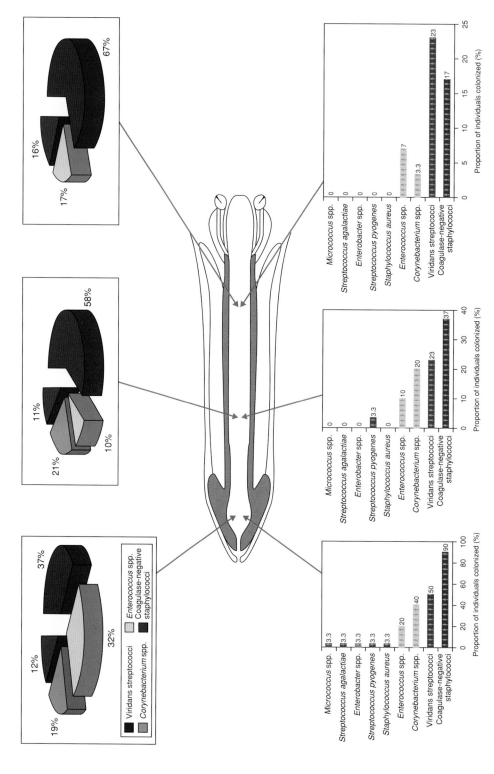

Fig. 7.5 Aerobic and facultatively anaerobic bacteria isolated from three different regions of the urethra of 30 adult males. No attempt was made to detect obligately anaerobic species in the study. The figure shows the composition of the cultivable microbiota at each site and the proportion of males colonized by each organism at each of the three sites (1).

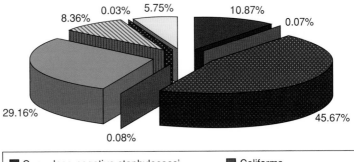
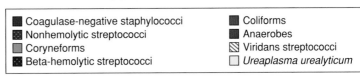

Fig. 7.6 Relative proportions of the various organisms comprising the cultivable urethral microbiota of adult males. Data are derived from an analysis of semen samples from 60 adult males (2).

Coagulase-negative staphylococci
Nonhemolytic streptococci
Coryneforms
Beta-hemolytic streptococci

Coliforms
Anaerobes
Viridans streptococci
Ureaplasma urealyticum

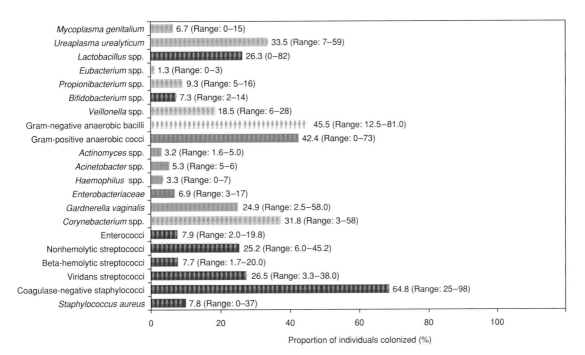

Fig. 7.7 Frequency of detection of various microbes in the urethra of healthy males. The data shown are mean values (and ranges) derived from the results of 11 studies involving 723 individuals (1–11).

in Table 7.2), only a number of *Streptococcus* spp. were frequently detected using this culture-independent approach.

A culture-independent assessment of the prevalence of various *Mollicutes* in the urethra has also been undertaken, and this study found that *Ureaplasma* spp. were considerably more prevalent than a number of *Mycoplasma* spp. (Fig. 7.10). Furthermore, the frequency of detection of the *Ureaplasma* spp. showed a significant correlation with increased sexual activity.

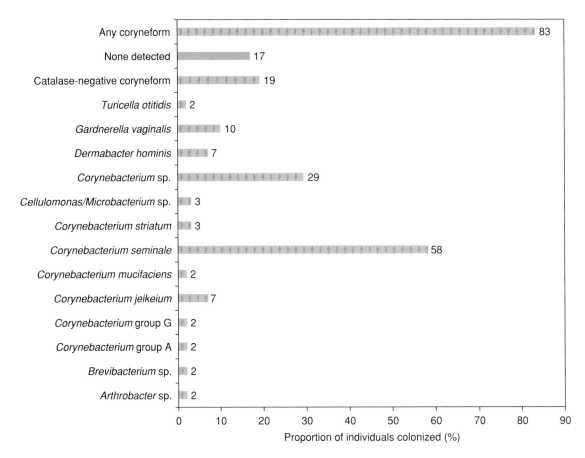

Fig. 7.8 Frequency of isolation of various coryneforms from the urethras of 59 healthy adult males (12)

7.4.3 Microbiota of the glans penis

The microbiota of the glans penis shows tremendous variation among individuals, depending on: (1) whether or not the individual is circumcised; (2) the level of hygiene practiced; and (3) the type and frequency of sexual activity (if any) in which the individual engages. In uncircumcised individuals, heat dissipation, evaporation of fluid, and epithelial desquamation are impaired due to the overlapping foreskin, resulting in the underlying region (the preputial space) having a damp and warm, primarily anaerobic, environment with a neutral to alkaline pH. An important source of nutrients in this region is lipid-rich smegma (see section 7.3). The population density of the microbiota

of the glans penis is higher in uncircumcised than in circumcised individuals, and Gram-negative anaerobes, enterococci, *Staph. aureus*, and facultatively anaerobic Gram-negative bacilli are more frequently detected in the former (Fig. 7.11). *Malassezia* spp. (mainly *Mal. sympodialis* and *Mal. globosa*) and *Can. albicans* are also frequently present. Three *Treponema* spp. are frequently found associated with the smegma of uncircumcised individuals – *T. phagedensis*, *T. refringens*, and *T. minutum*.

In circumcised individuals, the microbiota is less diverse and tends to resemble that of sebaceous gland-rich skin regions – *Propionibacterium acnes*, CNS, and coryneforms are frequently present.

A variety of yeasts may be recovered from the glans penis of uncircumcised individuals (Fig. 7.12). *Malassezia*

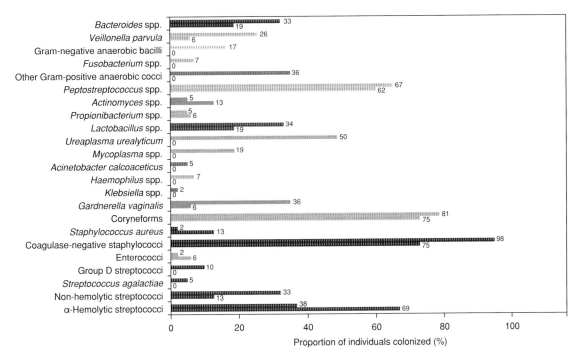

Fig. 7.9 Frequency of isolation of microbes from the urethra of sexually active (upper bars) and sexually inactive (lower bars) males. Figures represent mean values for the 58 individuals studied (7).

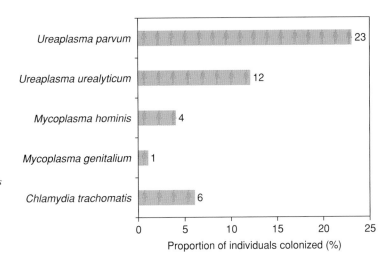

Fig. 7.10 Culture-independent assessment of the prevalence of *Mollicutes* in the urethras of 100 healthy adults. PCR amplification of 16S rRNA genes was carried out on DNA extracted from the samples using primers specific for a number of *Mollicutes* (14).

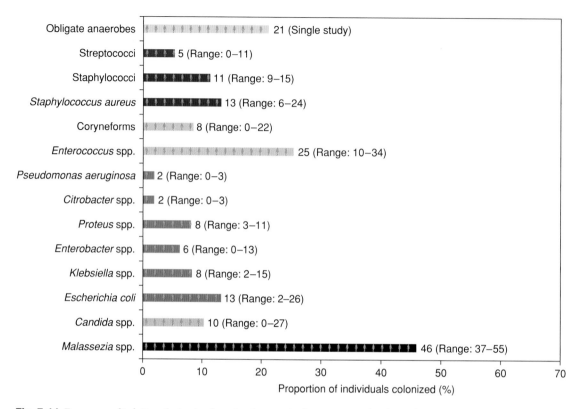

Fig. 7.11 Frequency of isolation of microbes from the glans penis of uncircumcised males. Values shown are the means (and ranges) derived from the results of seven studies involving 542 males of all ages (15–21).

spp. are the yeasts most frequently isolated, and this may be a consequence of the accumulation of smegma under the foreskin – *Malassezia* spp. have a preference for lipid-rich environments (see section 2.4.1.5).

7.4.4 Microbiota of the prostate

Whether or not the prostate of a healthy individual harbors a microbiota continues to be a controversial issue because of the difficulties in interpreting the results obtained from the studies performed. A major problem is obtaining samples free of contaminating urethral organisms. However, even when prostate biopsies have been investigated, the results have not been clear cut – studies using both molecular and culture approaches have reported both the presence and absence of bacteria.

7.5 OVERVIEW OF THE MICROBIOTA OF THE MALE URINARY AND REPRODUCTIVE SYSTEMS

Of the various regions of the urinary and reproductive tracts in males, only the distal portion of the urethra and the glans penis appear to be colonized by microbes. Unlike in females, where microbes are found along the whole length of the urethra, microbes can be detected only in the distal 6 cm of the male urethra. CNS, viridans streptococci, *Corynebacterium* spp., GPAC, and GNAB are the most frequently encountered organisms, and the composition of the microbiota varies along the urethra. Engaging in sexual activity has a dramatic effect on the urethral microbiota – it becomes more complex and contains organisms derived from the vagina. The microbiota of the glans penis differs substantially between circumcised and uncircumcised individuals – the population

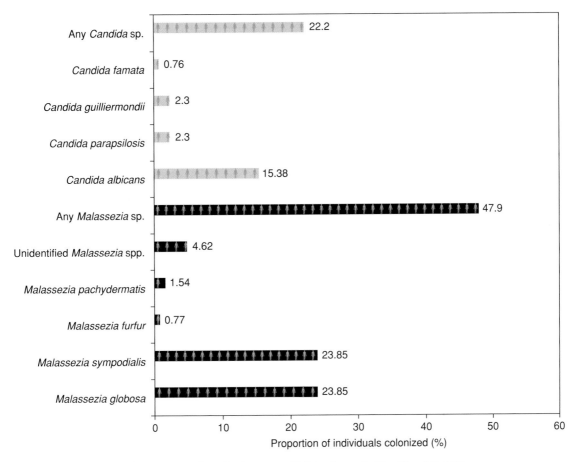

Fig. 7.12 Frequency of isolation of yeasts from the glans penis of 117 uncircumcised healthy adults (16).

density and species diversity being greater in the latter. *Malassezia* spp., anaerobes, and facultative Gram-negative bacilli are frequently encountered in uncircumcised individuals, whereas CNS, *Propionibacterium* spp., and *Corynebacterium* spp. dominate in circumcised individuals. Very few culture-independent studies of the urethral microbiota have been published.

7.6 SOURCES OF DATA USED TO COMPILE FIGURES

1 Montagnini Spaine, D., Mamizuka, E.M., Pereira Cedenho, A. and Srougi, M. (2000) *Urology* 56, 207–10.

2 Cottell, E., Harrison, R.F., McCaffrey, M., Walsh, T., Mallon, E. and Barry-Kinsella, C. (2000) *Fertil Steril* 74, 465–70.

3 Korrovits, P., Punab, M., Turk, S. and Mandar, R. (2006) *Eur Urol* 50, 1338–44.

4 Mazuecos, J., Aznar, J., Rodríguez-Pichardo, A. et al. (1998) *J Eur Acad Dermatol Venereol* 10, 237–42.

5 Bowie, W.R., Wang, S.P., Alexander, E.R., Floyd, J., Forsyth, P.S., Poilock, H.M., Lin, J.S., Buchanan, T.M. and Holmes, K.K. (1977) *J Clin Invest* 59, 735–42.

6 Furness, G., Kamat, K.L.H., Kaminski, Z. and Seebode, J.J. (1973) *Invest Urol* 10, 378–91.

7 Chambers, C.V., Shafer, M.A., Adger, H., Ohm-Smith, M., Millstein, S.G., Irwin, C.E. Jr., Schachter, J. and Sweet, R. (1987) *J Pediatr* 110, 314–21.

8 Iser, P., Read, T.R.H., Tabrizi, S., Bradshaw, C., Lee, D., Horvarth, L., Garland, S., Denham, I. and Fairley, C.K. (2005) *Sex Transm Inf* 81, 163–5.

9 Bradshaw, C.S., Tabrizi, S.N., Read, T.R.H., Garland, S.M., Hopkins, C.A., Moss, L.M. and Fairley, C.K. (2006) *J Infect Dis* 193, 336–45.

10 Levendoglu, F., Ugurlu, H., Ozerbil, O.M., Tuncer, I. and Ural, O. (2004) *Spinal Cord* 42, 106–9.

11 Eggert-Kruse, W., Rohr, G., Strock, W., Pohl, S., Schwalbach, B. and Runnebaum, B. (1995) *Human Reprod Update* 1, 462–78.

12 Turk, S., Korrovits, P., Punab, M. and Mandar, R. (2007) *Int J Androl* 30, 123–8.

13 Riemersma, W.A., van der Schee, C.J., van der Meijden, W.I., Verbrugh, H.A. and van Belkum, A. (2003) *J Clin Microbiol* 41, 1977–86.

14 Takahashi, S., Takeyama, K., Miyamoto, S., Ichihara, K., Maeda, T., Kunishima, Y., Matsukawa, M. and Tsukamoto, T. (2006) *J Infect Chemother* 12, 269–71.

15 Iskit, S., Ilkit, M., Turaç-Biçer, A., Demirhindi, H. and Turker, M. (2006) *Med Mycol* 44, 113–17.

16 Mayser, P., Schütz, M., Schuppe, H.C., Jung, A. and Schill, W.B. (2001) *BJU Int* 88, 554–8.

17 Savaş, C., Çakmak, M., Yorgancıgil, B. and Bezir M. (2000) *Int Urol Nephrol* 32, 85–7.

18 Tokgöz, H., Polat, F., Tan, M.O., Sipahi, B., Sultan, N. and Bozkirli, I. (2005) *Int Urol Nephrol* 37, 101–5.

19 Agartan, C.A., Kaya, D.A., Ozturk, C.E. and Gulcan, A. (2005) *Jpn J Infect Dis* 58, 276–8.

20 Gunsar, C., Kurutepe, S., Alparslan, O., Yilmaz, O., Daglar, Z., Sencan, A., Genc, A., Taneli, C. and Mir, E. (2004) *Urol Int* 72, 212–15.

21 Serour, F., Samra, Z., Kushel, Z., Gorenstein, A. and Dan, M. (1997) *Genitorin Med* 73, 288–90.

7.7 FURTHER READING

Agartan, C.A., Kaya, D.A., Ozturk, C.E. and Gulcan, A. (2005) Is aerobic preputial flora age dependent? *Jpn J Infect Dis* 58, 276–8.

Aridogăn, I.A., Ilkit, M., Izol, V. and Ates, A. (2005) *Malassezia* and *Candida* colonisation on glans penis of circumcised men. *Mycoses* 48, 352–6.

Balat, A., Karakok, M., Guler, E., Ucaner, N. and Kibar, Y. (2007) Local defense systems in the prepuce. *Scand J Urol Nephrol* 21, 1–3.

Bourgeon, F., Evrard, B., Brillard-Bourdet, M., Colleu, D., Jégou, B. and Pineau, C. (2004) Involvement of semenogelin-derived peptides in the antibacterial activity of human seminal plasma. *Biol Reprod* 70, 768–74.

Chambers, C.V., Shafer, M.A., Adger, H., Ohm-Smith, M., Millstein, S.G., Irwin, C.E. Jr., Schachter, J. and Sweet, R. (1987) Microflora of the urethra in adolescent boys: Relationships to sexual activity and nongonococcal urethritis. *J Pediatr* 110, 314–21.

Chowdhury, P., Sacks, S.H. and Sheerin, N.S. (2006) Toll-like receptors TLR2 and TLR4 initiate the innate immune response of the renal tubular epithelium to bacterial products. *Clin Exp Immunol* 145, 346–56.

Chromek, M., Slamová, Z., Bergman, P., Kovács, L., Podracká, L., Ehrén, I., Hökfelt, T., Gudmundsson, G.H., Gallo, R.L., Agerberth, B. and Brauner, A. (2006) The antimicrobial peptide cathelicidin protects the urinary tract against invasive bacterial infection. *Nat Med* 12, 636–41.

Cottell, E., Harrison, R.F., McCaffrey, M., Walsh, T., Mallon, E. and Barry-Kinsella, C. (2000) Are seminal fluid microorganisms of significance or merely contaminants? *Fertil Steril* 74, 465–70.

Deguchi, T., Yoshida, T., Miyazawa, T., Yasuda, M., Tamaki, M., Ishiko, H. and Maeda, S.I. (2004) Association of *Ureaplasma urealyticum* (biovar 2) with nongonococcal urethritis. *Sex Transm Dis* 31, 192–5.

Ganz, T. (2001) Defensins in the urinary tract and other tissues. *J Infect Dis* 183, S41–2.

Hochreiter, W.W., Duncan, J.L. and Schaeffer, A.J. (2000) Evaluation of the bacterial flora of the prostate using a 16S rRNA gene based polymerase chain reaction. *J Urol* 163, 127–30.

Iser, P., Read, T.R.H., Tabrizi, S., Bradshaw, C., Lee, D., Horvarth, L., Garland, S., Denham, I. and Fairley, C.K. (2005) Symptoms of non-gonococcal urethritis in heterosexual men: A case control study. *Sex Transm Inf* 81, 163–5

Iskit, S., Ilkit, M., Turaç-Biçer, A., Demirhindi, H. and Turker, M. (2006) Effect of circumcision on genital colonization of *Malassezia* spp. in a pediatric population. *Med Mycol* 44, 113–17.

Ivanov, Y.B. (2007) Microbiological features of persistent nonspecific urethritis in men. *J Microbiol Immunol Infect* 40, 157–61

Jensen, J.S. (2004) *Mycoplasma genitalium*: The aetiological agent of urethritis and other sexually transmitted diseases. *J Eur Acad Dermatol Venereol* 18, 1–11.

Jensen, J.S., Björnelius, E., Dohn, B. and Lidbrink, P. (2004) Use of TaqMan 5′ nuclease real-time PCR for quantitative detection of *Mycoplasma genitalium* DNA in males with and without urethritis who were attendees at a sexually transmitted disease clinic. *J Clin Microbiol* 42, 683–92.

Korrovits, P., Punab, M., Türk, S. and Mändar, R. (2006) Seminal microflora in asymptomatic inflammatory (NIH IV Category) prostatitis. *Eur Urol* 50, 1338–44.

Lee, J.C., Muller, C.H., Rothman, I., Agnew, K.J., Eschenbach, D., Ciol, M.A., Turner, J.A. and Berger, R.E. (2003) Prostate biopsy culture findings of men with chronic pelvic pain syndrome do not differ from those of healthy controls. *J Urol* 169, 584–8.

Levendoglu, F., Ugurlu, H., Ozerbil, O.M. Tuncer, I. and Ural, O. (2004) Urethral cultures in patients with spinal cord injury. *Spinal Cord* 42, 106–9.

Maeda, S.I., Deguchi, T., Ishiko, H., Matsumoto, T., Naito, S., Kumon, H., Tsukamoto, T., Onodera, S. and Kamidono, S. (2004) Detection of *Mycoplasma genitalium*, *Mycoplasma hominis*, *Ureaplasma parvum* (biovar 1) and *Ureaplasma urealyticum* (biovar 2) in patients with non-gonococcal urethritis using polymerase chain reaction-microtiter plate hybridization. *Int J Urol* 11, 750–4.

Mayser, P., Schütz, M., Schuppe, H.C., Jung, A. and Schill, W.B. (2001) Frequency and spectrum of *Malassezia* yeasts in the area of the prepuce and glans penis. *BJU Int* 88, 554–8.

Mazuecos, J., Aznar, J., Rodriguez-Pichardo, A., Marmesat, F., Borobio, M.V., Perea, E.J. and Camacho, F. (1998) Anaerobic bacteria in men with urethritis. *J Eur Acad Dermatol Venereol* 10, 237–42.

Montagnini Spaine, D., Mamizuka, E.M., Pereira Cedenho, A. and Srougi, M. (2000) Microbiologic aerobic studies on normal male urethra. *Urology* 56, 207–10.

Park, C.H., Valore, E.V., Waring, A.J. and Ganz, T. (2001) Hepcidin, a urinary antimicrobial peptide synthesized in the liver. *J Biol Chem* 276, 7806–10.

Pudney, J. and Anderson, D.J. (1995) Immunobiology of the human penile urethra. *Am J Pathol* 147, 155–65.

Riemersma, W.A., van der Schee, C.J., van der Meijden, W.I., Verbrugh, H.A., and van Belkum, A. (2003) Microbial population diversity in the urethras of healthy males and males suffering from nonchlamydial, nongonococcal urethritis. *J Clin Microbiol* 41, 1977–86.

Saemann, M.D., Weichhart, T., Horl, W.H. and Zlabinger, G.J. (2005) Tamm-Horsfall protein: A multilayered defence molecule against urinary tract infection. *Eur J Clin Invest* 35, 227–35.

Sanocka-Maciejewska, D., Ciupińska, M. and Kurpisz, M. (2005) Bacterial infection and semen quality. *J Reprod Immunol* 67, 51–6.

Serafini-Cessi, F., Monti, A. and Cavallone, D. (2005) *N*-Glycans carried by Tamm-Horsfall glycoprotein have a crucial role in the defense against urinary tract diseases. *Glycoconj J* 22, 383–94.

Sirigu, P., Perra, M.T. and Turno, F. (1995) Immunohistochemical study of secretory IgA in the human male reproductive tract. *Andrologia* 27, 335–9.

Spaine, D.M., Mamizuka, E.M., Cedenho, A.P. and Srougi, M. (2000) Microbiologic aerobic studies on normal male urethra. *Urol* 56, 207–10.

Svenstrup, H.F., Jensen, J.S., Björnelius, E., Lidbrink, P., Birkelund, S. and Christiansen, G. (2005) Development of a quantitative real-time PCR assay for detection of *Mycoplasma genitalium*. *J Clin Microbiol* 43, 3121–8.

Takahashi, S., Takeyama, K., Miyamoto, S., Ichihara, K., Maeda, T., Kunishima, Y., Matsukawa, M. and Tsukamoto, T. (2006) Detection of *Mycoplasma genitalium*, *Mycoplasma hominis*, *Ureaplasma urealyticum*, and *Ureaplasma parvum* DNAs in urine from asymptomatic healthy young Japanese men. *J Infect Chemother* 12, 269–71.

Taylor-Robinson, D., Gilroy, C.B. and Keane, F.E. (2003) Detection of several *Mycoplasma* species at various anatomical sites of homosexual men. *Eur J Clin Microbiol Infect Dis* 22, 291–3.

Tokgöz, H., Polat, F., Tan, M.O., Sipahi, B., Sultan, N. and Bozkirli, I. (2005) Preputial bacterial colonisation in preschool and primary school children. *Int Urol Nephrol* 37, 101–5.

Turk, S., Korrovits, P., Punab, M. and Mandar, R. (2007) Coryneform bacteria in semen of chronic prostatitis patients. *Int J Androl* 30, 123–8

Virecoulon, F., Wallet, F., Fruchart-Flamenbaum, A., Rigot, J.M., Peers, M.C., Mitchell, V. and Courcol, R.J. (2005) Bacterial flora of the low male genital tract in patients consulting for infertility. *Andrologia* 37, 160–5.

Volgmann, T., Ohlinger, R. and Panzig, B. (2005) *Ureaplasma urealyticum* – harmless commensal or underestimated enemy of human reproduction? A review. *Arch Gynecol Obstet* 273, 133–9.

Wijesinha, S., Atkins, B.L., Dudley, N.E. and Tam, P.K.H. (1998) Does circumcision alter the periurethral bacterial flora? *Pediatr Surg Int* 13, 146–8.

Wildeboer-Veloo, A.C., Harmsen, H.J., Welling, G.W. and Degener, J.E. (2007) Development of 16S rRNA-based probes for the identification of Gram-positive anaerobic cocci isolated from human clinical specimens. *Clin Microbiol Infect* 13, 985–92.

Willen, M., Holst, E., Myhre, E.B. and Olsson, A.M. (1996) The bacterial flora of the genitourinary tract in healthy fertile men. *Scand J Urol Nephrol* 30, 387–93.

Woolley, P.D. (2000) Anaerobic bacteria and non-gonococcal urethritis. *Int J STD AIDS* 11, 347–8.

Woolley, P.D., Kinghorn, G.R., Talbot, M.D. and Duerden, B.I. (1990) Microbiological flora in men with non-gonococcal urethritis with particular reference to anaerobic bacteria. *Int J STD AIDS* 1, 122–5.

Yoshida, T., Deguchi, T., Ito, M., Maeda, S., Tamaki, M. and Ishiko, H. (2002) Quantitative detection of *Mycoplasma genitalium* from first-pass urine of men with urethritis and asymptomatic men by real-time PCR. *J Clin Microbiol* 40, 1451–5.

Chapter 8

THE INDIGENOUS MICROBIOTA OF THE ORAL CAVITY

The oral cavity consists of a complex system of tissues and organs whose collective functions are to select food that is acceptable for intake and to process that food into a form suitable for digestion in the rest of the gastrointestinal tract (GIT) (see Chapter 9).

8.1 ANATOMY AND PHYSIOLOGY OF THE ORAL CAVITY

The oral cavity is formed from the cheeks, the hard and soft palates, and the tongue (Fig. 8.1), and is connected to the pharynx by an opening known as the fauces. Its total surface area, including the teeth, is approximately 200 cm², of which tooth surfaces account for approximately 20%.

The cheeks comprise the lateral walls of the oral cavity and terminate in fleshy folds (labia or lips) at the mouth's entrance. The hard and soft palates form the roof of the mouth. The hard palate separates the oral and nasal cavities, while the soft palate separates the oropharynx and nasopharynx. The tongue, which is a muscular structure important in mastication, forms the floor of the oral cavity. The sides and upper surfaces of the tongue are covered in projections (papillae) between which are crypts that can be several millimeters

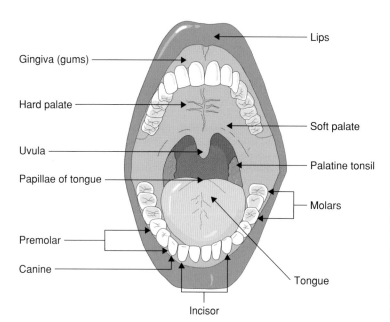

Fig. 8.1 The main anatomical features of the oral cavity. (From http://www.training.seer.cancer.gov/module_anatomy/anatomy_physiology_home.html; funded by the US National Cancer Institute's Surveillance, Epidemiology and End Results (SEER) Program with Emory University, Atlanta, Georgia, USA.)

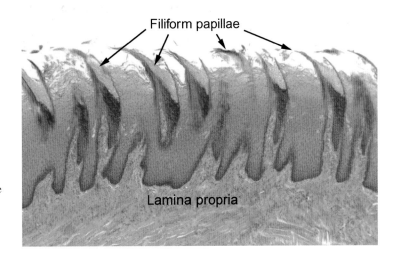

Fig. 8.2 Filiform papillae on the tongue surface. Image © 2006, David King, Southern Illinois University School of Medicine, Springfield and Carbondale, Illinois, USA, used with permission.

deep. Over the anterior two thirds of the tongue (known as the "body"), these projections ("filiform papillae") are conical and are covered by a keratinized epithelium (Fig. 8.2). The epithelium between the papillae is non-keratinized, which makes it flexible and extensible. Other types of papillae (fungiform and circumvallate) are also present on the surface of the tongue. Three

basic types of mucosa are present in the oral cavity – masticatory, lining, and specialized (Fig. 8.3).

The teeth are mineralized structures arising from sockets in the alveolar bone (Fig. 8.4). The upper regions of alveolar bone are covered by the gingivae (gums), which extend a short distance into the socket, resulting in a shallow depression (the gingival crevice

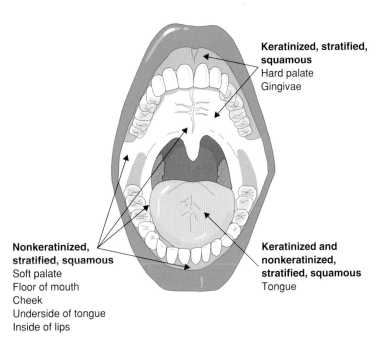

Fig. 8.3 The three types of oral mucosa: (1) masticatory – keratinized, stratified squamous; (2) lining – nonkeratinized, stratified, squamous; and (3) specialized – keratinized and nonkeratinized, stratified, squamous.

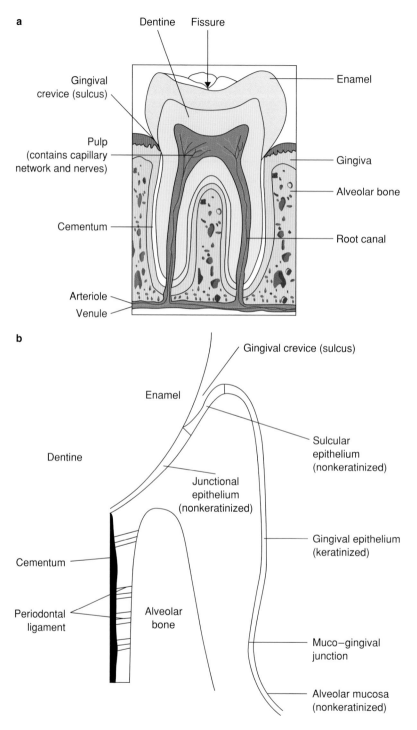

Fig. 8.4 Anatomy of the tooth, together with adjacent and supporting tissues. (a) The gross structure of a tooth and adjacent tissues. (b) Structure of the supporting tissues of the tooth.

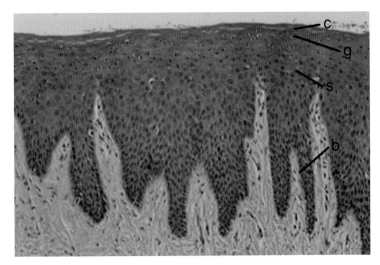

Fig. 8.5 Cross-section through the gingival epithelium (a keratinized, stratified, squamous epithelium), showing the various layers: b, basal layer; s, spinous layer; g, granular layer; c, cornified layer. Reproduced with the permission of Blackwell Publishing Ltd, Oxford, UK, from: Dale, B.A., Kimball, J.R., and Krisanaprakornkit, S. et al. (2001) Localized antimicrobial peptide expression in human gingiva. *J Periodont Res* 36, 285–94.

or sulcus) surrounding the tooth. The depth of the gingival crevice is usually no more than 2 mm. A serum-like exudate (gingival crevicular fluid [GCF]) continually enters the oral cavity from the gingival crevice. The gingivae are covered by a stratified, squamous, keratinized epithelium, which merges with the nonkeratinized "sulcular" epithelium in the gingival crevice (Fig. 8.5).

Teeth consist mainly of a bone-like material known as dentine, within which is a pulp cavity containing blood vessels, lymphatics, and nerves. The dentine is covered by other mineralized tissues – enamel on the tooth region that protrudes into the oral cavity (the crown) and cementum in the case of the root. Dentine consists mainly of a mineral, hydroxyapatite, together with organic matter (21%) and water (10%). Enamel is the hardest substance in the body and forms a layer of varying thickness, which protects the tooth from the wear associated with chewing and also from acids in the diet. It has a high mineral content (96–97%), the rest being water and organic matter. The mineral content consists mainly of hydroxyapatite and fluorapatite, with the latter rendering enamel more resistant to acid attack.

Saliva is a dilute, aqueous fluid which is involved in lubrication, digestion, temperature regulation, and host defense. It also acts as a buffer, thereby preventing extreme pH fluctuations and, because of its high content of calcium and phosphate ions, is involved in tooth remineralization. The rate of secretion of saliva and its composition are inter-related and are highly variable, being influenced by the time of day, the presence of food

or objects in the mouth, the smell or sight of food, as well as by a variety of other stimuli. Two types of saliva are recognized – that produced in the absence ("resting saliva") and in the presence ("stimulated saliva") of any stimulus. The resting rate of salivary flow is approximately 0.3 ml per minute, and this increases to 2.5–5.0 ml per minute during stimulation. The average daily production of saliva is 700–800 ml, and its composition is described in section 8.3.2.

8.2 ANTIMICROBIAL DEFENSE SYSTEMS OF THE ORAL CAVITY

Considerable mechanical and hydrodynamic forces are produced within the oral cavity as a result of biting, chewing, tongue movements, and salivary flow. These forces dislodge microbes from oral surfaces and so discourage colonization. Furthermore, mucins and various proteins (statherins, histadine-rich proteins, and proline-rich proteins [PRPs]) in saliva bind to microbes and/or encourage their co-aggregation, so that they remain in suspension in saliva and are eventually swallowed. Swallowing delivers the dislodged microbes to the stomach, where most are killed by the low pH. Approximately 8×10^{10} microbes are removed each day from the oral cavity in this way.

Microbial colonization of mucosal surfaces in the oral cavity is hindered by the mechanical and hydrodynamic forces mentioned above, by the presence of a mucous gel coating, and by the shedding of superficial

Table 8.1 Antimicrobial defense mechanisms operating in the oral cavity.

Defense mechanism	Function/effect
Mechanical forces due to chewing and talking	Microbes dislodged from oral surfaces and swallowed
Hydrodynamic forces due to salivary flow	Microbes dislodged from oral surfaces and swallowed
Mucous layer on mucosal surfaces	Reduces microbial colonization of mucosal surfaces
Shedding of epithelial cells	Shed cells carry adherent microbes with them and are swallowed
Lysozyme	Antibacterial; agglutinates bacteria, thereby keeping them suspended in saliva, which is then swallowed
Lactoferrin, transferrin	Deprive microbes of iron
Other antimicrobial peptides and proteins (apo-lactoferrin, lactoperoxidase, SLPI, HBD-1, HBD-2, HBD-3, HNP 1–4), LL-37, histatins, calprotectin, adrenomedullin	Kill or inhibit the growth of microbes; neutralize pro-inflammatory activities of bacterial components
TLR2, TLR4, NOD1, NOD2	Activation induces expression of pro-inflammatory cytokines, chemokines, and antimicrobial peptides
sIgA	Prevents microbial adhesion; agglutinates microbes, thereby keeping them suspended in saliva, which is then swallowed
IgG and IgM	Prevent microbial adhesion; activate complement; function as opsonins
Complement	Microbial lysis and opsonization
Phagocytes	Kill microbes

cells. Replacement of the outermost cells of the oral mucosa occurs every 24–48 h. Uniquely, however, the mouth also has nonshedding surfaces, the teeth, available for microbial colonization, and most of the microbes colonizing the oral cavity are found on tooth surfaces, thereby illustrating the effectiveness of epithelial cell shedding as a defense mechanism.

Saliva contains a range of antimicrobial compounds (Table 8.1) including lysozyme (11 mg/100 ml), lactoperoxidase, secretory leukocyte protease inhibitor (SLPI), lactoferrin, transferrin, and a variety of antimicrobial peptides produced by oral tissues (Fig. 8.6). The effect of some of these on one member of the oral microbiota is shown in Fig. 8.7. Oral epithelial cells also express a number of pattern-recognition molecules including Toll-like receptor (TLR) 2, TLR4, NOD1, and NOD2 (Fig. 8.8), which recognize bacterial components such as lipopolysaccharide (LPS), lipopeptides, and peptidoglycan fragments, thereby resulting in the upregulation of HBD-2 expression as well as in the production of pro-inflammatory cytokines (see section 1.2.4).

The continuous flow of GCF into the oral cavity brings with it lymphocytes, monocytes, polymorphonuclear leukocytes (PMNs), IgG, and IgM. These are present at relatively high levels in the gingival crevice but, because of dilution by saliva, only at low concentrations in the rest of the oral cavity. IgG is an important effector molecule of the acquired immune defense system in the gingival crevice and within gingival tissues where it is involved in opsonization, complement activation, neutralization of microbial enzymes and toxins, and the prevention of microbial adhesion. However, while opsonization and complement activation are effective at killing planktonic microbes, they are less effective against microbes in biofilms (see section 1.1.2.3). Once a biofilm has formed in the gingival crevice or on supragingival surfaces, access of complement components, antibodies, and PMNs to the organisms within the biofilm will be impaired by the biofilm matrix.

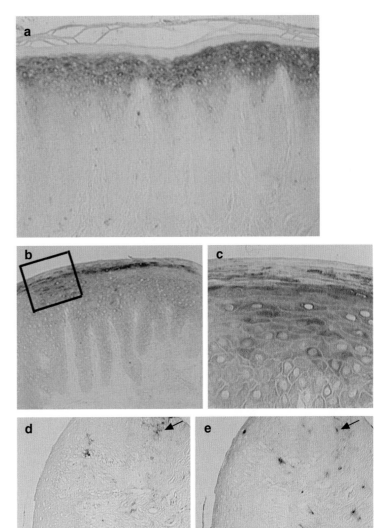

Fig. 8.6 Immunohistochemical detection of the expression of various antimicrobial peptides in gingival tissues. Brown staining areas denote the expression of (a) HBD-1; (b, c) HBD-2; (d) HNP 1–3 (human neutrophil peptides 1–3); and (e) LL-37. (c) Higher-magnification image of the boxed area in (b). The arrows in (d) and (e) indicate some of the cells that are expressing HNP 1–3 and LL-37, respectively. Original magnifications: (a) ×10, (b) ×10, (c) ×40, (d) ×4, and (e) ×4. Reproduced with the permission of Blackwell Publishing Ltd, Oxford, UK, from: Dale, B.A., Kimball, J.R., Krisanaprakornkit, S. et al. (2001) Localized antimicrobial peptide expression in human gingiva. *J Periodont Res* 36, 285–94.

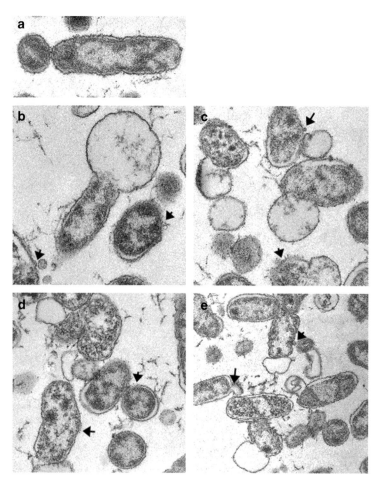

Fig. 8.7 Effect of antimicrobial peptides on the oral bacterium *Ag. actinomycetemcomitans*. Electron micrographs of thin sections of the bacterium following exposure to (a) phosphate-buffered saline, (b) HBD-1, (c) HBD-2, (d) HBD-3, or (e) LL-37. Typical membrane perforation is shown by arrowheads. Reprinted from Ouhara, K., Komatsuzawa, H., Yamada, S., Shiba, H., Fujiwara, T., Ohara, M., Sayama, K., Hashimoto, K., Kurihara, H. and Sugai, M. (2005) *J Antimicrob Chemother* 55, 888–96 by permission of the British Society for Antimicrobial Chemotherapy, Birmingham, UK.

8.3 ENVIRONMENTAL DETERMINANTS AT THE VARIOUS SITES WITHIN THE ORAL CAVITY

8.3.1 Mechanical determinants

Nowhere else in the body are microbes subjected to such extreme mechanical forces as those generated in the oral cavity as a result of biting, chewing, tongue movements, and salivary flow. The ability to adhere to some surface in the oral cavity in the face of such powerful removal forces is, therefore, of prime importance for any potential microbial colonizer. Nevertheless, certain sites are protected from these forces, e.g. the gaps between adjacent teeth (known as "approximal" regions), the naturally occurring fissures on the occlusal (i.e. biting) surfaces of pre-molar and molar teeth, the gingival crevice, and the gaps between papillae on the tongue surface. These regions tend to have higher population densities, as well as a more varied microbiota, than exposed sites such as the smooth buccal (i.e. facing the cheek) and lingual (i.e. facing the inside) surfaces of teeth.

8.3.2 Nutritional determinants

The main sources of nutrients for oral microbes are saliva, compounds produced by host cells, GCF, products of microbial metabolism, and constituents of the host's diet. Oral microbes can obtain a range of nutrients from food ingested by the host, and some

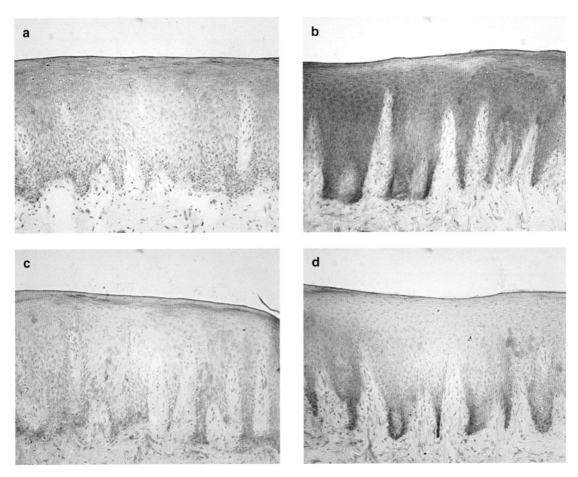

Fig. 8.8 Expression of TLR2, TLR4, NOD1, and NOD2 in human gingival epithelial tissues. Cryosections of healthy gingival tissues were stained brown with (a) anti-NOD1 antibody, (b) anti-NOD2 antibody, (c) anti-TLR2 antibody, and (d) anti-TLR4 antibody. The sections were counterstained blue with hematoxylin. Although all four pattern-recognition molecules were detected, expression of NOD1 and NOD2 was much greater than that of TLR2 and TLR4. Reproduced with the permission of the International Association for Dental Research, Alexandria, Virginia, USA, from: Sugawara, Y., Uehara, A., Fujimoto, Y., Kusumoto, S., Fukase, K., Shibata, K., Sugawara, S., Sasano, T. and Takada, H. (2006) Toll-like receptors, NOD1, and NOD2 in oral epithelial cells. *J Dent Res* 85, 524–9.

of these, such as easily fermentable carbohydrates, can have a profound effect on the composition of the oral microbiota.

All surfaces within the oral cavity are covered with a film of saliva, which is between 8 and 40 μm in thickness. Such a thin coating facilitates rapid transfer of nutrients, gases, etc. that are present in saliva to those microbes adhering to oral surfaces. Therefore, although saliva is continually being removed from the oral cavity, the mucosal and tooth surfaces over which it flows

are continually being supplied with a wide range of nutrients. The main constituents of saliva are proteins and glycoproteins (Table 8.2). Salivary mucins comprise approximately 25% of the total protein content, while amylase, IgA, and lysozyme are also major constituents.

The main salivary mucins are MG1 (MUC5B) and MG2 (MUC7). MG1 is highly glycosylated and forms a gel that covers and protects the mucosal surfaces from desiccation, mechanical damage, and microbes. MG1

Table 8.2 The main constituents of human saliva and GCF.

Substance	Concentration (mg/100 ml)	
	Saliva	Gingival crevicular fluid
Protein	140–640	Total: 6800–9200; albumin: 4000; transferrin: 300; complement components: 160; lactoferrin: 60; α_1-antitrypsin: 300; α_2-macroglobulin: 270; haptoglobin: 210
Amylase	38	NDA
Lysozyme	11	NDA
Glycoprotein sugars	110–300	NDA
Free carbohydrate	2	130
Amino acids	3.4–4.8	NDA
IgA	19	160
IgG	1.4	1230
IgM	0.2	120
Lipid	2–3	NDA
Peroxidase	0.3	NDA
Glucose	1.0	10
Urea	13	NDA
Uric acid	3	NDA
Citrate	0.2–2.0	NDA
cAMP	50	NDA
Potassium	80	0.02–0.27
Chloride	100	NDA
Bicarbonate	200	NDA
Phosphate	12	NDA
Sodium	60	0.24–0.51
Calcium	6	0.02–0.09
Thiocyanate	1–3	NDA
Ammonia	3	NDA
Vitamin C	NDA	NDA
B vitamins	NDA	NDA

NDA, no data available.

also binds to enamel and forms complexes with other salivary proteins, and therefore is a major constituent of the acquired enamel pellicle (AEP; section 8.4.2.1), a complex layer of adsorbed molecules which protects the enamel surface from acid dissolution and also serves as a substratum for microbial adhesion. The presence of calcium and phosphate ions at supersaturation concentrations is important for maintaining the integrity of tooth enamel. In addition to the mucins secreted by the salivary glands, the oral mucosa itself produces MUC1 and MUC4, which remain attached to the epithelial cell membrane where, like MG1, they have a protective function.

The presence of only low concentrations of free carbohydrates and amino acids in saliva means that, unless such compounds are available from other microbes or from the host's diet, oral microbes must obtain most of their carbon and energy from host proteins and glycoproteins. Collectively, oral microbes produce a wide range of proteases, peptidases, sialidases, and glycosidases, which enable them to hydrolyze a variety of proteins and glycoproteins (Table 8.3).

The urea, ammonia, and amino acids present in saliva, together with amino acids liberated from proteins and glycoproteins, constitute the main nitrogen sources for oral microbes.

Table 8.3 Oral species able to contribute to the degradation of glycoproteins.

Organism	Sialidase	Glycosidase	Protease/peptidase
Ag. actinomycetemcomitans	NDA	NDA	Yes
A. naeslundii genospecies 1	Yes	Yes	No
A. naeslundii genospecies 2	Yes	Yes	Yes
Bacteroides capillosus	Yes	NDA	NDA
Bifidobacterium spp.	Yes	Yes	Yes
Cap. sputigena	Yes	Yes	Yes
Cap. gingivalis	Yes	Yes	Yes
Cap. ochracea	Yes	Yes	Yes
Eub. nodatum	NDA	NDA	Yes
H. aphrophilus	NDA	Yes	NDA
P. acnes	Yes	NDA	Yes
Por. asaccharolytica	NDA	Yes	Yes
Por. endodontalis	NDA	No	Yes
Prev. denticola	Yes	Yes	Yes
Prev. loescheii	Yes	Yes	Yes
Prev. buccalis	Yes	Yes	NDA
Prev. buccae	Yes	Yes	NDA
Prevotella disiens	Yes	No	Yes
Prev. intermedia	NDA	Yes	Yes
Prev. nigrescens	NDA	NDA	Yes
Prev. oralis	Yes	Yes	Yes
Strep. oralis	Yes	Yes	Yes
Strep. intermedius	Yes	Yes	NDA
Strep. mitis	Yes	Yes	Yes
Strep. sanguinis	NDA	Yes	Yes
Streptococcus mitior	NDA	Yes	Yes
Strep. mutans	NDA	NDA	Yes
Streptococcus sobrinus	NDA	NDA	Yes
Streptococcus anginosus	NDA	NDA	Yes
Streptococcus pneumoniae	Yes	Yes	Yes
Sel. sputigena	NDA	Yes	NDA
T. denticola	Yes	NDA	Yes
T. vincentii	NDA	NDA	Yes
Tan. forsythensis	Yes	Yes	Yes

Yes, some strains produce the enzyme; no, most strains do not produce the enzyme; NDA, no data available.

GCF is a transudate that flows continuously from the gingival crevice at rates of between 3 and 137 µl/h depending on the oral health of the individual – higher rates being found in individuals with gingivitis or periodontitis. As most of the adult population has some degree of gingival inflammation, GCF constitutes a regular and abundant source of nutrients (see Table 8.2) to microbes inhabiting the gingival crevice. The fluid is usually alkaline due to the presence of urea, and its pH ranges from 7.5 to 8.0.

8.3.3 Physicochemical determinants

The temperature of the oral cavity varies from site to site, but falls within the range 33–37°C, and so is suitable for the growth of mesophiles. However, dramatic fluctuations in temperature occur during the consumption of cold or hot food and drinks, and the temperature of the subgingival regions can be as high as 39°C in individuals with periodontitis.

The main determinant of pH in the oral cavity is

saliva – except within the gingival crevice where GCF is more important. However, microbial activity can have a dramatic effect on the pH at individual oral sites. The pH of saliva varies with its rate of secretion and, in general, resting saliva (pH 6.5–6.9) is more acidic than stimulated saliva (pH 7.0–7.5). Because individuals produce stimulated saliva for most of the time, the pH of saliva is usually approximately neutral – and this is suitable for the growth of many types of microbe. Saliva has good buffering capacity because of the presence of proteins, amino acids, phosphates, and bicarbonates. However, owing to acid production by microbes when the host's diet is rich in easily fermentable carbohydrates such as sucrose, it is unable to prevent dramatic falls in pH. The frequent consumption of sucrose can select for colonization by acidophilic and/or aciduric species which, because of their continuous acid production, can give rise to dental caries. The fact that the bacteria on the tooth surface are growing as biofilms (section 1.1.2.3), which can partially exclude the buffering components of saliva and can also impede the removal of bacterially produced acids, exacerbates this problem.

The main determinant of the pH of the gingival crevice is GCF, which has a pH of between 7.5 and 8.0. In individuals with periodontitis, the pH of subgingival dental plaque ranges from 7.5 to 8.5.

Although all regions of the oral cavity are exposed to a plentiful supply of air, a number of anatomical and microbial factors render many sites suitable for the growth of microaerophiles and obligate anaerobes. Anatomical regions to which there is restricted access of oxygen-rich saliva, and which therefore have a low oxygen content and/or low redox potential, include gaps between adjacent teeth, regions between the tongue papillae, and the gingival crevice (the oxygen content of GCF is approximately 30 mmHg). The growth of obligate aerobes and/or facultative anaerobes at these sites can rapidly exhaust the available oxygen, resulting in the production of reducing compounds, thereby creating conditions suitable for colonization by capnophiles, microaerophiles, and obligate anaerobes. Another important factor that is responsible for a range of habitats with different oxygen and redox conditions is the biofilm mode of growth. Within a biofilm, gradients of oxygen and redox potential form as a result of microbial consumption of oxygen and inadequate replenishment by saliva. Hence, habitats suitable for the growth of aerobes, facultative anaerobes, capnophiles, microaerophiles, and obligate anaerobes are likely to be present at different

locations within the biofilm. Gradients are also produced within the biofilm with respect to the concentration of nutrients supplied by saliva and GCF. The concentration of endogenous nutrients (i.e. excreted products from microbes within the biofilm) as well as the pH (due to microbial acid production) will also vary at different locations within the biofilm (section 1.1.2.3).

The oxygen content and redox potential of various oral sites are shown in Fig. 8.9. Regions with the lowest oxygen concentrations are the tongue and the gingival crevice, and obligate anaerobes are particularly abundant within these regions. Sites with the lowest redox potentials are those at the gingiva–tooth interface, particularly between adjacent teeth, and these regions usually have high proportions of obligate anaerobes. All of the sites shown in Fig. 8.9b, other than saliva, have an Eh lower than the +200 mV found in aerobically metabolizing host tissues. Once plaque accumulates, oxygen consumption and the production of reducing compounds can result in dramatic reductions in Eh. Within 7 days, the Eh of plaque can be as low as −150 mV (Fig. 8.10). Even greater reducing conditions can exist in the subgingival plaques present in the periodontal pockets of individuals suffering from periodontitis, where the Eh has been reported to reach −300 mV.

The main features of the environment of each of the main oral sites are summarized in Fig. 8.11.

8.4 THE INDIGENOUS MICROBIOTA OF THE ORAL CAVITY

From the previous sections, it can be appreciated that the oral cavity provides a wide variety of habitats, each with a distinct set of environmental selection factors. Each of these sites supports a microbiota which, in many cases, is very complex, which means that the overall species diversity within the oral cavity is very high. Between 200 and 300 microbial species have been cultured from the oral cavity of humans, and a further 400 phylotypes have been detected by a variety of culture-independent analyses. Of these 700 possible colonizers of the oral cavity, a more limited number, between 100 and 200, are estimated to be present in the oral cavity of a healthy individual at any one time. The proportion of the microbiota that has been cultivated (approximately 50%) is higher for the oral cavity than for many other body sites. This is likely to be due to two main factors: (1) oral bacteria are responsible for two of the most common infections of humans – caries

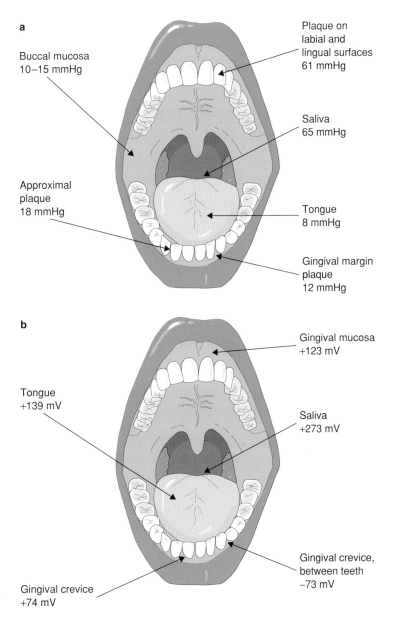

a

Buccal mucosa
10–15 mmHg

Plaque on
labial and
lingual surfaces
61 mmHg

Saliva
65 mmHg

Approximal
plaque
18 mmHg

Tongue
8 mmHg

Gingival margin
plaque
12 mmHg

b

Tongue
+139 mV

Gingival mucosa
+123 mV

Saliva
+273 mV

Gingival crevice
+74 mV

Gingival crevice,
between teeth
–73 mV

Fig. 8.9 Oxygen content and redox potential of various sites within the oral cavity. (a) Partial pressure of oxygen and (b) redox potential.

and periodontal diseases – and the etiologies of these diseases have been extensively investigated for many years; and (2) samples for analysis are very easy to obtain without causing discomfort or embarrassment to the individual. One unique feature of the oral cavity, which distinguishes it from other body sites, is the presence of nonshedding surfaces (i.e. the enamel and

cementum of the teeth) on which microbes can accumulate and form biofilms known as dental plaques. Biofilm formation is a characteristic feature of the organisms inhabiting the oral cavity, and the structure of dental plaques will be described later (section 8.4.2.1).

Because most regions within the oral cavity are subjected to substantial mechanical and hydrodynamic

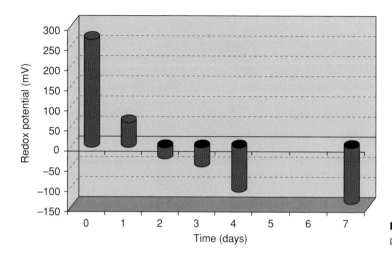

Fig. 8.10 Changes in redox potential as plaque develops on a tooth surface.

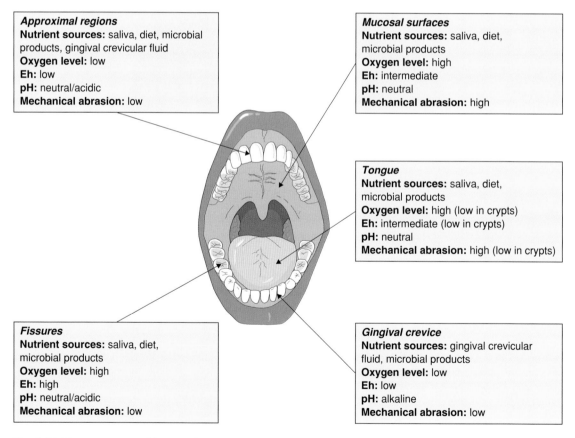

Approximal regions
Nutrient sources: saliva, diet, microbial products, gingival crevicular fluid
Oxygen level: low
Eh: low
pH: neutral/acidic
Mechanical abrasion: low

Mucosal surfaces
Nutrient sources: saliva, diet, microbial products
Oxygen level: high
Eh: intermediate
pH: neutral
Mechanical abrasion: high

Tongue
Nutrient sources: saliva, diet, microbial products
Oxygen level: high (low in crypts)
Eh: intermediate (low in crypts)
pH: neutral
Mechanical abrasion: high (low in crypts)

Fissures
Nutrient sources: saliva, diet, microbial products
Oxygen level: high
Eh: high
pH: neutral/acidic
Mechanical abrasion: low

Gingival crevice
Nutrient sources: gingival crevicular fluid, microbial products
Oxygen level: low
Eh: low
pH: alkaline
Mechanical abrasion: low

Fig. 8.11 Main environmental features of important habitats within the oral cavity.

forces, the ability to adhere to host tissues or to other microbes is an important prerequisite of any successful colonizer. Bacterial adhesion, therefore, will be an important theme in this chapter.

8.4.1 Members of the oral microbiota

The predominant genera detected in the oral cavity include *Streptococcus, Actinomyces, Neisseria, Haemophilus, Eubacterium, Lactobacillus, Fusobacterium, Abiotrophia, Atopobium, Granulicatella, Gemella, Rothia, Veillonella, Porphyromonas, Prevotella, Treponema, Bifidobacterium, Capnocytophaga, Eikenella, Leptotrichia, Peptostreptococcus, Mycoplasma, Bifidobacterium, Staphylococcus, Propionibacterium,* and *Kingella.* Most of these genera have been described in previous chapters, and this section will briefly outline the main characteristics of those that have not.

8.4.1.1 Oral streptococci and related Gram-positive cocci

The genus *Streptococcus* has been described in section 4.4.1.1, and this section will deal only with the oral streptococci. These organisms can be divided into four main groups (Table 8.4). Oral streptococci are among the dominant members of the oral microbiota and are found on mucosal and tooth surfaces. Table 8.5 lists some of the many adhesins that have been identified in these organisms.

Many species synthesize polysaccharides, which form the principal components of the matrix of dental plaque and can also act as carbon and energy sources for other oral bacteria. They are acidogenic, and many species are acidophilic. The acid produced by these organisms can demineralize enamel, cementum, and dentine, resulting in the formation of a caries lesion. Some species (e.g. *Streptococcus salivarius* and *Streptococcus vestibularis*) are ureolytic, and the ammonia produced, as well as serving as a nitrogen source for other oral species, can also neutralize the acids produced by glycolysis, thereby raising the pH and reducing the risk of dental caries. *Streptococcus mitis* secretes a surfactant that can inhibit the adhesion of some organisms to the AEP. Many oral streptococci secrete bacteriocins that are active mainly against Gram-positive species. Members of the mutans group can cause caries, while mitis group streptococci are frequently associated with endocarditis, and anginosus group species can cause abscesses. The 2.03-Mbp genome of *Streptococcus mutans* has been sequenced (http://cmr.tigr.org/tigr-scripts/CMR/GenomePage.cgi?org=ntsm02).

Organisms with complex nutritional requirements (e.g. a requirement for vitamin B_6), previously known as nutritionally variant streptococci, have been assigned to two new genera – *Abiotrophia* and *Granulicatella*. The main species found in the oral cavity are *Abiotrophia defective, Granulicatella elegans,* and *Granulicatella adiacens* – the latter being encountered most frequently.

8.4.1.2 *Gemella* spp.

Gemella spp. are facultatively anaerobic, nonsporing, nonmotile Gram-positive cocci that occur in pairs, tetrads, small clusters, and small chains. They are catalase-negative. The genus consists of four species, two of which are found in the human oral cavity – *Gemella morbillorum* and *Gemella haemolysans.* The G+C content of their DNA is 31–33.5 mol%. They produce acid from glucose and from some other carbohydrates. Under anaerobic conditions, the main end products are lactic,

Table 8.4 The four main groups of viridans streptococci found in the oral cavity.

Group	Main habitat	Examples
Mitis group	Dental plaque	*Strep. mitis, Strep. gordonii, Strep. oralis, Strep. sanguinis* (formerly *Strep. sanguis*), *Strep. parasanguinis* (formerly *Strep. parasanguis*), *Strep. crista, Strep. infantis, Strep. peroris*
Mutans group	Dental plaque; caries lesions	*Strep. mutans, Strep. sobrinus*
Salivarius group	Mucosal surfaces	*Strep. salivarius, Strep. vestibularis*
Anginosus group	Gingival crevice	*Strep. anginosus, Strep. constellatus, Strep. intermedius*

Table 8.5 Oral bacterial adhesins and their receptors. Reprinted from: Rosan, B. and Lamont, R.J. (2000) Dental plaque formation. *Microbes Infect* 2, 1599–607. Copyright 2000, with permission from Elsevier, Amsterdam, The Netherlands.

AEP receptor(s)	Adhesin(s)	Approximate size (kDa)	Species
Parotid salivary agglutinin, salivary glycoproteins, proline-rich proteins, collagen	Antigen I/II family (Ag I/II, AgB, P1 [SpaP], PAc, Sr, SpaA, PAg, SspA, SspB, SoaA)	160–175	*Strep. mutans, Strep. sobrinus, Strep. gordonii, Strep. oralis, Strep. intermedius*
Salivary components in pellicle	Lral family (FimA, SsaB)	35	*Strep. parasanguinis, Strep. sanguinis*
Salivary components in pellicle	Fap1	200	*Strep. parasanguinis*
α-Amylase	Amylase-binding proteins	82, 65, 20, 15, 12,	*Strep. gordonii, Strep. mitis, Strep. crista*
Submandibular salivary protein (73 kDa)	Antigen complex	80, 62, 52	*Strep. gordonii*
Salivary glycoprotein presenting *N*-acetylneuraminic acid	Surface lectins	96, 70, 65	*Strep. oralis, Strep. mitis*
Proline-rich proteins, statherin	Type 1 fimbriae-associated protein	—	*A. naeslundii*

acetic, and formic acids, while aerobic growth results in the production of acetate and CO_2.

8.4.1.3 *Actinomyces* spp.

The genus *Actinomyces* consists of Gram-positive rods, which may be short, filamentous, pleomorphic, or branching. The main characteristics of *Actinomyces* spp. are as follows:
- Gram-positive bacilli;
- Nonmotile;
- Nonsporing;
- G+C content of their DNA is 55–71 mol%;
- Most are facultative anaerobes, some are obligate anaerobes;
- Growth is stimulated by high concentrations of CO_2;
- Growth over the range 20–45°C;
- Optimum growth temperature is 35–37°C;
- Aciduric;
- Nutritionally exacting;
- Use of carbohydrates as a carbon and energy source;
- Can contribute to mucin degradation.

At least 24 species have been isolated from humans, of which eight are found in the oral cavity – *A. naeslundii* genospecies 1, *A. naeslundii* genospecies 2 (includes strains formerly classified as *A. viscosus*), *A. israelii, A. gerencseriae, A. georgiae, A. odontolyticus, A. graevenitzii,* and *A. meyeri*. Fermentable carbohydrates are their preferred source of carbon and energy, the main end products being a variety of acids (succinic, lactic, formic, and acetic acids), which vary depending on whether growth is aerobic or anaerobic. They are generally not proteolytic but do produce sialidases and glycosidases (α- and β-galactosidases, β-glucuronidases, α- and β-glucosidases, β-xylosidases, and α-fucosidases), and so can contribute to mucin degradation. *A. naeslundii* genospecies 1 and 2 and *A. meyeri* are ureolytic. The 3.04-Mbp genome of *A. naeslundii* has been sequenced (http://cmr.tigr.org/tigr-scripts/CMR/GenomePage.cgi?org=gan).

A. naeslundii genospecies 1 and 2 are found mainly on teeth and on the buccal mucosa, whereas *A. odontolyticus* preferentially colonizes the tongue. *Actinomyces* spp. can adhere to both tooth and mucosal surfaces, mainly by means of fimbriae. Two types of fimbriae have been identified – type 1 and 2: the first of these mediates binding to PRPs present in the AEP and is therefore involved in dental plaque formation. Type 2 fimbriae bind to glycoproteins present on epithelial cells, enabling adhesion of the organism to the oral mucosa. They also recognize carbohydrate receptors on the surfaces of many viridans streptococci and other oral bacteria and are responsible for co-aggregation with such species. Many *Actinomyces* spp. also produce polysaccharides (e.g. levan, dextran, or glycogen) which mediate adhesion to hydroxyapatite and contribute to the plaque matrix.

A. israelii is responsible for actinomycosis, while other *Actinomyces* spp. are associated with caries (particularly root surface caries), gingivitis, and periodontitis.

8.4.1.4 *Rothia dentocariosa*

R. dentocariosa is a non-spore-forming, nonhemolytic, nonmotile, facultatively anaerobic Gram-positive bacillus. Its morphology varies from coccoid to diphtheroid to filamentous. The G+C content of its DNA is 47–53 mol%. The organism is fermentative, with the main end products of carbohydrate metabolism being lactic and acetic acids. Most strains are catalase-positive and urease-negative.

8.4.1.5 *Veillonella* spp.

Veillonella spp. are Gram-negative cocci, which usually occur in pairs, clusters, or chains. The main characteristics of *Veillonella* spp. are as follows:
- Gram-negative cocci;
- Obligate anaerobes;
- Nonmotile;
- G+C content of their DNA is 40.3–44.4 mol%;
- Growth over the temperature range 24–40°C;
- Optimum growth at 30–37°C;
- Aciduric, grow well at a pH of 4.5;
- Unable to use carbohydrates as an energy source;
- Co-aggregation with a wide variety of oral species.

Seven species are recognized, of which *V. parvula*, *V. atypica*, and *V. dispar* are members of the oral microbiota. They cannot use carbohydrates as an energy source but use lactate, pyruvate, fumarate, malate, and some purines for this purpose and generate propionate, acetate, and hydrogen as metabolic end products. *V. atypica* produces a soluble chemical that induces amylase expression in *Streptococcus gordonii*, thereby increasing carbohydrate fermentation and lactic acid production by the organism. While they demonstrate only weak adhesion to hydroxyapatite or to clean enamel, *Veillonella* spp. are able to co-aggregate with a variety of oral bacteria including species belonging to the genera *Streptococcus*, *Fusobacterium*, *Gemella*, *Eubacterium*, *Actinomyces*, *Neisseria*, *Rothia*, and *Propionibacterium*. Co-aggregation appears to be mediated by either a 45-kDa surface protein or another, as yet unidentified, molecule. Because of their ability to co-aggregate, *Veillonella* spp. are key organisms in dental plaque formation and constitute important "secondary colonizers" subsequent to initial adhesion of "primary colonizers" (e.g. certain streptococci) to the AEP coating on tooth surfaces. They are found at all sites within the oral cavity, but *V. atypica* and *V. dispar* tend to be found on the tongue and in saliva, while *V. parvula* is encountered more frequently in dental plaque. They are rarely responsible for infections in humans.

8.4.1.6 Anaerobic and microaerophilic Gram-negative rods

8.4.1.6.1 Fusobacterium *spp.*

Fusobacterium spp. are Gram-negative rods which are often pleomorphic and/or filamentous. The main characteristics of *Fusobacterium* spp. are as follows:
- Gram-negative bacilli;
- G+C content is 26–34 mol%;
- Nonsporing;
- Nonmotile;
- Anaerobic;
- Optimum growth temperature is 35–37°C;
- Optimum pH for growth is 7.0;
- Generally unable to ferment sugars;
- Energy derived from fermentation of amino acids and peptides;
- Co-aggregation with a wide variety of oral species.

Many species are spindle-shaped with tapering ends, i.e. they have a "fusiform" morphology. Seventeen species are recognized, and those most commonly isolated from the oral cavity are *F. nucleatum*, *F. periodonticum*, *F. sulci*, and *F. alocis*, all of which require rich media for growth. *F. nucleatum* is the species isolated most frequently, and six subspecies are recognized: subsp. *nucleatum*, subsp. *polymorphum*, subsp. *vincentii*, subsp. *fusiforme*, subsp. *animalis*, and subsp. *canifelium*. They obtain their energy by the fermentation of peptides and amino acids to butyric and acetic acids. Generally, fusobacteria are only weakly proteolytic, although a number of metallo-proteases have been detected in *F. nucleatum*. Most *Fusobacterium* spp.

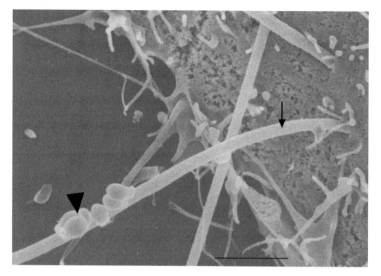

Fig. 8.12 Invasion of oral epithelial cells by *F. nucleatum* or *Strep. cristatus*. *F. nucleatum* readily invades the cells (arrow). *Strep. cristatus* can be seen adhering to *F. nucleatum* (arrowhead) and can be transported into the cell by the invading *F. nucleatum*. Scale bar, 2 μm. Image kindly supplied by Joel Rudney, School of Dentistry, University of Minnesota, Minneapolis, Minnesota, USA.

produce a DNase. Some strains of *F. nucleatum* can hydrolyze dextran. The 2.17-Mbp genome of *F. nucleatum* has been sequenced (http://cmr.tigr.org/tigr-scripts/CMR/GenomePage.cgi?org=ntfn01).

Fusobacterium spp. can co-aggregate with many species from most genera of oral bacteria (see section 8.4.2.1), and so are important in plaque formation, where they act as "bridging" organisms between the primary colonizers of the tooth surface and later colonizers. In the case of *F. nucleatum*, two galactose-binding adhesins (proteins with molecular masses of 30 and 42 kDa, respectively) in the outer membrane are important in mediating co-aggregation with other oral bacteria. However, co-aggregation with some oral bacteria appears to involve protein–protein interactions that are inhibited by arginine. Co-aggregation with streptococci leads to characteristic "corn-cob" structures, which are a common feature of dental plaque. Many oral fusobacteria produce bacteriocins with activity against a number of oral species, including, most frequently, *Eggerthella lenta* and *Peptostreptococcus anaerobius*. As well as being able to adhere to oral epithelial cells, *F. nucleatum* can also invade them (Fig. 8.12).

F. nucleatum is invariably present in dental plaque, but is found less frequently on the tongue or in saliva. The main habitat of *F. sulci* and *F. alocis* is the gingival sulcus.

8.4.1.6.2 Porphyromonas spp.
Porphyromonas spp. are anaerobic Gram-negative rods or coccobacilli, and their main characteristics are as follows:

• Gram-negative bacilli;
• Nonmotile;
• Nonsporing;
• Obligate anaerobes;
• G+C content of their DNA is 46–54 mol%;
• Optimum growth temperature is 37°C;
• Nutritionally fastidious – require hemin and vitamin K for growth;
• Asaccharolytic or weakly saccharolytic;
• Energy derived from fermentation of amino acids and peptides;
• Butyrate, acetate, and propionate produced.

Twelve species are recognized, of which *Por. gingivalis*, *Por. endodontalis*, and *Por. asaccharolytica* are the ones most frequently detected in the human oral cavity. They produce large quantities of cell-associated protohemin, resulting in black-pigmented colonies on blood agar. Peptides and amino acids are their main sources of energy, with butyrate, acetate, and propionate being the principal metabolic end products. *Por. gingivalis* is highly proteolytic, but most of the other species have little or no proteolytic activity. The proteinases (known as gingipains) of *Por. gingivalis* are cysteine proteinases and can degrade a wide range of proteins, including many host structural and defense proteins. The organism also produces nucleases, sialidases, and glycosaminoglycan-degrading enzymes. The main habitat of *Por. gingivalis* is the subgingival region, but it has also been detected on the tongue, tonsils, in saliva, and in supragingival plaque. The organism has fimbriae which, along with its hemagglutinins, mediate adhesion to host cells and to other bacteria. It can also invade epithelial cells. The 2.34-Mbp genome of *Por. gingivalis* has been sequenced (http://cmr.tigr.org/tigr-scripts/CMR/GenomePage.cgi?org=gpg).

Por. gingivalis is one of the main etiological agents of periodontitis, while *Por. endodontalis* is one of several organisms associated with endodontic infections.

8.4.1.6.3 Prevotella *spp.*

Prevotella spp. are morphologically similar to *Porphyromonas* spp. but, unlike the latter, are saccharolytic. The G+C content of their DNA is 40–50 mol%. Their optimum growth temperature is 37°C, but they can grow over the range 25–42°C. They ferment carbohydrates, producing mainly acetate and succinate, and most species are proteolytic. Some species (e.g. *Prev. intermedia* and *Prev. nigrescens*) produce lipases. At least 20 species are recognized, and they can be classified into two main groups – pigmented and nonpigmented – on the basis of their ability to convert hemoglobin to protohemin. Pigmented species found in the oral cavity include *Prev. intermedia*, *Prev. nigrescens*, *Prev. melaninogenica*, *Prev. loescheii*, *Prev. corporis*, and *Prev. denticola*. Nonpigmented oral species include *Prev. buccae*, *Prev. buccalis*, *Prev. oralis*, and *Prev. oulora*. Many strains of *Prev. intermedia* produce a bacteriocin active against *F. nucleatum*, *Por. gingivalis*, and *Prev. nigrescens*. The 2.69-Mbp genome of *Prev. intermedia* has been sequenced (http://cmr.tigr.org/tigr-scripts/CMR/GenomePage.cgi?org=gpi).

Some *Prevotella* spp. (particularly *Prev. intermedia* and *Prev. nigrescens*) have been implicated in the pathogenesis of periodontitis, while others have been isolated from infections at other body sites.

8.4.1.6.4 *Spirochaetes*

Spirochaetes are large, spiral-shaped organisms which are motile despite not having external flagella. They are helical and tightly coiled and range in length from 5 to 20 μm and in diameter from 0.1 to 0.4 μm. At least ten species have been isolated from the human oral cavity, but a large number of additional taxa (approximately 70) have been detected using molecular techniques. Their main habitat is the gingival crevice and subgingival plaque. They are nutritionally fastidious, are very oxygen-sensitive, and are difficult to grow in the laboratory. All species require fatty acids for growth. They prefer a slightly alkaline pH (approximately 7.5) and grow best at a temperature of 37°C. The main species isolated from healthy individuals are *Treponema*

socranskii, *Treponema denticola*, *Treponema vincentii*, and *Treponema pectinovorum*. Elevated proportions of these species are also present in patients with periodontitis.

T. denticola produces chymotrypsin-like and trypsin-like proteinases as well as peptidases, and so can liberate amino acids from proteins present in the gingival sulcus. It can invade epithelial cells and excretes ammonia and H_2S. The organism also produces glycine and pyruvate from the glutathione present in the subgingival region, and these are valuable nutrients for other species colonizing this region. Its 2.84-Mbp genome has been sequenced (http://cmr.tigr.org/tigr-scripts/CMR/GenomePage.cgi?org=gtd).

8.4.1.6.5 *Other anaerobic species*

Tannerella forsythensis (formerly *Bacteroides forsythus*) is a nonmotile, fusiform bacillus, the DNA of which has a G+C content of 44–48 mol%. It is nutritionally fastidious, and most strains require N-acetyl muramic acid for growth. It uses peptides and amino acids as energy sources and produces mainly acetic, butyric, isovaleric, and propionic acids as end products. It produces proteases, a sialidase, and a number of glycosidases and so can contribute to the degradation of mucins. Its main habitat is subgingival plaque, and it is one of several organisms associated with periodontitis.

Leptotrichia spp. are nonsporing, nonmotile, large Gram-negative fusiforms. *Lep. buccalis* is the type species. Other species in the genus are *Lep. goodfellowii*, *Lep. hofstadii*, *Lep. wadei*, and *Lep. trevisanii*. The G+C content of their DNA is 28–30 mol%. They are catalase-positive, and their optimum temperature for growth is 37°C. They use carbohydrates as an energy source and produce lactic acid as the main end product. The primary habitat of *Leptotrichia* spp. is the oral cavity, but some have been found in the female genito-urinary tract. In the oral cavity, they have been detected in plaque, saliva, the gingival sulcus, and on most mucosal surfaces – *Lep. buccalis* is the species most frequently detected.

Selenomonas spp. are motile, anaerobic Gram-negative curved rods which are bile-sensitive. The G+C content of their DNA is 48–57 mol%. They are saccharolytic and produce mainly acetate and propionate from glucose. Six species have been detected in the oral cavity – *Sel. artemidis*, *Sel. noxia*, *Sel. flueggei*, *Sel. infelix*, *Sel. sputigena*, and *Sel. dianae*. They are found mainly in dental plaque, particularly in the gingival crevice, and on the tongue.

Campylobacter spp. are motile (apart from *Camp. gracilis*), anaerobic, or microaerophilic, nonsporing

Gram-negative slender, helical, or curved rods. The G+C content of their DNA is 29–47 mol%. Their optimum temperature for growth is between 30 and 42°C. They are asaccharolytic and use amino acids or tricarboxylic acid cycle intermediates as carbon and energy sources, producing mainly succinate as a metabolic end product. Most species require formate, fumarate, and hydrogen for growth. Fifteen species are known, of which six are found in the oral cavity of humans – *Camp. sputorum* subsp. *sputorum*, *Camp. concisus*, *Camp. rectus* (formerly *Wolinella recta*), *Camp. curvus*, *Camp. gracilis* (formerly *Bacteroides gracilis*), and *Camp. showae*. Their main habitat is the gingival crevice and subgingival plaque, and some species are associated with periodontitis.

Filifactor alocis is an obligately anaerobic, Gram-negative, non-spore-forming rod with rounded or tapered ends. Some strains show twitching motility, although no flagella have been detected. The G+C content of its DNA is 34 mol%. The organism ferments glucose to produce butyrate and acetate. Other metabolic end products include hydrogen and ammonia. The principal habitat of *Fil. alocis* is the human gingival sulcus.

Dialister spp. are obligately anaerobic, nonmotile, nonsporing, nonfermentative Gram-negative coccobacilli. The G+C content of their DNA ranges from 35 to 46 mol%. Two species have been detected in the oral cavity – *Dial. pneumosintes* and *Dial. invisus*. They are unreactive in commonly used biochemical tests.

8.4.1.7 Facultatively anaerobic Gram-negative bacilli

Aggregatibacter (formerly *Actinobacillus*) *actinomycetemcomitans* is a coccobacillus, and its main characteristics are as follows:

- Gram-negative coccobacillus;
- Nonmotile;
- Facultative anaerobe;
- Growth improved by 5–10% CO_2;
- G+C content of its DNA is 42.7 mol%;
- Growth over the range 25–42°C;
- Optimum temperature for growth is 37°C;
- Optimum pH for growth is 7.0–8.0;
- Nutritionally fastidious.

It is fermentative and can utilize a number of carbohydrates for energy generation, including glucose, fructose, and mannose. Its growth is stimulated by certain steroid hormones including estrogen, progesterone, and testosterone. The organism produces a bacteriocin, actinobacillin, which is active against some streptococci and *Actinomyces* spp. It can invade epithelial cells and has collagenolytic activity. The organism is found in plaque and in saliva, as well as on a number of mucosal surfaces, including the tongue. It is an important etiological agent of localized aggressive periodontitis, and is also responsible for diseases at extra-oral sites, including endocarditis, septicemia, and meningitis. Its 2.2-Mb genome has been sequenced (http://www.genome.ou.edu/act.html).

Eikenella corrodens is a nonsporing coccobacillus, the growth of which is enhanced by the presence of CO_2. It is the sole member of the genus *Eikenella* and its DNA has a G+C content of 56–58 mol%. It exhibits twitching motility although it does not have any flagella. It grows best at 35–37°C, but can grow over the range 27–43°C. Its optimum pH for growth is 7.0–8.0. The organism characteristically produces pitting on agar surfaces. As well as being implicated in the pathogenesis of periodontal diseases, the organism is responsible for a number of infections at extra-oral sites, e.g. endocarditis and meningitis.

Capnocytophaga spp. are slender rods that exhibit gliding motility. Growth is enhanced aerobically and anaerobically by CO_2 and is optimal at 35–37°C. The G+C content of their DNA is 37.6–39.1 mol%. Seven species are recognized, of which *Cap. ochracea*, *Cap. sputigena*, *Cap. gingivalis*, *Cap. granulosa*, and *Cap. haemolytica* have been detected in the oral cavity – mainly in dental plaque. They ferment carbohydrates to produce acetic and succinic acids and produce a range of extracellular enzymes, including proteases, peptidases, acid and alkaline phosphatases, phospholipase A_2, sialidase, and glycosidases. *Cap. ochracea* produces a bacteriocin that is active against a number of oral streptococci and *Propionibacterium acnes*. In addition to being associated with periodontal diseases, *Capnocytophaga* spp. have also been isolated from cases of septicemia, endocarditis, and abscesses at a variety of body sites.

8.4.1.8 *Mycoplasma* spp.

Mycoplasmas are common in the oral cavity, and a number of species have been found on the oral mucosa (*Myc. buccale*, *Myc. orale*, and *Myc. pneumoniae*), in dental plaque (*Myc. peumoniae*, *Myc. buccale*, and *Myc. orale*) and in saliva (*Myc. salivarium*, *Myc. pneumoniae*, and *Myc. hominis*).

8.4.1.9 *Megasphaera* spp.

Megasphaera spp. are anaerobic Gram-negative cocci which occur in pairs and short chains. They are catalase-negative and can utilize a range of sugars and lactate as carbon sources. Glucose is fermented mainly to caproate and formate, while lactate is converted to short-chain fatty acids, mainly butyrate and propionate, and hydrogen. They require a number of amino acids for growth, and acetate is stimulatory.

8.4.2 Community composition at different sites

The oral cavity has two main types of surfaces for microbial colonization – shedding (mucosa) and non-shedding (teeth). However, because of anatomical factors, each of these types of surfaces provides a range of habitats, each of which has a characteristic microbiota. Although saliva contains large numbers of bacteria (approximately 10^8 per ml), it is considered not to have an indigenous microbiota because its high flow rate and generally low content of easily assimilable nutrients do not enable bacterial multiplication to occur in vivo. The organisms in saliva are those shed by, or dislodged from, oral surfaces, and the proportions and types of organisms found are similar to those present on the dorsal surface of the tongue.

8.4.2.1 Supragingival plaque

Supragingival plaques are those biofilms on the tooth surface above the level of the gingival margin. The main types of supragingival plaque are: smooth-surface plaque (found on the lingual and buccal surfaces), interproximal plaque (found between adjacent teeth), and fissure plaque (found in the fissures on the biting surfaces of premolars and molars). Although technically "supragingival", plaque present at the gingival margin has a composition that is distinct from the make-up of other supragingival plaques because of the very different environment of this region; the microbiota of the gingival crevice will be described separately in section 8.4.2.2). The formation of plaque on the supragingival tooth surface will be discussed in general terms and then the composition of the microbial communities at different anatomical sites will be described.

Consisting mainly of proteins and, to a lesser extent, lipids and glycolipids, the AEP layer of adsorbed molecules that covers tooth surfaces is a tenacious layer 0.1–1.0 µm thick, in which the adsorbed molecules form distinct globular and fibrillar structures. PRPs, cystatins, lysozyme, IgA, mucins, lactoferrin, and statherin are consistently detected in the AEP, and a range of other molecules are usually also present including bacterial components and their products, e.g. glucosyltransferases and glycans. As well as acting as a substratum for bacterial adhesion, the AEP protects enamel against demineralization by acids and enhances its remineralization.

The first organisms to adhere to the AEP will be those with adhesins that are able to bind to receptors on molecules present in the pellicle. A wide range of receptors for bacterial adhesins have been identified in the AEP (Table 8.5). Within a few hours, between 12 and 32% of the enamel surface is covered by bacteria, and those species that can grow and reproduce under the prevailing environmental conditions become established and constitute the pioneer community. Initially, most of these organisms are present as microcolonies on the tooth surface (Fig. 8.13).

The supragingival surface of the tooth is an aerobic environment with a relatively high redox potential and a neutral pH, and the main cultivable pioneer species include viridans streptococci (mainly *Strep. oralis*, *Strep. mitis* biovar 1, *Strep. sanguinis*, and *Strep. intermedius*), *Actinomyces* spp. (mainly *A. naeslundii* genospecies 1 and 2, *A. gerencseriae*, and *A. odontolyticus*), *Neisseria* spp., and *Haemophilus* spp.

A culture-independent analysis of supragingival plaque formation in vivo has revealed that a wide variety of organisms are able to adhere to the clean tooth surface. However, not all of these are capable of surviving and growing there as, 6 h later, a more restricted range of organisms is present in substantial proportions (Fig. 8.14). *Actinomyces* spp. were found to be present in the greatest proportions initially, but after 4 h streptococci began to dominate the microbiota. *A. naeslundii* genospecies 1 and genospecies 2 together comprised 19% of the microbiota after 2 h, but after 4 h they comprised only 13%. *Strep. mitis* and *Strep. oralis*, on the other hand, together comprised 23%. During the 6-h period, the population density on the tooth surface increased approximately five-fold.

Once the cell density reaches $2.5–4.0 \times 10^6$ cells per mm^2, the growth rate increases dramatically, which implies that quorum-sensing mechanisms are operating. Indeed, studies have shown that the quorum-sensing signaling molecule of *Strep. mutans*,

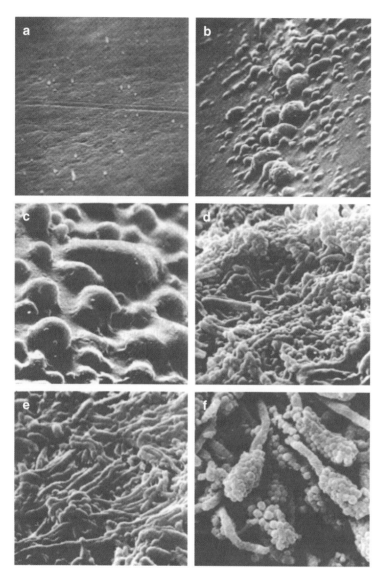

Fig. 8.13 Electron micrographs showing the development of dental plaque. (a) A recently (<1 h) cleaned tooth surface coated with AEP. (b) Attachment of individual bacteria (1–4 h). (c) Formation of microcolonies (4–24 h). (d) Increasing species diversity due to microbial succession (>24 h). (e) Mature plaque (>24 h). (f) An example of a frequently encountered structure in mature plaque – a "corn-cob". Reprinted with the permission of Cambridge University Press, Cambridge, UK, from: Bradshaw, D.J., Marsh, P.D. (2003) Novel microscopic methods to study the structure and metabolism of oral biofilms. In: Wilson, M. and Devine, D. (eds). *Medical Implications of Biofilms.*

competence stimulating peptide (CSP), is involved in biofilm formation by the organism. Attached bacteria grow exponentially for the first few days, and the growth rate then decreases. Growth of the pioneer community alters the environment in a number of ways (Table 8.6).

This altered environment enables colonization by organisms (known as "secondary colonizers"), which are unable to either adhere to, or survive on, the original tooth surface. This autogenic succession results in a more diverse community that includes microaerophiles and obligate anaerobes (Fig. 8.15). Figure 8.15 shows the results of a culture-independent analysis of supragingival plaque in healthy adults who refrained from oral hygiene measures for a 4-day period. After 1 day, the total number of organisms remained constant at approximately 2.1×10^7 per tooth surface, and the microbiota was dominated by *Actinomyces* spp. By day 4, organisms such as *Capnocytophaga* spp., *Campylobacter* spp., and *Eik. corrodens*, together with

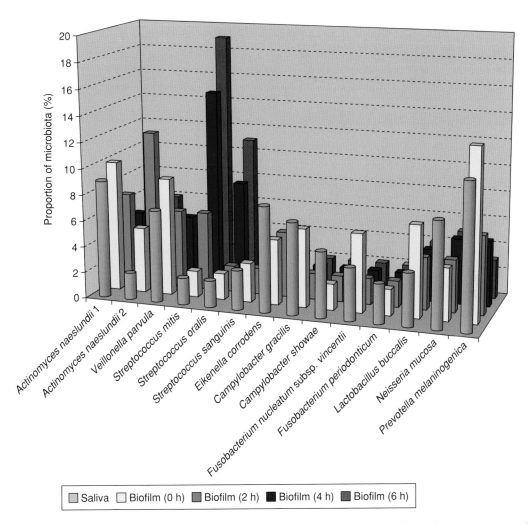

Fig. 8.14 Dental plaque formation on supragingival tooth surfaces over a 6-h period in 15 healthy adults. DNA was extracted from the saliva and plaque samples and was assayed for the presence of 40 oral species using whole-genome DNA probes. Only those organisms comprising at least 2% of the microbiota are shown (1).

Table 8.6 Environmental impact of growth of the pioneer community on the tooth surface.

Microbial process	Consequences
Oxygen utilization and CO_2 generation	Creates adverse conditions for growth of obligate aerobes; provides conditions suitable for capnophiles, microaerophiles, and anaerobes
Production of reducing compounds	Lowers the redox potential, thereby providing conditions suitable for growth of anaerobes
Excretion of metabolic end products	Increases the range of nutrients available to microbes
Degradation of host macromolecules	Increases the range of nutrients available to microbes; possibility of loss of receptors complementary to adhesins of pioneer species
Production of exopolymers	Increases the range of nutrients available to microbes; provides additional receptors for bacterial adhesion; may impede access of nutrients to underlying organisms and/or dispersal of metabolic end products
Biofilm formation	Results in the production of gradients with respect to physicochemical parameters, nutrients, and waste products

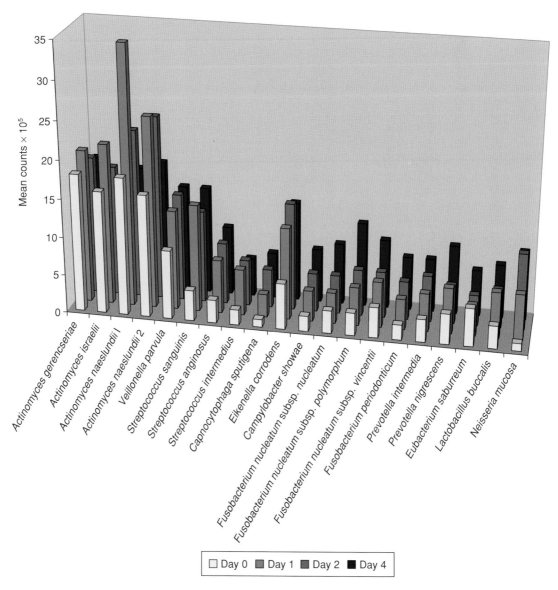

Fig. 8.15 Composition of supragingival plaques obtained from ten volunteers on various occasions over a 4-day period. Counts of the various species were determined using a checkerboard DNA–DNA hybridization technique. Only those species that had counts of more than 5×10^5 on at least one of the sampling occasions are shown (2).

obligately anaerobic species such as *Fusobacterium* spp. and *Prevotella* spp., had begun to comprise appreciable proportions of the microbiota.

Colonization of the growing and diversifying plaque by additional bacterial species present in saliva occurs by two main mechanisms: (1) co-adhesion – this involves the adhesion of a planktonic cell in saliva to a microbe already present in the biofilm; and (2) co-aggregation – this involves the adhesion of a microbial aggregate in saliva to a microbe that is already present in the biofilm. The adhesin–receptor interactions enabling both co-adhesion and co-aggregation have been determined for many oral microbes, and examples are shown in Table 8.7.

Table 8.7 Adhesins and receptors involved in co-adhesion and co-aggregation of oral bacteria. Reprinted from: Rosan, B. and Lamont, R.J. Dental plaque formation. *Microbes Infect* (2000) 2, 1599-607. Copyright 2000, with permission from Elsevier, Amsterdam, The Netherlands.

Species	Adhesin	Approximate size (kDa)	Receptor(s)	Partner species
Strep. gordonii, Strep. mitis, Strep. oralis	Antigen I/II family (SspA, SspB)	171 (SspA), 160 (SspB)	Bacterial surface proteins, yeast mannoproteins	*Por. gingivalis, Strep. mutans, A. naeslundii, Candida albicans*
Strep. gordonii	CshA, CshB	259, 245	Bacterial surface proteins, yeast mannoproteins	*A. naeslundii, Can. albicans, Strep. oralis*
Strep. gordonii	LraI family (ScaA)	35	—	*A. naeslundii*
Strep. gordonii	Co-aggregation-mediating adhesin	100	Carbohydrate containing lactose or lactose-like moieties	*Streptococcus* spp.
Strep. salivarius	Fibrillar antigen B (VBP)	320	—	*V. parvula*
A. naeslundii	Type 2 fimbriae-associated protein	95	Cell-wall polysaccharide containing Galβ1 → 3GalNAc and GalNAcβ1 → 3Gal glycosidic linkages	*Streptococcus* spp.
Por. gingivalis	Fimbrillin	43	Surface proteins	*Strep. gordonii, Strep. oralis, A. naeslundii*
Por. gingivalis	Outer-membrane protein	35	Surface protein	*Strep. gordonii*
Por. gingivalis	Outer-membrane protein	40	—	*A. naeslundii*
Prev. loescheii	PlaA	75	Cell-wall polysaccharide containing Galβ1 → 3GalNAc and GalNAcβ1 → 3Gal glycosidic linkages	*Streptococcus* spp.
Prev. loescheii	Fimbria associated protein	45	—	*A. israelii*
F. nucleatum	Outer-membrane proteins	42, 30	Galactose-containing carbohydrate	*Por. gingivalis*
F. nucleatum	"Corn-cob" receptor	39	—	*Strep. cristatus*
V. atypica	Outer-membrane protein	45	Carbohydrate containing lactose or lactose-like moieties	*Streptococcus* spp.
Cap. gingivalis	Outer-membrane protein	140	Cell-wall carbohydrate	*A. israelii*
Cap. ochracea	Outer-membrane protein	155	Cell-wall carbohydrate	*Strep. oralis*
Treponema medium	Outer-membrane protein	37	fimbriae	*Por. gingivalis*

The secondary colonizer *F. nucleatum* is particularly important in both co-adhesion and co-aggregation as it has been shown to bind to virtually all oral bacterial species (Fig. 8.16). Other substrata available for adhesion in growing dental plaque are the exopolymers produced by plaque bacteria. Furthermore, degradation of host macromolecules in the AEP by bacterial enzymes can reveal previously unavailable receptors (known as "cryptitopes") for bacterial adhesins.

The secondary colonizers in the biofilm will, of course, also alter the local environment, thereby enabling colonization by yet other species. Microbial succession eventually results in the formation of a stable, climax community with a high species diversity. The members of this climax community exist in a state of dynamic equilibrium because of complex spatial, physiological, and nutritional interactions, both beneficial (Fig. 8.17) and antagonistic (Table 8.8), and their relative proportions will remain constant provided that there is no change in any of the environmental determinants at the site. Any changes that do occur will result in a shift in the relative proportions of the constituent members, with the possible elimination of some members and recruitment of others, until a new climax community is established. Studies of the composition of dental plaques over several years have shown that the relative proportions of the predominant organisms remain remarkably stable despite the fact that allochthonous

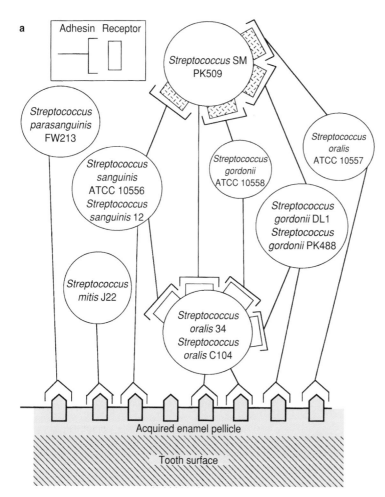

Fig. 8.16 Adhesive interactions between bacteria in dental plaque. (a) Adhesins on early colonizers bind to receptors on the AEP coating the tooth surface.

species are continually being introduced into the oral cavity and autochthonous microbes are repeatedly being removed by oral hygiene procedures.

As the biofilm grows, it becomes increasingly susceptible to mechanical and hydrodynamic shear forces, which will tend to decrease its size. Even in the absence of tooth brushing and other oral hygiene measures, plaque thickness reaches a maximum after 3 to 4 days. The effects of mechanical abrasion are obviously greater for biofilms on exposed tooth surfaces and can limit their size. In contrast, biofilms growing in anatomically protected regions will be largely unaffected until they protrude beyond the confines of their sheltered site. Detachment of bacteria and biofilm sections from the

"parent" biofilm provides a means by which biofilm formation can be initiated at other sites once re-attachment has taken place.

The main constituents of dental plaque are microbes, their exopolymers, and host macromolecules. A variety of exopolymers are produced, and these function as adhesins and as receptors, provide protection against host defenses, can be utilized as a carbon and energy source, and can help to maintain the integrity of the plaque. The main exopolymers present are the glucans and fructans synthesized by streptococci. These polymers are produced from dietary sucrose by the enzymes glucosyltransferase and fructosyltransferase, respectively. Both water-soluble polymers and those that are

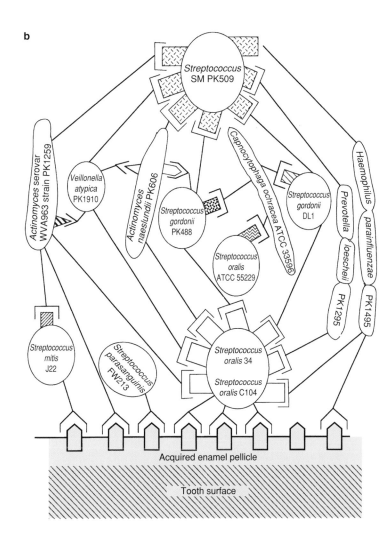

Fig. 8.16 (*Cont'd*) Adhesive interactions between bacteria in dental plaque. (b) Secondary colonizers then adhere to the early colonizers.

c

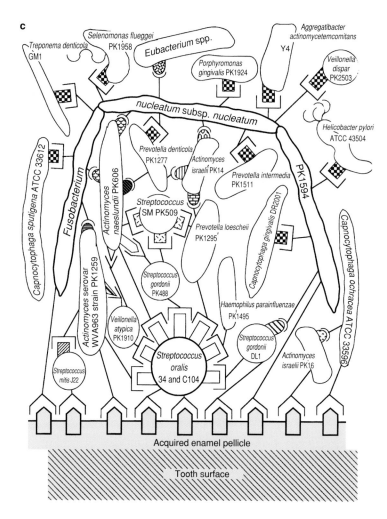

Fig. 8.16 (*Cont'd*) Adhesive interactions between bacteria in dental plaque. (c) Interactions occurring in mature plaque. Several kinds of co-aggregations are depicted as complementary sets of symbols of different shapes. One set is depicted in the box at the top of (a). Proposed adhesins (symbols with a stem) represent surface components that are sensitive to heat and to proteases; their complementary receptors (symbols without a stem) are not usually affected by heat or proteases as they are often a carbohydrate. Identical symbols represent components that are functionally similar but may not be structurally identical. Reprinted with the permission of Cambridge University Press, Cambridge, UK, from: Wilson, M. (2005) *Microbial Inhabitants of Humans: Their Ecology and Role in Health and Disease.*

insoluble in water are present in plaque, with the latter being the most important from the point of view of maintaining plaque structure, and these consist mainly of α-1, 3-linked glucose residues.

Electron microscopy of plaques reveals tightly packed bacteria near the tooth surface with a more open structure within the outermost layers (Fig. 8.18).

Characteristic arrangements of bacteria can often be seen, e.g. microcolonies and "corn-cobs" consisting of cocci (*Streptococcus crista*) surrounding a central filamentous organism such as *F. nucleatum* or *Corynebacterium matruchotii* (see Figs 8.13f and 8.18). However, as the biofilm matrix is composed mainly of water (often as much as 97%), the specimen preparation

(dehydration, fixation, and staining) that is essential for electron microscopy alters the native plaque structure. In contrast, confocal laser scanning microscopy (CLSM) enables the biofilms to be viewed in their hydrated state, thereby preserving their native structure. These have revealed that some dental plaques have a more open structure than that implied by electron microscopy, and this is in keeping with the structures found in biofilms from other environments (Fig. 8.19). Hence, water channels permeating "stacks" of bacteria that are enclosed in an exopolymeric matrix have been observed. A microcolony forms at the particular location within a stack that has the appropriate combination of environmental factors that

Table 8.8 Antagonistic interactions among oral bacteria.

Organism	Mechanism	Target organism(s)
Strep. mutans	Production of a bacteriocin (mutacin)	Other strains of *Strep. mutans*, other streptococci, *Actinomyces* spp.
Streptococci; *Actinomyces* spp.; *Lactobacillus* spp.; *Veillonella* spp.	Generation of low pH	Wide range of nonaciduric spp., particularly Gram-negatives
Streptococci (many species)	Hydrogen peroxide	*Ag. actinomycetemcomitans, Cap. sputigena, F. nucleatum, Camp. rectus, Bacteroides* spp.
C. matruchotii	Bacteriocins	*Actinomyces* spp., *L. casei, L. acidophilus, L. fermentum, F. nucleatum, Strep. salivarius*
Ag. actinomycetemcomitans	Bacteriocin (actinobacillicin)	Other strains of the organism, *Strep. sanguinis, A. viscosus, Strep. uberis*
Por. gingivalis	Hematin	Streptococci, *Actinomyces* spp., *C. matruchotii, P. acnes*
Strep. sanguinis	Bacteriocin (sanguicin)	*Por. gingivalis*
Prev. nigrescens	Bacteriocin (nigrescin)	*Por. gingivalis, Prev. intermedia, Tan. forsythensis, Actinomyces* spp.
Prev. intermedia	Bacteriocin TH14	*Prev. intermedia, F. nucleatum*
Streptococcus spp., *Actinomyces* spp., *Bifidobacterium* spp., *Gemella* spp., *Leptotrichia* spp.	Lactic acid	*T. denticola* and other susceptible species
Strep. salivarius	Bacteriocin (salivaricin)	Other streptococci
Cap. ochracea	Bacteriocin	*Strep. sanguis, Strep. mitis, Strep. mutans, P. acnes*
Strep. mitis	Surfactant	Inhibits adhesion of organisms to the AEP

are suitable for the survival and growth of that particular organism (Fig. 8.20). The appropriate combination of environmental factors may arise because of gradients in nutrient concentration, pH, oxygen content, etc.

The above description of biofilm formation on tooth surfaces will, in general terms, apply to biofilm formation on any nonshedding oral surface, including enamel, dentine, and cementum as well as oral appliances and prostheses such as dentures, implants, and artificial crowns. However, the composition of the climax community will be affected by the nature of the environmental determinants operating at the particular site.

As can be seen from Fig. 8.21, the predominant cultivable bacteria in supragingival plaques on smooth surfaces, in approximal regions, and in fissures are streptococci and *Actinomyces* spp. These organisms are consistently present at all three sites in most individuals. *Veillonella* spp., *Neisseria* spp., and lactobacilli are often (but not always) present, and may also comprise appreciable proportions of the cultivable microbiota.

Smooth-surface plaques tend to be thinner, and have a lower species diversity, than other supragingival regions because they are more exposed to mechanical abrasion. Owing to autogenic succession, this limits their growth and, consequently, the range of microhabitats that can develop within the plaque. The cultivable microbiota of approximal plaque tends to be dominated by *Actinomyces* spp. rather than streptococci. Their anatomical location means that they are largely protected from mechanical abrasion, and they

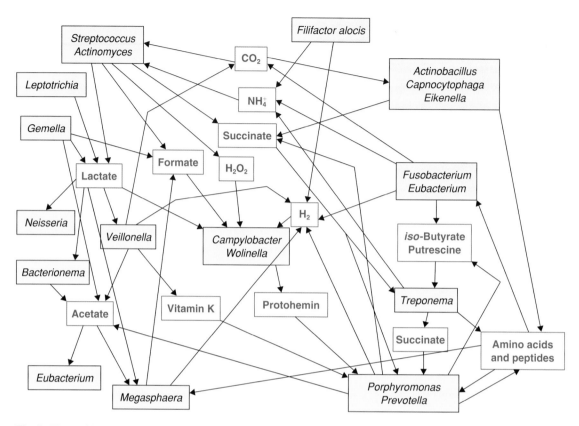

Fig. 8.17 Possible nutritional interactions that may occur between organisms present in the oral cavity.

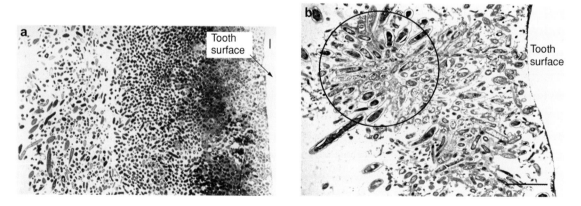

Fig. 8.18 Transmission electron micrographs of supragingival plaque from a healthy volunteer. (a) Bacteria are tightly packed near the tooth surface, but the packing is less dense in the outermost regions. (b) A microcolony (circled) is apparent within the plaque. Scale bar, 3 μm. Images kindly supplied by Mrs Nicola Mordan, UCL Eastman Dental Institute, University College London, London, UK.

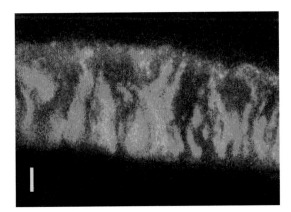

Fig. 8.19 Confocal laser scanning micrograph of supragingival dental plaque. Plaque was allowed to accumulate on dentine discs attached to oral appliances that were worn by volunteers for up to 48 h. The specimens were then stained to reveal live (green) and dead (red) bacteria in the plaque. The figure shows that after 48 h, the plaque has a complex structure consisting of clusters of live and dead cells with interspersed cell-free regions. The dentine surface is at the bottom. Scale bar, 10 μm. Reproduced with the permission of the International Association for Dental Research, Alexandria, Virginia, USA, from: Zaura-Aritel, E., van Marle, J. and ten Cate, J.M. (2001) Confocal microscopy study of undisturbed and chlorhexidine-treated dental biofilm. *J Dent Res* 80, 1436–40.

Fig. 8.20 pH variation at different locations within a multi-species oral biofilm resolved using two-photon excitation microscopy-fluorescence lifetime imaging microscopy (TPEM-FLIM). Reproduced with the permission of Cambridge University Press, Cambridge, UK, from: Bradshaw, D.J. and Marsh, P.D. (2003) Novel microscopic methods to study the structure and metabolism of oral biofilms. In: Wilson, M. and Devine, D. (eds), *Medical Implications of Biofilms*.

have a lower oxygen content because of poor penetration by saliva. Gram-negative anaerobes (mainly *Prevotella* spp.) are frequently present, although their proportions are low. Occlusal fissures often contain food that has become impacted, and this abundance of nutrients may explain the diversity of the cultivable microbiota at these sites. Mutans streptococci are found more often, and in higher proportions, at these sites, and this correlates with occlusal fissures being the most likely sites of dental caries. Staphylococci are also frequently present, although their proportions are low.

Although the predominant cultivable organisms in supragingival plaques are streptococci, *Actinomyces* spp., and *Veillonella* spp., a wide variety of other organisms are often isolated, and these are listed in Table 8.9.

Only a limited number of culture-independent studies of the supragingival plaque microbiota have been carried out. In one of these, the most frequently detected species in smooth-surface plaque from 22 healthy adults were *Actinomyces* spp., *Lep. buccalis*, *Gem. morbillorum*,

F. nucleatum, *Strep. sanguinis*, *Strep. oralis*, *Eubacterium nodatum*, and *Parvimonas micra* (formerly *Micromonas micros* and *Peptostreptococcus micros*), all of which were present in at least 50% of individuals (Fig. 8.22a). *Actinomyces* spp. were the dominant organisms in the plaque community, with streptococci and *Fusobacterium* spp. also accounting for appreciable proportions of the microbiota (Fig. 8.22b).

Using a similar approach (extraction of DNA from plaque followed by hybridization with whole-genome

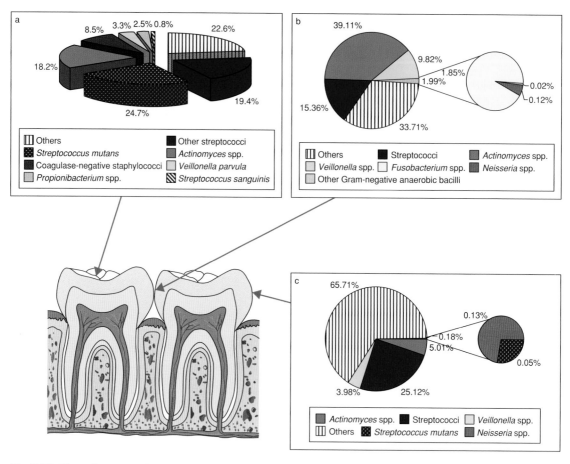

Fig. 8.21 The predominant cultivable microbiota of the three main types of supragingival plaque: (a) fissure, (b) approximal, and (c) smooth surface. Data are derived from three studies involving a total of 40 healthy adults (3–5).

DNA probes for 40 oral species), similar results were obtained in a study of smooth-surface plaque from 44 adults – the predominant organisms were found to be *A. naeslundii* genospecies 1 and 2, *A. israelii*, *A. gerencseriae*, *V. parvula*, and *Lep. buccalis*.

The limitations of the two above-mentioned studies are that the results obtained relate only to the prevalence and proportions of the 40 species for which probes were available, and do not take account of the presence of not-yet-cultivated species which would almost certainly be present. In a study without these limitations – involving the amplification, cloning, and sequencing of 16S rRNA genes present in supragingival plaque – the number of species detected in each of five healthy adults ranged from 12 to 27 (Table 8.10).

Strep. mitis, *Strep. sanguinis*, and *Strep. gordonii* were detected in at least four of the five individuals, while *Streptococcus* sp. clone EK048, *R. dentocariosa*, *Gem. haemolysans*, *Gran. adiacens*, and *Abiotrophia defectiva* were found in three of the five subjects.

The diet of the host can have a profound effect on the composition of plaque communities. Easily fermentable carbohydrates (e.g. sucrose) are metabolized by many members of the plaque microbiota to organic acids (mainly lactic and acetic acids), and this lowered pH selects for aciduric species such as mutans streptococci and lactobacilli. Such species are also acidogenic, and exposure of enamel to low pHs for long periods of time can result in its demineralization and the formation of a caries lesion. On the other hand, consumption of milk

Table 8.9 Bacteria that have been isolated from supragingival plaque. This is not an exhaustive list but gives an indication of the species diversity of supragingival plaques.

Genus	Examples
Actinomyces	*A. naeslundii* genospecies 1, *A. naeslundii* genospecies 2, *A. odontolyticus*, *A. israelii*, *A. gerencseriae*, *A. meyeri*
Streptococcus	*Strep. gordonii*, *Strep. sanguinis*, *Strep. intermedius*, *Strep. oralis*, *Strep. mitis*, *Strep. mutans*, *Strep. anginosus*, *Strep. salivarius*
Abiotrophia	*Ab. defectiva*, *Ab. adiacens*, *Ab. elegans*
Lactobacillus	*L. casei*, *L. plantarum*, *L. fermentum*, *L. acidophilus*, *L. rhamnosus*, *L. buccalis*
Enterococcus	*Ent. faecalis*
Haemophilus	*H. parainfluenzae*, *H. segnis*, *H. paraphrophilus*, *H. haemolyticus*
Leptotrichia	*Lep. buccalis*
Propionibacterium	*P. acnes*
Neisseria	*N. subflava*, *N. sicca*, *N. mucosa*, *N. perflava*
Veillonella	*V. parvula*, *V. dispar*, *V. atypica*
Eubacterium	*Eub. nodatum*, *Eub. brachy*, *Eub. saburreum*
Rothia	*R. dentocariosa*, *R. mucilaginosa*
Porphyromonas	*Por. gingivalis*, *Por. endodontalis*
Selenomonas	*Sel. sputigena*
Bifidobacterium	*Bif. dentium*, *Bif. denticolens*, *Bif. adolescentis*, *Bif. inopinatum*
Capnocytophaga	*Cap. ochracea*, *Cap. gingivalis*
Fusobacterium	*F. nucleatum* subsp. *nucleatum*, *F. nucleatum* subsp. *vincentii*, *F. nucleatum* subsp. *polymorphum*
Prevotella	*Prev. intermedia*, *Prev. nigrescens*
Aggregatibacter	*Ag. actinomycetemcomitans*
Parvimonas	*Parvimonas micra* (formerly *Peptostreptococcus micros*)
Gemella	*Gem. morbillorum*, *Gem. haemolysans*
Arachnia	*Arachnia propionica*

and cheese can protect against dental caries as the proteins present exert a buffering action and can become incorporated into the AEP, where they enhance remineralization by sequestering calcium phosphate. Their presence in the salivary pellicle reduces adhesion of mutans streptococci, and their degradation by plaque bacteria can release pH-raising amines and ammonia.

8.4.2.2 Gingival crevice

The gingival crevice constitutes a very different habitat from those existing on the supragingival tooth surfaces for three main reasons: (1) the predominant source of nutrients in the gingival crevice is not saliva, but is GCF, which has a high protein content; (2) because of its anatomical location, it is less exposed to the mechanical abrasion that occurs in supragingival regions; and (3) penetration of oxygen-rich saliva into the region is hindered by the outward flow of GCF, so that the oxygen content and redox potential will be lower than those found supragingivally. Consequently, the microbiota of the gingival crevice differs substantially from those found in supragingival plaques. Although culture-based studies have shown that the microbiota is dominated by streptococci and *Actinomyces* spp., the gingival microbiota has a greater species diversity than supragingival plaques, with higher proportions of obligate anaerobes including spirochaetes, which are very rarely found in supragingival plaques. The relative proportions of the main cultivable groups of organisms in the gingival crevice are shown in Fig. 8.23.

Culture-independent studies have shown that fewer than half of the organisms present in the gingival crevice have been cultivated. In one such investigation,

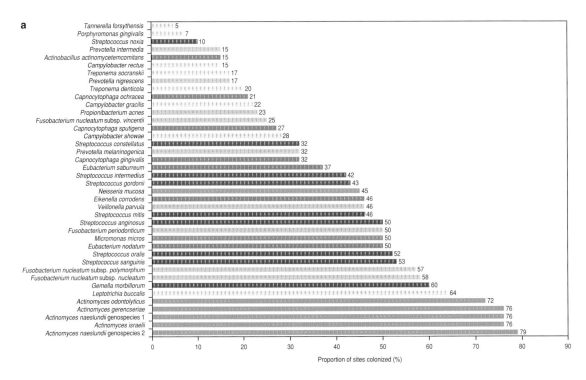

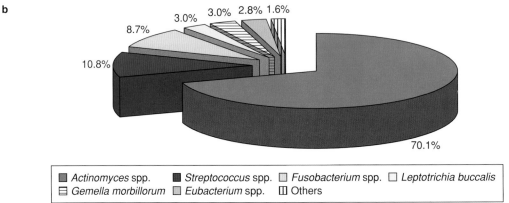

Fig. 8.22 Culture-independent analysis of smooth-surface supragingival plaque in 22 healthy adults. DNA was extracted from plaque samples taken from multiple sites in each individual and was assayed for the presence of 40 oral species using whole-genome DNA probes. (a) Prevalence of the various species detected. The bars indicate the proportion of sites colonized by each of the 40 species. (b) Relative proportions (%) of the organisms present in the 22 individuals (6).

187 phylotypes were detected in the gingival sulci of 15 periodontally healthy adults (Fig. 8.24). Most of the phylotypes belonging to the genera *Selenomonas*, *Veil-lonella*, and *Peptostreptococcus* were not-yet-cultivated organisms, while all of the phylotypes belonging to *Deferribacteres*, *Megasphaera*, *Desulfobulbus*, and *Lachno-spira* were not yet cultivated. Streptococci and *Veillonella* spp. comprised the largest proportions of the microbiota. Streptococci are also among the most frequently detected species in the gingival crevice, while

Table 8.10 Culture-independent analysis of pooled supragingival plaque samples from five healthy adults. DNA was extracted from the samples and 16S rRNA genes were amplified, cloned and sequenced. The distribution and levels of bacterial species/phylotypes are indicated as not detected, less than 15% of the total number of clones assayed, or more than 15% of the total number of clones assayed. A novel phylotype is indicated in bold (7).

Organism	Subject				
	1	**2**	**3**	**4**	**5**
Strep. mitis	>	<	<	>	<
Streptococcus clone EK048	<	>	ND	ND	<
Strep. infantis	ND	<	ND	ND	ND
Streptococcus clone DP009	<	<	ND	ND	ND
Streptococcus clone DN025	ND	ND	<	ND	ND
Streptococcus strain H6	ND	<	ND	ND	ND
Strep. parasanguinis	<	ND	ND	ND	ND
Streptococcus clone FN051	<	ND	ND	ND	<
Strep. cristatus	<	ND	ND	ND	ND
Streptococcus clone FN042	ND	ND	ND	ND	<
Strep. peroris	<	ND	ND	ND	ND
Strep. sanguinis	<	<	<	<	<
Streptococcus clone AY020	<	ND	ND	ND	ND
Strep. gordonii	>	>	ND	<	<
Strep. intermedius	ND	ND	<	ND	ND
Ab. defectiva	ND	<	ND	<	<
Gran. adiacens	<	<	ND	<	ND
Gem. haemolysans	ND	<	<	<	ND
Gem. morbillorum	<	ND	<	ND	ND
C. matruchotii	ND	ND	ND	ND	<
Corynebacterium clone AK153	ND	ND	ND	ND	<
C. durum	ND	ND	<	ND	<
Rothia strain CCUG 25688	ND	ND	ND	<	ND
R. dentocariosa	<	ND	<	ND	>
A. odontolyticus	<	ND	ND	ND	ND
A. naeslundii genospecies 2	ND	ND	ND	<	<
Actinomyces clone AG004	ND	ND	ND	ND	<
Actinomyces clone BL008	ND	ND	<	ND	>
Actinomyces clone AP064	ND	ND	ND	ND	<
Veillonella clone AA050	<	<	ND	ND	ND
V. parvula or *V. dispar*	ND	ND	ND	ND	<
Leptotrichia clone DT031	ND	ND	<	ND	ND
Leptotrichia clone E1022	ND	<	ND	ND	ND
Fusobacterium clone BS011	<	ND	ND	ND	ND
Fusobacterium clone CZ006	ND	ND	<	ND	ND
F. nucleatum subsp. *animalis*	<	ND	<	ND	ND
Eubacterium clone BU014	>	ND	ND	ND	ND
TM7 clone HD027	ND	ND	<	ND	ND
Camp. gracilis	ND	ND	<	ND	ND
Lautropia clone FX006	ND	ND	ND	<	ND
N. bacilliformis	ND	ND	<	ND	ND
Neisseria clone AP132	ND	ND	ND	<	ND
N. mucosa	ND	ND	ND	ND	<
King. denitrificans	ND	ND	ND	<	ND
N. elongata	<	ND	ND	ND	ND
Cap. gingivalis	<	ND	<	ND	ND
Porphyromonas clone AW032	ND	ND	<	ND	ND
Porphyromonas clone CW034	<	ND	ND	<	ND
Prevotella clone BU035	<	ND	ND	ND	ND
Prev. nigrescens	<	ND	ND	ND	ND
Prevotella clone BE073	<	ND	ND	ND	ND
Prev. melaninogenica	<	ND	ND	ND	ND

ND, not detected; <, <15% of the total number of clones assayed; >, >15% of the total number of clones assayed.

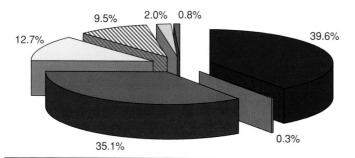

■ Gram-positive facultatively anaerobic cocci
(mainly streptococci)
☐ Gram-negative aerobic/facultatively anaerobic cocci
(mainly *Neisseria* spp.)
■ Gram-positive facultatively anaerobic rods (mainly *Actinomyces* spp.)
☐ Gram-negative anaerobic rods
◩ Gram-positive anaerobic rods
☐ Gram-negative anaerobic cocci (mainly *Veillonella* spp.)
■ Gram-positive anaerobic cocci

Fig. 8.23 Relative proportions of organisms comprising the cultivable microbiota of the gingival crevice. Data are derived from a study involving seven healthy adults (8).

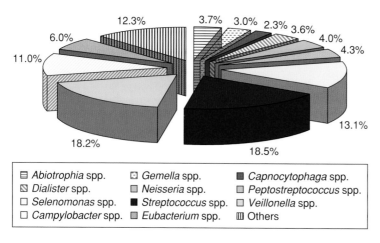

☐ *Abiotrophia* spp.   ☐ *Gemella* spp.   ■ *Capnocytophaga* spp.
◩ *Dialister* spp.   ☐ *Neisseria* spp.   ☐ *Peptostreptococcus* spp.
☐ *Selenomonas* spp.   ■ *Streptococcus* spp.   ☐ *Veillonella* spp.
☐ *Campylobacter* spp.   ☐ *Eubacterium* spp.   ▥ Others

Fig. 8.24 Culture-independent analysis of plaque from the gingival sulci of 15 periodontally healthy adults. DNA was extracted from the samples, and 16S rRNA genes were amplified, cloned and sequenced. The data shown are mean values of the proportions of only the 11 most frequently detected genera – these comprised 48 species and accounted for 87.7% of the microbiota (9).

Fusobacterium spp., *Granulicatella* spp., and *Atopobium* spp. are also frequently detected (Fig. 8.25).

8.4.2.3 Tongue

The epithelium of the tongue, unlike other oral mucosal surfaces, is highly convoluted due to the presence of numerous papillae (see Fig. 8.2). Owing to limited access to oxygen-rich saliva, the crypts between the papillae are oxygen-depleted habitats. The crypts also protect the resident microbial communities from

mechanical abrasion, thereby enabling biofilm formation. The environment within the crypts is nutrient-rich due to the retention of appreciable quantities of food, desquamated epithelial cells, and other debris. Consequently, the population density on the tongue is higher than that of other oral mucosal surfaces (each epithelial cell has approximately 100 attached bacteria), and the microbiota is more diverse. The population density of cultivable microbes on the tongue ranges from 10^7 to 10^9 cfu/cm^2, with the greater densities being found towards the back of the tongue.

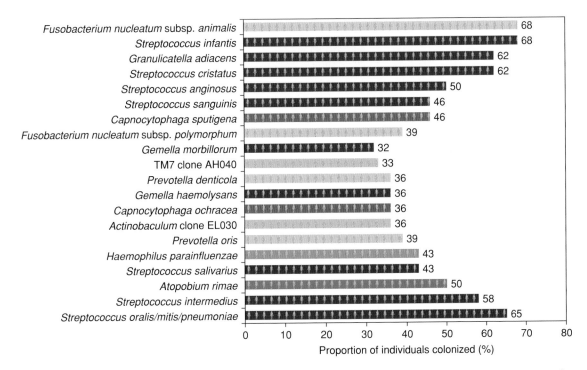

Fig. 8.25 Frequency of detection of organisms in the gingival sulci of 28 periodontally healthy individuals. DNA was extracted from the samples, and 16S rRNA genes were amplified and identified by hybridization with complementary oligonucleotide probes that recognize various oral bacterial species. The frequency of detection is shown only for those species found to be present in more than 30% of individuals (10).

Culture-dependent studies have revealed a diverse microbiota, with streptococci, *Veillonella* spp., *Actinomyces* spp., and Gram-negative anaerobic rods being frequently isolated and comprising appreciable proportions of the microbiota (Fig. 8.26).

In a culture-independent study of the communities on the tongue dorsum of five healthy individuals, a total of 55 phylotypes were detected (Fig. 8.27). Thirty-two percent of the phylotypes are unique to the tongue as they have not been found among over 6000 sequences obtained from other oral sites. The number of phylotypes in each individual ranged from 16 to 22. Streptococci were detected in all individuals, with *Strep. salivarius* and *Strep. parasanguinis* comprising the predominant organisms in most cases. *R. mucilaginosa* and *Strep. infantis* were also present in all individuals but comprised lower proportions of the microbiota.

The results of a study involving a larger number of individuals are shown in Fig. 8.28. The major findings of this culture-independent study differ from those

shown in Fig. 8.27 as obligate anaerobes rather than streptococci were found to dominate the tongue microbiota. This disparity may be due to the fact that in this study the microbiotas were probed for only 40 common oral species and these, unfortunately, did not include those organisms (*Strep. salivarius* and *Strep. parasanguinis*) found to predominate in the study referred to in Fig. 8.27.

8.4.2.4 Other mucosal surfaces

Mucosal surfaces in the oral cavity are predominantly aerobic regions with the major sources of nutrients being saliva and the mucous gel covering the epithelium. The ability to adhere to the epithelium or the mucous gel is an important requirement for species colonizing these regions, although recently it has been shown that bacteria are also present within oral epithelial cells in most individuals. Because of desquamation and mechanical abrasion, mucosal surfaces other than

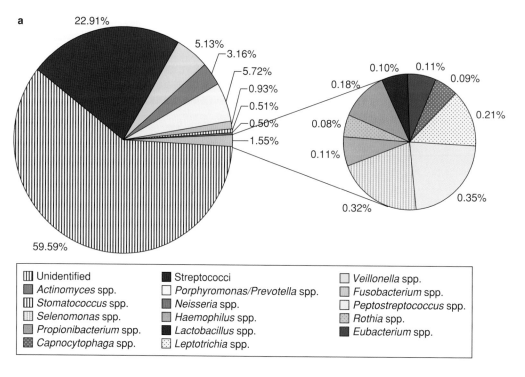

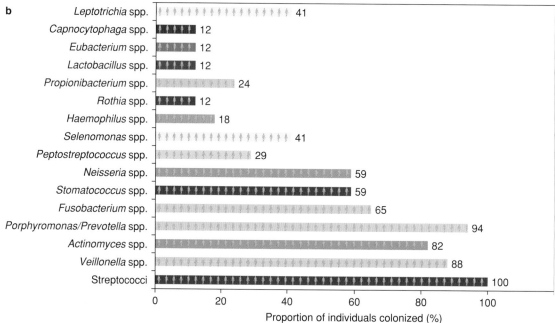

Fig. 8.26 Culture-dependent analysis of the tongue microbiota. (a) Relative proportions of the various organisms comprising the cultivable microbiota of the tongue. (b) Frequency of detection of cultivable organisms on the tongue. Data are derived from a study involving 17 healthy adults (11).

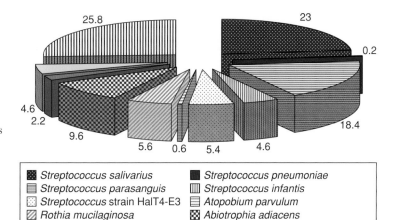

Fig. 8.27 Culture-independent analysis of the microbiota of the dorsal surface of the tongue. DNA was extracted from samples taken from five healthy individuals and 16S rRNA genes were amplified, cloned and sequenced. The figures represent the proportions (%) of the ten most frequently detected species which together comprised 74.2% of the tongue microbiota (12).

Legend for Fig. 8.27:
- *Streptococcus salivarius*
- *Streptococcus parasanguis*
- *Streptococcus* strain HalT4-E3
- *Rothia mucilaginosa*
- *Neisseria flavescens*
- Others
- *Streptococcus pneumoniae*
- *Streptococcus infantis*
- *Atopobium parvulum*
- *Abiotrophia adiacens*
- *Veillonella parvula/Veillonella dispar*

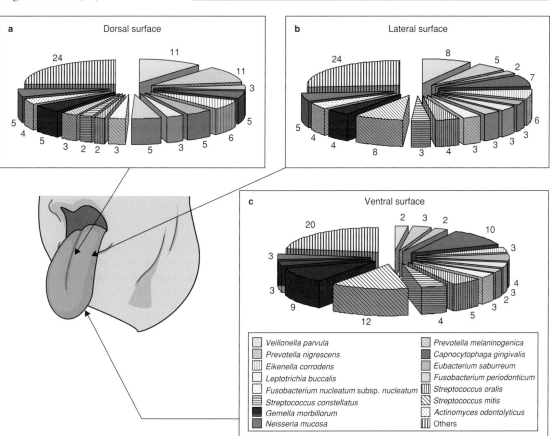

Fig. 8.28 Culture-independent analysis of the microbiotas of the (a) dorsal, (b) lateral, and (c) ventral surfaces of the tongue in 225 healthy adults. DNA was extracted from the samples and assayed for the presence of 40 oral species using whole-genome DNA probes. The data show the relative proportions (%) of the predominant 15 species. The other 25 species were all detected at the three sites although each species always comprised less than 5% of the microbiota of the site (13).

Legend for Fig. 8.28:
- *Veillonella parvula*
- *Prevotella nigrescens*
- *Eikenella corrodens*
- *Leptotrichia buccalis*
- *Fusobacterium nucleatum* subsp. *nucleatum*
- *Streptococcus constellatus*
- *Gemella morbillorum*
- *Neisseria mucosa*
- *Prevotella melaninogenica*
- *Capnocytophaga gingivalis*
- *Eubacterium saburreum*
- *Fusobacterium periodonticum*
- *Streptococcus oralis*
- *Streptococcus mitis*
- *Actinomyces odontolyticus*
- Others

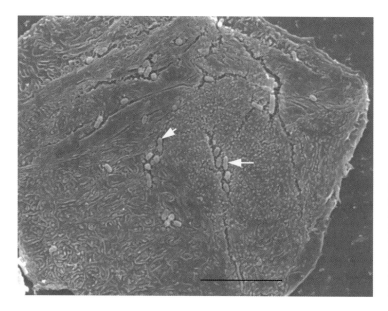

Fig. 8.29 Scanning electron micrograph of an epithelial cell from the cheek. A limited number of bacteria are present, and these consist mainly of pairs of oval-shaped cocci – probably streptococci (arrowed). Scale bar, 10 μm. Image kindly supplied by Mrs Nicola Mordan, UCL Eastman Dental Institute, University College London, London, UK.

the tongue have a relatively sparse microbiota compared to those found on the tooth surface. The average number of bacteria per epithelial cell is usually between 5 and 25 (Fig. 8.29).

The cultivable microbiota of the buccal mucosa is dominated by facultatively anaerobic and capnophilic species. Viridans streptococci (mainly *Strep. oralis*, *Strep. mitis* biovar 1, *Strep. sanguinis*, *Strep. salivarius* and *Strep. vestibularis*) are the dominant organisms, with *Neisseria* spp. and *Haemophilus* spp. also invariably present in high numbers. The main *Haemophilus* spp. found are *H. parainfluenzae*, *H. segnis* and *H. aphrophilus*. Organisms also often present, but in lower proportions, include staphylococci, *Veillonella* spp., lactobacilli, *Actinomyces* spp., and *Propionibacterium* spp. A morphologically unique member of the buccal microbiota is *Simonsiella muelleri* which is a motile organism consisting of between eight and 12 flat, wide cells joined together to form a filament.

The cultivable microbiota of the hard palate is again dominated by viridans streptococci with high proportions of *Actinomyces* spp. but lower proportions of *Neisseria* spp. and *Haemophilus* spp. *Corynebacterium* spp., *Prevotella* spp., and *Veillonella* spp. are also usually isolated but in much lower proportions.

In a culture-independent study of the microbial communities residing on the hard palate, soft palate, buccal epithelium, and anterior vestibule, the numbers

of phylotypes (*n*) detected at each site were in the range of *n* = 4 to 21, *n* = 6 to 20, *n* = 4 to 20, and *n* = 3 to 9, respectively (Fig. 8.30). The microbiotas of the hard and soft palates were generally more diverse than those on the buccal epithelium and anterior vestibule. *Strep. mitis* was the only organism that was detected at every site in all five individuals, and was the only species comprising more than 15% of the clones analyzed in each sample. *Gem. haemolysans* was detected at each site in most individuals. Obligate anaerobes were not frequently detected at any of the sites.

With regard to the relative proportions of the various species comprising mucosal communities, the results of an extensive culture-independent study (Fig. 8.31) agree in some respects with those obtained from culture-based studies, in that streptococci (mainly *Strep. mitis* and *Strep. oralis*) predominate at all of the mucosal sites. Another frequently detected species was *Gem. morbillorum*. This is a Gram-positive coccus which is difficult to identify and has many physiological and biochemical properties similar to those of viridans streptococci – it is possible that culture-based studies have mis-identified this organism as a viridans streptococcus. Unfortunately, the probes used in the culture-independent study included only one for a *Neisseria* species – *N. mucosa* – and no probes were included for any *Haemophilus* spp., although *Haemophilus* spp. are frequently detected in culture-based studies. Surprisingly, a number of

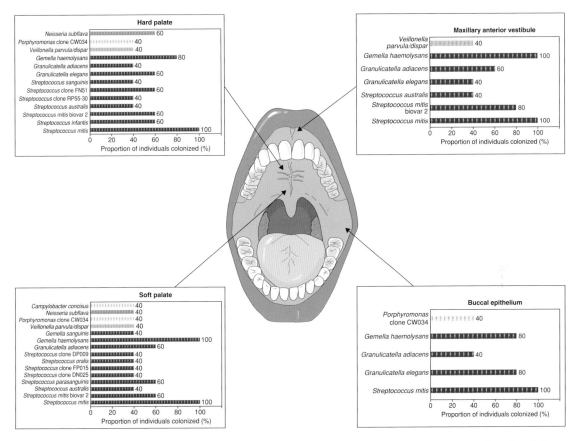

Fig. 8.30 Culture-independent analysis of the microbial communities residing on the hard palate, soft palate, buccal epithelium and maxillary anterior vestibule of five healthy individuals. Identification of the resident microbes involved the amplification, cloning, and sequencing of the 16S rRNA genes present in DNA extracted from the communities. Results are shown only for those taxa that were detected in at least two individuals (7).

anaerobic species were detected at all sites, although each of these tended to comprise a low (≤5%) proportion of the microbiota of each site – these included *V. parvula*, *Lep. buccalis*, *Eubacterium saburreum*, *F. periodonticum*, and *Prev. intermedia*. Presumably, their survival at these sites is dependent on oxygen utilization by aerobic and facultative anaerobes.

Recently, it has become apparent that oral epithelial cells are a significant habitat for a number of oral bacteria. In one study, every epithelial cell obtained from 71 adults was found to contain some bacteria, with streptococci being the most frequently detected. Other species found were *Gem. adiacens*, *Gem. haemolysans*, *Camp. rectus*, *F. nucleatum*, *Prev. intermedia*, *Por. gingivalis*, *Tan. forsythensis*, *Ag. actinomycetemcomitans*, and

E. corrodens (Fig. 8.32). In general, each epithelial cell contained more than 100 streptococci, whereas fewer than ten cells of each of the other species detected were present.

8.5 OVERVIEW OF THE ORAL MICROBIOTA

Because of its complex anatomy, the oral cavity has a large variety of habitats available for microbial colonization. Uniquely, it also has nonshedding surfaces, the teeth, which enable the formation of substantial and complex biofilms. Mechanical forces (due to chewing and tongue and jaw movements) and salivary

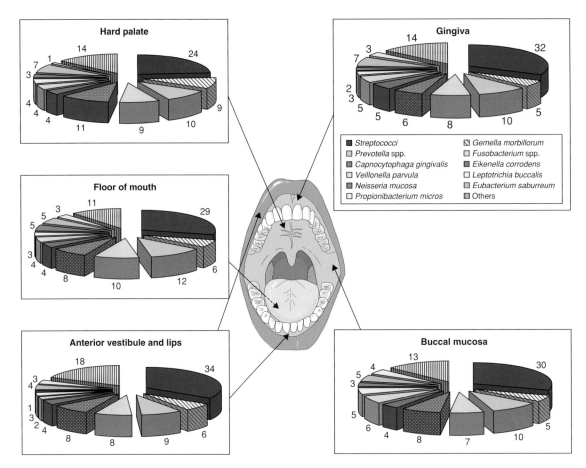

Fig. 8.31 Results of a culture-independent analysis of the microbiotas of various mucosal sites (the floor of the mouth, buccal mucosa, hard palate, lip, and the attached gingiva) in 225 adults. DNA was extracted from samples from each site and probed with whole-genome DNA probes specific for 40 oral bacterial taxa. Results are shown for only the 11 most plentiful species present in the communities. Each figure represents the percentage of the total DNA probe count, and so does not necessarily represent its proportion of the microbiota of the site because of the presence of organisms other than those recognized by the probes used (13).

flow hinder microbial colonization of exposed regions within the oral cavity, which means that successful microbial colonizers need to be able to adhere strongly to a surface or else to colonize sites protected from these powerful removal forces. The oral cavity harbors a large number of microbial communities, most of which have a high species diversity. As many as 700 phylotypes have been detected in the oral cavity, and approximately 50% of these have not yet been cultivated. However, culture-independent approaches are now being widely used to characterize the various microbial

communities that are present in the oral cavity, and these techniques are beginning to reveal the complexity of such communities.

Although mucosal surfaces comprise 80% of the total surface area of the oral cavity, most of the bacteria present in the mouth are found on tooth surfaces in biofilms known as dental plaques. The microbial composition of these plaques is complex and is dependent on their anatomical location. In supragingival plaques, viridans streptococci and *Actinomyces* spp. are usually the dominant organisms, but anaerobes

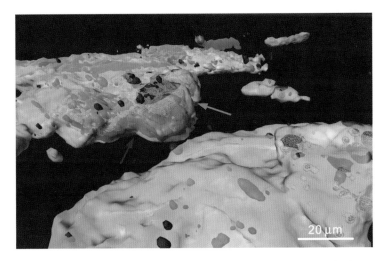

Fig. 8.32 The presence of intracellular bacteria in human oral epithelial cells as revealed by confocal microscopy combined with fluorescence in situ hybridization (FISH). Bacteria recognizing a FISH probe directed toward a 16S rRNA sequence present in all bacterial species are shown in red. A 16S rRNA probe recognizing *Ag. actinomycetemcomitans* was also used. Co-localization of the *Ag. actinomycetemcomitans* and universal probes is depicted by a green wire-frame over a red interior. The surfaces of the oral epithelial cell are shown in blue. The red and green colors are muted when bacterial masses are intracellular, and are brighter when bacteria project out of the surface. The large mass, which appeared to have a lobular structure, was seen to be a cohesive unit containing *Ag. actinomycetemcomitans* in direct proximity to other species (red and green arrows). Reproduced with the permission of the International Association for Dental Research, Alexandria, Virginia, USA, from: Rudney, J.D., Chen, R. and Sedgewick, G.J. (2005) *Actinobacillus actinomycetemcomitans, Porphyromonas gingivalis*, and *Tannerella forsythensis* are components of a polymicrobial intracellular flora within human buccal cells. *J Dent Res* 84, 59–63.

such as *Veillonella* spp. and *Fusobacterium* spp. are also invariably present. The microbial composition of plaque alters with time and is also affected by the host's diet. The microbiota of the plaque found in the gingival crevice is more diverse than that of supragingival plaques and, although streptococci are usually the dominant organisms, the proportion of anaerobes is greater than in supragingival plaques. Anaerobic organisms frequently detected include *Veillonella* spp., Gram-positive anaerobic cocci, *Prevotella* spp., *Fusobacterium* spp., *Selenomonas* spp., *Eubacterium* spp., and spirochaetes.

The tongue is densely colonized by microbes, and the composition of the resident communities varies with the anatomical location. Streptococci, again, are generally the dominant organisms, and a variety of anaerobes are frequently present, including species belonging to the genera *Prevotella, Veillonella, Eubacterium*, and *Fusobacterium*. Other mucosal surfaces are relatively sparsely populated compared with the tongue. The community composition varies with the anatomical

location, but facultative anaerobes and capnophiles are usually the dominant organisms, e.g. streptococci, *Gemella* spp., *Neisseria* spp., *Haemophilus* spp., and *Capnocytophaga* spp. However, anaerobes such as *Fusobacterium* spp., *Veillonella* spp., and *Prevotella* spp. are also often present.

8.6 SOURCES OF DATA USED TO COMPILE FIGURES

1 Li, J., Helmerhorst, E.J., Leone, C.W., Troxler, R.F., Yaskell, T., Haffajee, A.D., Socransky, S.S. and Oppenheim, F.G. (2004) *J Appl Microbiol* 97, 1311–18.

2 Ramberg, P., Sekino, S., Uzel, N.G., Socransky, S. and Lindhe, J. (2003) *J Clin Periodontol* 30, 990–5.

3 Beighton, D., Brailsford, S.R., Lynch, E., Chen, H.Y. and Clark, D.T. (1999) *Caries Res* 33, 349–56.

4 Bowden, G.H., Hardie, J.M. and Slack, G.L. (1975) *Caries Res* 9, 253–77.

5 Theilade, E., Fejerskov, O., Karring, T. and Theilade, J. (1982) *Infect Immun* 36, 977–82.

6 Ximénez-Fyvie, L.A., Haffajee, A.D. and Socransky, S.S. (2000) *J Clin Periodontol* 27, 648–57.

7 Aas, J.A., Paster, B.J., Stokes, L.N., Olsen, I. and Dewhirst, F.E. (2005) *J Clin Microbiol* 43, 5721–32.

8 Slots, J. (1977) *Scand J Dent Res* 85, 247–54

9 Kumar, P.S., Griffen, A.L., Moeschberger, M.L. and Leys, E.J. (2005) *J Clin Microbiol* 43, 3944–55.

10 Tanner, A.C.R., Paster, B.J., Lu, S.C., Kanasi, E., Kent, R. Jr., Van Dyke, T. and Sonis, S.T. (2006) *J Dent Res* 85, 318–23.

11 Hartley, M.G., El-Maaytah, M.A., McKenzie, C. and Greenman, J. (1996) *Micr Ecol Hlth Dis* 9, 215–23

12 Kazor, C.E., Mitchell, P.M., Lee, A.M., Stokes, L.N., Loesche, W.J., Dewhirst, F.E. and Paster, B.J. (2003) *J Clin Microbiol* 41, 558–63.

13 Mager, D.L., Ximenez-Fyvie, L.A., Haffajee, A.D. and Socransky, S.S. (2003) *J Clin Periodontol* 30, 644–54.

8.7 FURTHER READING

8.7.1 Books

Kuramitsu, H.K. and Ellen, R.P. (eds) (2000) *Oral Bacterial Ecology: The Molecular Basis*. Horizon Scientific Press, Wymondham, UK.

Lamont, R., Burne, R., Lantz, M. and Leblanc, D. (eds). (2006) *Oral Microbiology and Immunology*. ASM Press, Washington, DC, USA.

Marsh, P. and Martin, M.V. (1999) *Oral Microbiology*. Wright, Oxford, UK.

Newman, H.N. and Wilson, M. (eds) (1999) *Dental Plaque Revisited: Oral Biofilms in Health and Disease*. Bioline, Cardiff, UK.

8.7.2 Reviews and papers

Aas, J.A., Paster, B.J., Stokes, L.N., Olsen, I. and Dewhirst, F.E. (2005) Defining the normal bacterial flora of the oral cavity. *J Clin Microbiol* 43, 5721–32.

Al-Ahmad, A., Wunder, A., Auschill, T.M., Follo, M., Braun, G., Hellwig, E. and Arweiler, N.B. (2007) The in vivo dynamics of *Streptococcus* spp., *Actinomyces naeslundii*, *Fusobacterium nucleatum* and *Veillonella* spp. in dental plaque biofilm as analysed by five-colour multiplex fluorescence in situ hybridization. *J Med Microbiol* 56, 681–7.

Corby, P.M., Lyons-Weiler, J., Bretz, W.A., Hart, T.C., Aas, J.A., Boumenna, T., Goss, J., Corby, A.L., Junior, H.M., Weyant, R.J. and Paster, B.J. (2005) Microbial risk indicators of early childhood caries. *J Clin Microbiol* 43, 5753–9.

Dale, B.A. and Fredericks, L.P. (2005) Antimicrobial peptides in the oral environment: Expression and function in health and disease. *Curr Issues Mol Biol* 7, 119–33.

Dale, B.A., Tao, R., Kimball, J.R. and Jurevic, R.J. (2006) Oral antimicrobial peptides and biological control of caries. *BMC Oral Health* 6 (Suppl 1), S13.

Diaz, P.I., Chalmers, N.I., Rickard, A.H., Kong, C., Milburn, C.L., Palmer, R.J. Jr. and Kolenbrander, P.E. (2006) Molecular characterization of subject-specific oral microflora during initial colonization of enamel. *Appl Environ Microbiol* 72, 2837–48.

Dodds, M.W.J., Johnson, D.A. and Yeh, C.K. (2005) Health benefits of saliva: A review. *J Dentistry* 33, 223–33.

Duncan, M.J. (2005) Oral microbiology and genomics. *Periodontology 2000* 38, 63–71.

Edwards, A.M., Grossman, T.J. and Rudney, J.D. (2006) *Fusobacterium nucleatum* transports noninvasive *Streptococcus cristatus* into human epithelial cells. *Infect Immun* 74, 654–62.

Goodson, J.M. (2003) Gingival crevice fluid flow. *Periodontology 2000* 31, 43–54

Helmerhorst, E.J. and Oppenheim, F.G. (2007) Saliva: A dynamic proteome. *J Dent Res* 86, 680–93.

Hintao, J., Teanpaisan, R., Chongsuvivatwong, V., Ratarasan, C. and Dahlen, G. (2007) The microbiological profiles of saliva, supragingival and subgingival plaque and dental caries in adults with and without type 2 diabetes mellitus. *Oral Microbiol Immunol* 22, 175–81.

Hu, S., Loo, J.A. and Wong, D.T. (2007) Human saliva proteome analysis. *Ann N Y Acad Sci* 1098, 323–19.

Hull, M.W. and Chow, A.W. (2007) Indigenous microflora and innate immunity of the head and neck. *Infect Dis Clin North Am* 21, 265–82.

Ji, S., Hyun, J., Park, E., Lee, B.L., Kim, K.K. and Choi, Y. (2007) Susceptibility of various oral bacteria to antimicrobial peptides and to phagocytosis by neutrophils. *J Periodontal Res* 42, 410–19.

Kazor, C.E., Mitchell, P.M., Lee, A.M., Stokes, Loesche, W.J., Dewhirst, F.E. and Paster, B.J. (2003) Diversity of bacterial populations on the tongue dorsa of patients with halitosis and healthy patients. *J Clin Microbiol* 41, 558–63.

Kolenbrander, P.E., Andersen, R.N., Blehert, D.S., Egland, P.G., Foster, J.S. and Palmer, R.J. Jr. (2002) Communication among oral bacteria. *Microbiol Mol Biol Rev* 66, 486–505.

Kreth, J., Merritt, J., Shi, W. and Qi, F. (2005) Competition and coexistence between *Streptococcus mutans* and *Streptococcus sanguinis* in the dental biofilm. *J Bacteriol* 187, 7193–203.

Kumar, P.S., Griffen, A.L., Moeschberger, M.L. and Leys, E.J. (2005) Identification of candidate periodontal pathogens and beneficial species by quantitative 16S clonal analysis. *J Clin Microbiol* 43, 3944–55.

Ledder, R.G., Gilbert, P., Huws, S.A., Aarons, L., Ashley, M.P., Hul, l.P.S. and McBain, A.J. (2007) Molecular analysis of the subgingival microbiota in health and disease. *Appl Environ Microbiol* 73, 516–23.

Li, J., Helmerhorst, E.J., Leone, C.W., Troxler, R.F., Yaskel, T., Haffajee, A.D., Socransky, S.S. and Oppenheim, F.G. (2004)

Identification of early microbial colonizers in human dental biofilm. *J Appl Microbiol* 97, 1311–18.

Li, Y., Saxena, D., Barnes, V.M., Trivedi, H.M., Ge, Y. and Xu, T. (2006) Polymerase chain reaction-based denaturing gradient gel electrophoresis in the evaluation of oral microbiota. *Oral Microbiol Immunol* 21, 333–9.

Mager, D.L., Ximenez-Fyvie, L.A., Haffajee, A.D. and Socransky, S.S. (2003) Distribution of selected bacterial species on intraoral surfaces. *J Clin Periodontol* 30, 644–54.

Marcy, Y., Ouverney, C., Bik, E.M., Lösekann, T., Ivanova, N., Martin, H.G., Szeto, E., Platt, D., Hugenholtz, P., Relman, D.A. and Quake, S.R. (2007) Dissecting biological "dark matter" with single-cell genetic analysis of rare and uncultivated TM7 microbes from the human mouth. *Proc Natl Acad Sci U S A* 104, 11889–94.

Marsh, P.D. (2005) Dental plaque: Biological significance of a biofilm and community life-style. *J Clin Periodontol* 32 (Suppl 6), 7–15.

Marsh, P.D. (2006) Dental plaque as a biofilm and a microbial community – implications for health and disease. *BMC Oral Health* 6, S14.

Marshall, R.I. (2004) Gingival defensins: Linking the innate and adaptive immune responses to dental plaque. *Periodontology 2000* 35, 14–20.

Ouhara, K., Komatsuzawa, H., Yamada, S., Shiba, H., Fujiwara, T., Ohara, M., Sayama, K., Hashimoto, K., Kurihara, H. and Sugai, M. (2005) Susceptibilities of periodontopathogenic and cariogenic bacteria to antibacterial peptides, β-defensins and LL37, produced by human epithelial cells. *J Antimicrob Chemother* 55, 888–96.

Palmer, R.J. Jr., Gordon, S.M., Cisar, J.O. and Kolenbrander, P.E. (2003) Coaggregation-mediated interactions of streptococci and actinomyces detected in initial human dental plaque. *J Bacteriol* 185, 3400–9.

Paster, B.J., Olsen, I., Aas, J.A. and Dewhirst, F.E. (2006) The breadth of bacterial diversity in the human periodontal pocket and other oral sites. *Periodontology 2000* 42, 80–7.

Robinson, C., Strafford, S., Rees, G., Brookes, S.J., Kirkham, J., Shore, R.C., Watson, P.S. and Wood, S. (2006) Plaque biofilms: The effect of chemical environment on natural human plaque biofilm architecture. *Arch Oral Biol* 51, 1006–14.

Rudney, J.D., Chen, R. and Sedgewick, G.J. (2005) *Actinobacillus actinomycetemcomitans, Porphyromonas gingivalis,* and *Tannerella forsythensis* are components of a polymicrobial intracellular flora within human buccal cells. *J Dent Res* 84, 59–63.

Ruhl, S., Sandberg, A.L. and Cisar, J.O. (2004) Salivary receptors for the proline-rich protein-binding and lectin-like adhesins of oral actinomyces and streptococci. *J Dent Res* 83, 505–10.

Sakamoto, M., Takeuchi, Y., Umeda, M., Ishikawa, I. and Benno, Y. (2003) Application of terminal RFLP analysis to characterize oral bacterial flora in saliva of healthy subjects and patients with periodontitis. *J Med Microbiol* 52, 79–89.

Sakamoto, M., Umeda, M. and Benno, Y. (2005) Molecular analysis of human oral microbiota. *J Periodont Res* 40, 277–85.

Sugawara, Y., Uehara, A., Fujimoto, Y., Kusumoto, S., Fukase, K., Shibata, K., Sugawara S., Sasano, T. and Takada, H. (2006) Toll-like receptors, NOD1 and NOD2 in oral epithelial cells. *J Dent Res* 85, 524–9.

Suntharalingam, P. and Cvitkovitch, D.G. (2005) Quorum sensing in streptococcal biofilm formation. *Trends Microbiol* 13, 3–6.

Suzuki, N., Yoshida, A. and Nakano, Y. (2005) Quantitative analysis of multi-species oral biofilms by TaqMan real-time PCR. *Clin Medicine Res* 3, 176–85.

Tabak, L.A. (2006) In defense of the oral cavity: The protective role of the salivary secretions. *Pediatr Dent* 28, 110–17.

Takamatsu, D., Bensing, Prakobphol, A., Fisher, S.J. and Sullam, P.M. (2006) Binding of the streptococcal surface glycoproteins GspB and Hsa to human salivary proteins. *Infect Immun* 74, 1933–40.

Tanner, A.C.R., Paster, B.J., Lu, S.C., Kanasi, E., Kent, R. Jr., Van Dyke, T. and Sonis, S.T. (2006) Subgingival and tongue microbiota during early periodontitis. *J Dent Res* 85, 318–23.

ten Cate, J.M. (2006) Biofilms, a new approach to the microbiology of dental plaque. *Odontology* 94, 1–9.

Teng, Y.T. (2006) Protective and destructive immunity in the periodontium: Part 1 - Innate and humoral immunity and the periodontium. *J Dent Res* 85, 198–208.

Walker, D.M. (2004) Oral mucosal immunology: An Overview. *Ann Acad Med Singapore* 33, 27S–30S.

Wang, H., Wang, Y., Chen, J., Zhan, Z., Li, Y. and Xu, J. (2007) Oral yeast flora and its ITS sequence diversity among a large cohort of medical students in Hainan, China. *Mycopathologia* 164, 65–72.

Yoshida, Y., Palmer, R.J., Yang, J., Kolenbrander, P.E. and Cisar, J.O. (2006) Streptococcal receptor polysaccharides: Recognition molecules for oral biofilm formation. *BMC Oral Health* 6 (Suppl 1), S12.

Chapter 9

THE INDIGENOUS MICROBIOTA OF THE GASTROINTESTINAL TRACT

The gastrointestinal tract (GIT), together with the accessory digestive organs (i.e. teeth, tongue, salivary glands, liver, gallbladder, and pancreas), constitutes the digestive system, whose function it is to break down dietary constituents into small molecules and then to absorb these for subsequent distribution throughout the body. The GIT consists of several anatomically and functionally distinct regions – the oral cavity, the pharynx (which is also part of the respiratory tract), esophagus, stomach, small intestine (duodenum, jejunum, and ileum) and the large intestine (cecum, colon, and rectum). Essentially, however, it can be considered to be a continuous tube extending from the mouth to the anus (Fig. 9.1).

The environmental determinants within each region of the GIT are very different, and each region, therefore, has a distinctive microbiota. Nowhere is this more apparent than in the oral cavity where the complex anatomy, the presence of shedding and nonshedding surfaces, and the existence of large mechanical forces combine to provide a group of very different habitats. In contrast, each of the other regions of the GIT has a simpler structure, is not subjected to large mechanical forces, and does not contain nonshedding surfaces;

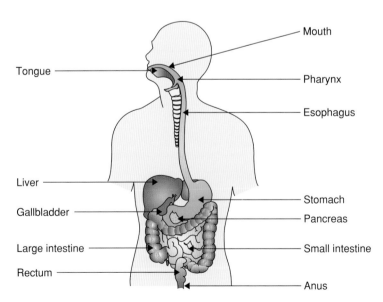

Fig. 9.1 Diagram showing the main organs of the digestive system, which consists of the intestinal tract (mouth, pharynx, esophagus, stomach, small and large intestines, rectum, and anus) and the accessory digestive organs (teeth, tongue, salivary glands, liver, gallbladder, and pancreas). (From http://www.training.seer.cancer.gov/module_anatomy/anatomy_physiology_home.html; funded by the US National Cancer Institute's Surveillance, Epidemiology and End Results (SEER) Program with Emory University, Atlanta, Georgia, USA.)

consequently, each has a lower habitat diversity than that of the oral cavity. Because of these important differences, it is convenient to describe the oral cavity and its various microbial communities separately, and this has been done in Chapter 8.

The microbiota of the pharynx has already been described in Chapter 4. The present chapter, therefore, will be concerned only with that part of the GIT extending from the esophagus to the anus.

9.1 ANATOMY AND PHYSIOLOGY OF THE GASTROINTESTINAL TRACT

Apart from the oral cavity (Chapter 8) and pharynx (Chapter 4), the GIT essentially consists of a tube comprised of four layers of tissue (Fig. 9.2). The innermost layer is the mucosa, which consists of an epithelium surrounded by connective tissue and a thin layer of muscle known as the muscularis mucosae. Mucosa-associated lymphoid tissue (MALT) is also present. Muscular contractions cause folding of the mucosa, which increases its surface area, thereby aiding digestion and absorption of the resulting products – the total surface area of the mucosa is 200–300 m^2, which makes it the largest body surface in contact with the external environment. The remaining layers consist of

the submucosa (connective tissue), the muscularis (sheets of circular and longitudinal muscles), and the serosa (connective tissue covered by squamous epithelium). Contraction of the muscles comprising the muscularis causes mixing of the food with digestive fluids, and so contributes to the physical breakdown of food. The muscles contract in sequence, thereby producing a peristaltic wave which propels the resulting mixture along the GIT. Although the above general description applies to those regions of the GIT from the esophagus to the anus, each region shows characteristic variations on this theme, and these will now be described briefly.

The esophagus connects the laryngopharynx to the stomach, and its function is to transport food between these regions. It is a tubular structure approximately 25 cm long and is lined by a stratified, squamous epithelium and has numerous mucous glands. Food entering the esophagus is propelled towards the stomach by peristalsis, and movement of the food bolus is aided by the presence of saliva and by mucus.

The stomach is a J-shaped structure (Fig. 9.3), linking the esophagus to the duodenum and has a total capacity of approximately 1500 ml. Food enters the stomach through the cardiac orifice, which is situated below the uppermost region of the stomach (the fundus). The pyloric sphincter separates the stomach from

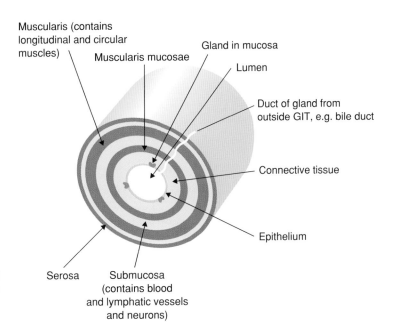

Fig. 9.2 Cross-section through the wall of the gastrointestinal tract, showing the arrangement of the four basic layers of tissues.

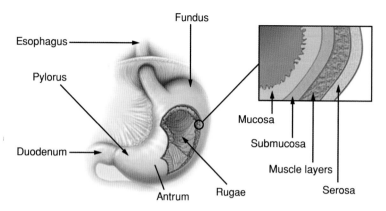

Fig. 9.3 Diagram showing the main regions of the stomach and a cross-section through the stomach wall. (From http://www.training.seer.cancer.gov; funded by the US National Cancer Institute's Surveillance, Epidemiology and End Results (SEER) Program with Emory University, Atlanta, Georgia, USA.)

the duodenum. When the stomach is empty, its mucosa forms folds known as rugae.

The gastric epithelium consists of a layer of non-ciliated simple columnar epithelial cells and has a large number of shallow involutions, known as gastric pits, into the base of which the gastric glands open (Fig. 9.4). The gastric glands secrete gastric juice, which is a mixture of mucus, pepsinogen, hydrochloric acid, intrinsic

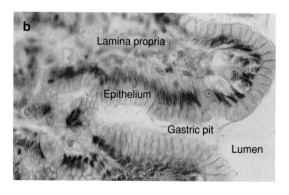

Fig. 9.4 Human gastric mucosa. (a) Diagram of a cross-section through the gastric mucosa showing gastric pits and glands. (b) Cross-section through the human gastric mucosa. The surface is indented to form numerous short gastric pits, which are open to the lumen. The epithelium consists mainly of mucus-secreting cells. Image © 2006, David King, Southern Illinois University School of Medicine, Springfield and Carbondale, Illinois, USA; used with permission.

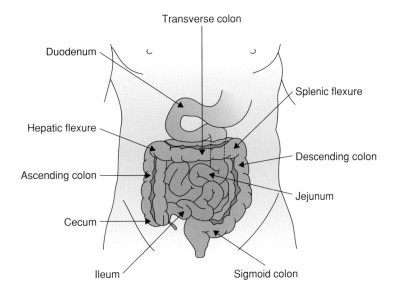

Fig. 9.5 Diagram showing the main regions of the small and large intestines. (From http://www.training.seer.cancer.gov; funded by the US National Cancer Institute's Surveillance, Epidemiology and End Results (SEER) Program with Emory University, Atlanta, Georgia, USA.)

factor, and gastrin. The HCl converts the enzymically inactive pepsinogen to the active enzyme pepsin, which hydrolyzes dietary proteins to peptides and has its optimum activity at the low pH (usually less than 2.0) of the stomach. Gastrin stimulates secretion of gastric juice and increases gut motility. The epithelium is protected from the low pH by a layer of mucus 0.2–0.6 mm thick. Peristaltic movements of the stomach mix food with the gastric juice to produce a watery fluid known as chyme, small quantities of which pass through the pyloric sphincter into the small intestine.

The small intestine is the main site of digestion and absorption (Fig. 9.5). It is a tubular structure approximately 3 m long and 2.5–3.5 cm in diameter and consists of three regions – the duodenum (approximately 25 cm in length), the jejunum (approximately 1.0 m), and the ileum (approximately 2.0 m).

The mucosa is highly folded to produce permanent ridges, which project into the lumen, and has numerous finger-like projections (0.5–1.0 mm in length), known as villi, of which there are between 20 and 40 per mm^2 (Fig. 9.6). By increasing the surface area, these features enhance digestion and absorption.

The epithelial surface is covered in a layer of mucus 10–250 μm thick. The epithelium consists mainly of simple columnar epithelial cells, absorptive cells (enterocytes), and mucus-secreting goblet cells (Fig. 9.7). The surface of each enterocyte has numerous projections (microvilli), which provide a very large surface area (known as a "brush border") for absorption of low

molecular mass digestion products. In the mucosa between the bases of the villi are a number of crypts (between six and ten per villus), known as the crypts of Lieberkuhn. These contain undifferentiated stem cells, which exit the crypt and move from the base of the villus to its tip, undergoing transformation to an enterocyte, a goblet cell, or an enteroendocrine cell. On reaching the tip of the villus, each cell is shed into the lumen. The whole process takes approximately 3 days. The crypts also produce Paneth cells, which remain within the crypts and produce a range of antimicrobial compounds (section 9.2.1). The combined secretions of the cells of the small intestine are known as intestinal juice. MALT is also abundant in the small intestine, and is present as solitary lymphatic nodules and as aggregated lymphatic follicles known as Peyer's patches (see section 9.2.2). In adults, the latter are localized predominantly in the ileum.

After each meal, approximately 2.0 liters of chyme enters the duodenum, and localized contractions cause mixing of the chyme with enzymes produced by epithelial cells, intestinal juice, pancreatic juice, and bile. Enzymatic hydrolysis of food macromolecules takes place, resulting in small molecules which are taken up by the absorptive cells – approximately 90% of all absorption occurs in the small intestine. This absorptive phase is followed by peristalsis, which expels the chyme into the large intestine.

The large intestine is about 1.5 m long and 6.5 cm in diameter, with a surface area of approximately

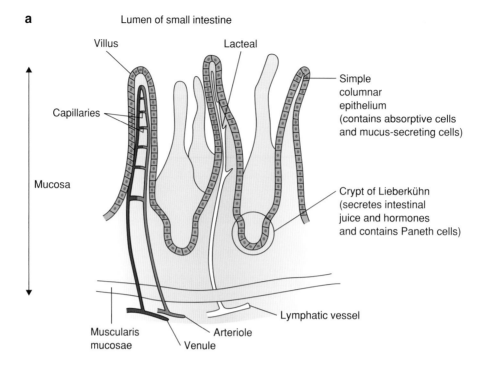

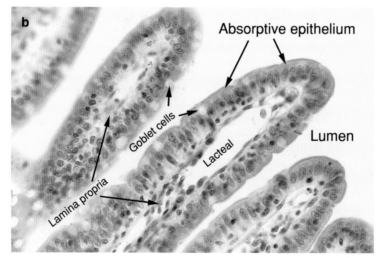

Fig. 9.6 Small intestine. (a) Diagram showing the structure of the mucosal surface of the small intestine. (b) Cross-section through the mucosa of the human duodenum, showing numerous villi. Image © 2006, David King, Southern Illinois University School of Medicine, Springfield, Illinois, USA; used with permission.

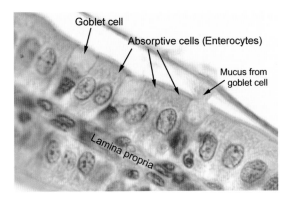

Fig. 9.7 Surface of the epithelium of the small intestine, showing goblet and absorptive cells. Mucus can be seen discharging from a goblet cell. Image © 2006, David King, Southern Illinois University School of Medicine, Springfield, Illinois, USA; used with permission.

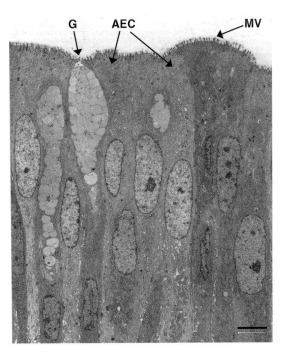

Fig. 9.9 Transmission electron micrograph of intestinal epithelium, showing a goblet cell (G) and absorptive epithelial cells (AEC) with associated microvilli (MV). Scale bar, 3 μm. Reproduced with kind permission of Springer Science and Business Media, Berlin, Germany, from: Mayhew, T.M. et al. (1992) Structural and enzymatic studies on the plasma membrane domains and sodium pump enzymes of absorptive epithelial cells in the avian lower intestine. *Cell Tissue Res* 270, 577–85.

1250 cm^2. It consists of several regions – the cecum, colon (ascending, transverse, descending, and sigmoid), rectum, and anal canal. Its mucosal surface is very different from that of the small intestine in that it has no permanent folds or villi and has a large number of narrow invaginations (known as intestinal glands or crypts) lined by absorptive and goblet cells (Fig. 9.8). The absorptive cells, which have many microvilli, are involved primarily in water absorption, while the

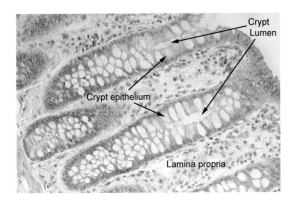

Fig. 9.8 Cross-section through the mucosa of the human colon. No villi are present, but there are many narrow invaginations lined with absorptive and goblet cells. Image © 2006, David King, Southern Illinois University School of Medicine, Springfield, Illinois, USA; used with permission.

goblet cells secrete mucus which provides lubrication to facilitate the peristaltic movement of the luminal contents (Fig. 9.9).

The mucous layer coating the epithelial surface increases in thickness from the proximal colon (107 ± 48 μm) to the rectum (155 ± 54 μm) – it is continually being synthesized by goblet cells and degraded by microbial action. No digestive enzymes are secreted by the mucosa of the large intestine – further breakdown of dietary constituents in this region is carried out by the resident microbiota. Between 3 and 10 h after entering the large intestine, the chyme becomes solid as a result of water absorption and is known as feces. Feces are transferred into the rectum from the lower, S-shaped region of the colon known as the sigmoid colon.

9.2 ANTIMICROBIAL DEFENSE SYSTEMS OF THE GASTROINTESTINAL TRACT

9.2.1 Innate defense systems

The intestinal epithelium, like all mucosae, is a shedding surface. The lifespan of an intestinal epithelial cell is between 2 and 5 days, after which time it is shed from the mucosal surface, taking with it any attached microbes. Approximately 2×10^{11} epithelial cells are shed into the lumen of the GIT each day. This constant shedding, coupled with the rapid transit of material through the upper regions of the GIT (esophagus, stomach, and small intestine), tends to limit colonization of the mucosal surfaces in these regions.

The mucous layer (Fig. 9.10) that covers the whole of the intestinal epithelium is an important component of the innate defense system. It consists mainly of water and mucins (principally MUC2), with a high oligosaccharide content (>80%), which polymerize to form a viscous gel. Other mucins produced by the intestinal mucosa remain anchored to the surface of epithelial cells. The mucin content of the mucus gradually decreases from the stomach (50 mg/ml) to the colon (20 mg/ml). The high carbohydrate content of the mucins endows the gel with lectin-binding properties, which enable it to bind and trap bacteria, thereby preventing their access to the underlying epithelium. In addition to mucins, the mucous layer contains various effector molecules of the innate and acquired host defense systems. The main functions of the mucous layer are to: (1) protect the underlying epithelium from microbial colonization; (2) provide lubrication to facilitate movement of the luminal contents; (3) protect the underlying mucosa (mainly in the stomach and duodenum) from acid and digestive enzymes; and (4) protect the mucosa from the shearing forces generated by movement of material through the GIT.

The intestinal mucosa produces a range of molecules with antimicrobial activities – lysozyme, lactoferrin, lactoperoxidase, secretory phospholipase A_2, bacterial permeability-inducing protein, collectins, adrenomedullin, histone H1, RegIIIγ and several antimicrobial peptides (Fig. 9.11).

With regard to the antimicrobial peptides produced, the α-defensins human defensin-5 (HD-5) and human defensin-6 (HD-6) are secreted by Paneth cells in the crypts of the small intestine (Fig. 9.12). HD-5 is also produced by intermediate cells of the villi. In contrast, the β-defensins human β-defensin-1 (HBD-1), HBD-2, and HBD-3 are secreted by a variety of epithelial cells in the stomach, small intestine, and colon. HBD-1 is constitutively expressed, although increased quantities are produced in response to IL-1α, whereas HBD-2 is produced by the mucosa only in response to IL-1α, bacterial infection, or during inflammation. HBD-4 is produced constitutively by epithelial cells of the gastric

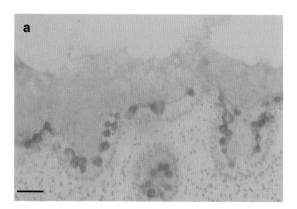
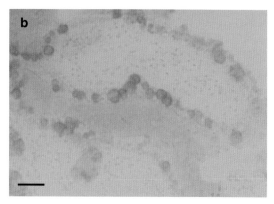

Fig. 9.10 Photomicrographs showing the mucous layer coating (a) the colon and (b) terminal ileum. Stained with Alcian blue. The epithelium of the colon is covered with a thick flat mucous layer (blue), and the crypts are filled with mucus. In the terminal ileum, the mucus fills most of the space between the villi. The luminal side of the villi is covered with a thin mucous layer. Scale bar, 50 μm. Reproduced with the permission of Lippincott, Williams & Wilkins, Philadelphia, Pennsylvania, USA, from: van der Waaij, L.A., Harmsen, H.J., Madjipour, M., Kroese, F.G., Zwiers, M., van Dullemen, H.M., de Boer, N.K., Welling, G.W. and Jansen, P.L. (2005) Bacterial population analysis of human colon and terminal ileum biopsies with 16S rRNA-based fluorescent probes: Commensal bacteria live in suspension and have no direct contact with epithelial cells. *Inflamm Bowel Dis* 11, 865–71.

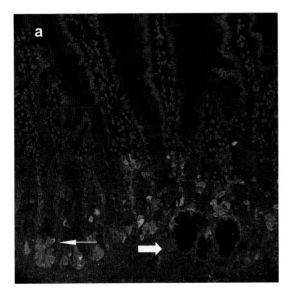
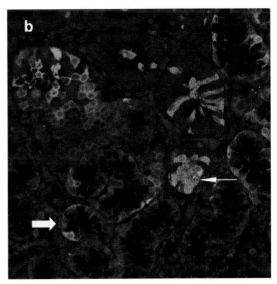

Fig. 9.11 Localization of lysozyme and HD-5 in human duodenal mucosa. Lysozyme (green staining) in Brunner's glands is denoted by the broad arrows. Thin arrows denote Paneth cells stained red, indicating co-localization of HD-5 and lysozyme. The image was kindly supplied by Dr Bo Shen. Reprinted with the permission of the Federation of the European Biochemical Societies (University College London, London, UK), from: Wehkamp, J., Chu, H., Shen, B., Feathers, R.W., Kays, R.J., Lee, S.K. and Bevins, C.L. (2006) Paneth cell antimicrobial peptides: Topographical distribution and quantification in human gastrointestinal tissues *FEBS Lett* 580, 5125–8, ©2006.

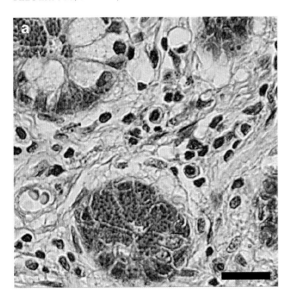
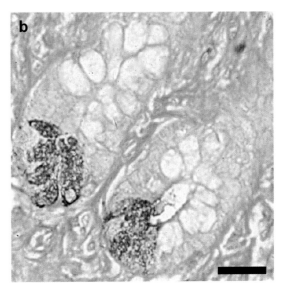

Fig. 9.12 Secretion of antimicrobal peptides by Paneth cells in the crypts of the small intestine. Phloxine tartrazine staining of human ileal mucosa (left), showing antimicrobial peptide-rich granules (stained red). Immunohistochemical localization (brown) of HD-5 in Paneth cells (right). Scale bars, 25 µm. Reproduced with permission from: Wehkamp, J., Salzman, N.H., Porter, E., Nuding, S., Weichenthal, M., Petras, R.E., Shen, B., Schaeffeler, E., Schwab, M., Linzmeier, R., Feathers, R.W., Chu, H., Lima, H. Jr., Fellermann, K., Ganz, T., Stange, E.F. and Bevins, C.L. (2005) Reduced Paneth cell α-defensins in ileal Crohn's disease. *Proc Natl Acad Sci U S A* 102, 18129–34. © 2005, National Academy of Sciences, Washington, DC, USA.

antrum and exhibits synergy with lysozyme. Human cathelicidin LL-37 is produced in the stomach and colon but not in the small intestine. The antimicrobial properties of these antimicrobial peptides are described in section 1.5.4.

In the large intestine, expression of all known Toll-like receptors (TLRs) other than TLR10 has been detected, whereas in the small intestine neither TLR6 nor TLR10 appear to be expressed (Fig. 9.13). Expression of NOD1 and NOD2 by intestinal epithelial cells

has also been reported. Recognition of microbe-associated molecular patterns (MAMPs) by TLRs in Paneth cells and intestinal epithelial cells has been shown to result in the production of antimicrobial peptides such as HBD-2, HD-5, and HD-6.

Excessive secretion of fluids (i.e. diarrhea) may be considered to be a host defense mechanism as it flushes out organisms that are resident in the GIT.

The low pH of the stomach (pH < 2) protects it from colonization by a wide range of microbes that are unable

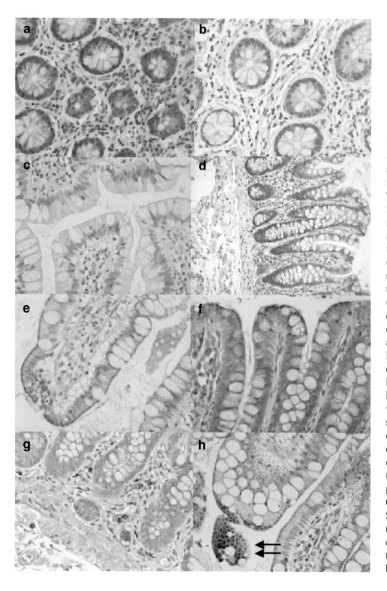

Fig. 9.13 Immunohistochemical analysis of normal healthy human colon stained with either (a–d) a monoclonal mouse anti-human TLR2 IgG2a, or (e–h) TLR3 IgG1 antibody, visualized using immunoperoxidase (dark-staining regions). (a) Transverse section (×40) across the crypt region, with staining in the epithelium and lamina propria. (b) Transverse and (c) longitudinal section of more mature epithelial cells that interface with the gut lumen (×40). (d) A longitudinal section (×20) from crypt to lumen. TLR2 expression is confined to crypt cells with no expression at the luminal surface. (e and f) Longitudinal sections (×40) through the epithelium at the luminal surface. (g) A longitudinal section (×20) from crypt to lumen, and (h) a longitudinal section with a transverse region (×40) showing staining seen from the gut lumen. TLR3 expression is localized in the most mature columnar epithelial cells exiting the crypts and forming the luminal surface of the epithelium. Reproduced with the permission of Blackwells Publishing Ltd, Oxford, UK, from: Furrie, E., Macfarlane, S., Thomson, G. and Macfarlane, G.T. (2005) Toll-like receptors-2, -3 and -4 expression patterns on human colon and their regulation by mucosal-associated bacteria. *Immunol* 115, 565–74.

to tolerate such acidic conditions. This low pH also prevents access of acid-sensitive organisms to the small and large intestines. Both the small and large intestines contain bile and proteolytic enzymes, which can kill a range of microbes. Bile is a mixture of cholesterol, phospholipids, bile acids, and immunoglobulins, which is produced in the liver and secreted into the duodenum. The bile acids are amphipathic, surface-active compounds with potent, but selective, antimicrobial activity in vitro. In general, they are more active against Gram-positive than Gram-negative species. The different bile acids vary in their antimicrobial potency, with the unconjugated compounds having greater activity than the conjugated forms. However, many organisms present in the colon (e.g. *Bacteroides* spp., *Bifidobacterium* spp., *Clostridium* spp., *Lactobacillus* spp., and *Streptococcus* spp.) are able to convert conjugated bile acids to their unconjugated form.

9.2.2 Acquired immune defense system

An immune response is generated following uptake of an antigen by the M cells of Peyer's patches. Secretory IgA is able to block adhesion of microbes to the epithelium and causes microbial aggregation, resulting in aggregates that are more easily expelled from the system (Fig. 9.14). However, the acquired immune response appears to have little effect on the composition of the microbiota of the GIT. This may be due to the inability of antibodies to function effectively at the low pH of the stomach and duodenum. In the colon and ileum, IgA may not be able to exert a detectable effect because of the vast numbers of organisms present. Nevertheless, IgA may play a role in the initial colonization of the GIT, and could exert an effect on the composition of the mucosa-associated microbiota of these regions.

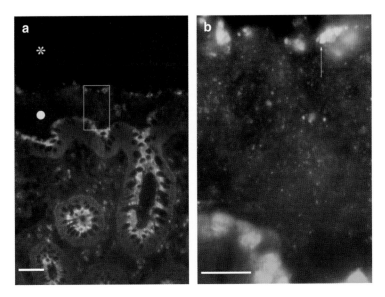

Fig. 9.14 Photomicrographs showing the colon hybridized with rhodamine-BACT338 (which hybridizes with all bacteria and stains orange-red) and stained with FITC-anti IgA (green). (b) Enlargement of the rectangular area shown in (a). Large amounts of IgA (green) are present in the mucus (denoted by the white circle in [a]), at the apical side of the epithelial cells, and in plasma cells within the lamina propria. The remaining cells of the colon are dark red because of autofluorescence. Bacteria (orange-yellow, arrow in [b]) are present within the mucous layer, and many of them are coated with IgA. The lumen is denoted by an asterisk. Scale bars: (a) 50 μm and (b) 10 μm. Reproduced with the permission of Lippincott, Williams & Wilkins, Philadelphia, Pennsylvania, USA, from: van der Waaij, L.A., Harmsen, H.J., Madjipour, M., Kroese, F.G., Zwiers, M., van Dullemen, H.M., de Boer, N.K., Welling, G.W. and Jansen, P.L. (2005) Bacterial population analysis of human colon and terminal ileum biopsies with 16S rRNA-based fluorescent probes: Commensal bacteria live in suspension and have no direct contact with epithelial cells. *Inflamm Bowel Dis* 11, 865–71.

Table 9.1 Host defense mechanisms in the gastrointestinal tract.

Mechanism	Effect
Production of mucus	Prevents microbial adhesion to epithelial cells; traps microbes; facilitates movement of gut contents, thereby expelling microbes; provides receptors for adhesion of members of indigenous microbiota, thereby preventing colonization by pathogens
Rapid transit of luminal contents in upper regions	Hinders colonization in esophagus, stomach, and small intestine
Desquamation	Removes microbes attached to exfoliated cells
Production of antimicrobial proteins and peptides: lysozyme, lactoferrin, lactoperoxidase, secretory phospholipase A_2, bacterial permeability-inducing protein, collectins, adrenomedullin, RegIIIγ, HD-5, HD-6, HBD-1, HBD-2, HBD-3, and HBD-4	Compounds exert a microbicidal or microbistatic effect
Release of histone H1 from apoptotic epithelial cells	Kills or inhibits microbes
Expression of TLRs and NODs	TLRs and NODs induce release of antimicrobial peptides by Paneth and epithelial cells
Release of cytokines by epithelial cells	Cytokines attract and activates phagocytes
Low pH of stomach	Kills or inhibits wide range of microbes
Bile acids	Antimicrobial – Gram-positive species generally more susceptible than Gram-negative species
Proteolytic enzymes	Microbicidal
Stimulation of excessive fluid secretion	Flushes out microbes from GIT
Production of IgA	Blocks adhesion of microbes to epithelial cells

The host defense mechanisms operating in the GIT are summarized in Table 9.1.

9.3 ENVIRONMENTAL DETERMINANTS WITHIN DIFFERENT REGIONS OF THE GASTROINTESTINAL TRACT

In general, many of the environmental determinants operating in the GIT remain unknown, because most of the regions of the GIT are difficult to access and/or involve sampling procedures that are uncomfortable or embarrassing for the individual. One of the distinguishing features of the GIT, which sets it apart from other body sites, is that the microbes colonizing some regions have access to nutrients present in food ingested by the host, and therefore are not dependent wholly on host

secretions or on other microbes for their nutrients. The vast numbers of epithelial cells shed into the lumen of the GIT constitute another important source of nutrients.

9.3.1 Esophagus

The esophagus is an aerobic region with a temperature of 37°C. The pH of the mucous layer covering the epithelium is approximately 6.8. The esophagus receives saliva continuously, and food and beverages intermittently, and is therefore subjected to hydrodynamic and mechanical forces. The ability of an organism to adhere to the epithelium, therefore, is an important requisite for any potential colonizer of this region. Although nutrient-rich, host-ingested food and fluids pass regularly through the esophagus, their transit time is short, and they are therefore unlikely to contribute significantly to the nutrients available to resident microbes. The only reliable source of nutrients is the mucous layer, which contains mucins as well as substances excreted and secreted by epithelial cells.

9.3.2 Stomach

The temperature of gastric fluid is normally 37°C, but will be affected temporarily by the intake of hot and cold foods and beverages. It is an aerobic environment, and the partial pressure of oxygen at the luminal surface of the mucosa is 46.3 ± 15.4 mmHg, which corresponds to 29% of the oxygen content of air. Although the median 24 h intragastric pH is 1.4, the pH is influenced by many factors, including age, diet, and whether or not food and/or drinks have recently been ingested. Food has a strong buffering action, and the pH of gastric juice over a 24-h period ranges from approximately 1 to 5, with the higher pHs corresponding to mealtimes (section 9.4.2.2). The main sources of nutrients for microbes in the gastric lumen are shown in Table 9.2.

Although a number of molecules with antimicrobial properties are present in the stomach (e.g. IgA, pepsin, antimicrobial peptides, lysozyme, etc.), many of these, apart from pepsin, are not effective at the low pH of the gastric lumen. However, they are likely to exert an antimicrobial effect at the mucosal surface or during periods of elevated pH when food is present. Recently, it has been shown that O-glycans with terminal α-1,4-linked N-acetylglucosamine residues inhibit the growth of *Helicobacter pylori*. These glycans are produced by cells present in human gastric glands, and this is likely to account, in part, for the failure of the organism to colonize these structures.

9.3.3 Small intestine

Rapid peristalsis in the duodenum, jejunum, and proximal region of the ileum results in a short transit time (3–5 h) for the luminal contents of these regions. However, peristalsis in the distal section of the ileum is much slower. In an adult, a total of approximately 9.0 liters of fluid per day enters the small intestine, and this exerts a flushing action which hinders microbial colonization, particularly in the upper regions. This fluid consists of 2.0 liters of drinks, 1.5 liters of saliva, 2.5 liters of gastric juices, 0.5 liters of bile, 1.5 liters of pancreatic juice, and 1.0 liters of intestinal fluid. More than 80% of this is absorbed by the small intestine. Other factors limiting microbial colonization are the high concentrations of bile salts and proteolytic enzymes, other host

Table 9.2 The principal nutrient sources for microbes in the gastric lumen.

Nutrient source	Fate in the stomach
Food ingested by the host	Transit time of 2–6 h; salivary amylases convert polysaccharides to oligosaccharides and disaccharides, which are further degraded by oral microbes (present in saliva) and by the gastric microbiota; gastric proteases hydrolyze proteins to oligopeptides, which are degraded to amino acids by microbes
Local host secretions	Mucins in the mucous layer are degraded to carbohydrates and amino acids by the concerted action of salivary organisms and *Hel. pylori*
Interstitial tissue fluid released into the lumen as a result of damage to the epithelium caused by the vacuolating cytotoxin of *Hel. pylori*	Interstitial fluid contains a variety of high and low molecular mass constituents, which can be utilized by the resident microbiota
Salivary constituents	Proteins and mucins are degraded by salivary and gastric microbes
Microbes	Many microbes are killed by the low pH, and their components are degraded by acid and enzymes in gastric fluid; metabolic end products are produced by organisms that are able to survive in the stomach

antimicrobial defenses (see section 9.2), and the low pH of the upper regions. Although chyme entering the small intestine is very acidic, the pH soon increases because of the alkaline fluids produced by, and delivered to, the small intestine. These include intestinal juice (pH = 7.6), pancreatic juice (pH = 7.1–8.2), and bile (pH = 7.6–8.6). The pH gradually increases along the small intestine, and ranges from 5.7 to 6.4 in the duodenum, from 5.9 to 6.8 in the jejunum, and from 7.3 to 7.7 in the ileum. The mucosa itself also secretes bicarbonate ions, which raise the pH.

Chyme provides an abundance of a variety of nutrients for resident microbes. Additional nutrient sources include mucins, exfoliated intestinal cells, and microbial end products of metabolism.

Oxygen is present in the various fluids secreted into the lumen and diffuses into the lumen from the tissues underlying the muosa. Nevertheless, its concentration in the small intestine is low, and its partial pressure at the lumenal surface of the mucosa ranges from 34 to 36 mmHg, which corresponds to approximately 22% of the oxygen content of air. There is very little information regarding the oxygen concentration of the lumenal contents. Microbial activity, however, will reduce the oxygen content as the chyme passes along the small intestine, and the redox potential of the ileum has been reported to be as low as −150 mV.

9.3.4 Large intestine

The human colon is approximately 150 cm long and has a volume of approximately 540 ml. Approximately 1.5 kg of material enters the colon each day, but much of this is water and is rapidly absorbed. The fluid material entering the first section of the large intestine, the cecum, is mixed thoroughly, and readily digestible compounds are rapidly utilized by resident bacteria. Digested matter remains in the cecum for approximately 18 h, from where it passes into the ascending colon. Portions of the digested material are periodically transferred to the transverse colon, and water is absorbed as it moves through the rest of the large intestine. The quantity of material within the colon of an adult averages 220 g, of which 35 g is dry matter. The mean transit time of material through the colon is between 30 and 60 h, and during this time the water content of the material decreases from 86% in the cecum to 77% in the rectum. Approximately 120 g of feces is produced each day, and bacteria comprise 55%

of the fecal solids. Microbial activity continually reduces the nutrient content and alters the composition of the material as it passes along the colon. This, in turn, dictates which organisms can survive and grow in a particular region and, therefore, the type of microbial metabolic end products produced and the nature of the local environment.

The pH of the cecum (approximately 5.7) is much lower than that of the ileum due to the rapid bacterial fermentation of carbohydrates to short-chain fatty acids (SCFAs), i.e. butyrate, acetate, and propionate. The pH remains low in the ascending colon (mean pH = 5.6) and in the traverse colon (mean pH = 5.7), but then increases to 6.6 in the descending and sigmoid colons due to the absorption of SCFAs and the secretion of bicarbonate by the mucosa. The pH in the rectum ranges from 6.6 to 6.8.

The lumen of the colon is an anaerobic region and has a very low redox potential with values in the range −200 to −300 mV. In contrast, the mucosal surface has a relatively high oxygen content because it is supplied with this gas by the underlying tissue. The partial pressure of oxygen on the lumenal side of the mucosa is approximately 30 mmHg in the cecum, 39 mmHg in the transverse colon, 29 mmHg in the descending colon, and 39 mmHg in the sigmoid colon. These values correspond to 19, 25, 18, and 25%, respectively, of the oxygen content of air.

A wide range of nutrients are available to colonic microbes including those present in, or derived from, the individual's diet, host secretions, cells shed from the intestinal mucosa, and products of microbial metabolism (Table 9.3).

The host's diet has a marked effect on the quantity and composition of the material entering the colon. A wide range of carbohydrates may be present, including unabsorbed monosaccharides, oligosaccharides, and polysaccharides (e.g. starch, cellulose, pectins, xylan, inulin, and hemicelluloses), as well as various gums and mucopolysaccharides (e.g. hyaluronic acid and chondroitin sulfate). Oligosaccharides and/or monosaccharides may also be liberated from host glycoproteins and glycosphingolipids by bacterial enzymes. A wide range of bacterial genera are able to degrade the various polysaccharides that may be present in the human colon (Table 9.4). However, complete degradation of the more complex molecules (e.g. mucins and glycosphingolipids) usually requires the concerted actions of several different organisms (section 1.5.3). In fact, only four species – *Ruminococcus*

Table 9.3 Nutrients present in the colon and their origins. End products of microbial metabolism constitute an additional source of a variety of nutrients.

Nutrient	Source	Amount reaching colon (g per day)
Nonstarch polysaccharide	Diet	8–18
Starch	Diet	8–40
Oligosaccharides	Diet	2–8
Unabsorbed sugars	Diet	2–10
Proteins and peptides	Diet	10–15
Fats (as fatty acids and glycerol)	Diet	6–8
Digestive enzymes	Host	5–8
Bile acids	Host	0.5–1.0
Mucins	Host	2–3
Epithelial cells	Host	20–30

Table 9.4 Polysaccharide and glycoprotein degradation by representative colonic genera.

Polysaccharide	Polysaccharide-degrading ability				
	Bacteroides	*Bifidobacterium*	*Eubacterium*	*Ruminococcus*	*Clostridium*
Pectin	Yes	Yes	Yes	No	Yes
Cellulose	Yes	No	No	Yes	Yes
Mucins	Yes	Yes	No	Yes	Yes
Heparin	Yes	No	No	No	No
Chondroitin sulfate	Yes	No	No	No	Yes
Xylan	Yes	Yes	No	No	Yes
Guar gum	Yes	No	No	Yes	NDA
Amylose	Yes	Yes	Yes	Yes	Yes
Amylopectin	Yes	Yes	Yes	Yes	Yes
Arabinogalactan	Yes	Yes	No	No	NDA
Galactomannans	Yes	No	No	Yes	NDA
Hyaluronic acid	Yes	NDA	NDA	NDA	Yes
Starch	Yes	Yes	Yes	NDA	Yes
Inulin	Yes	Yes	Yes	NDA	Yes

Yes, some species are able to degrade the polysaccharide; No, most species are unable to degrade the polysaccharide; NDA, no data available.

torques, *Ruminococcus gnavus*, a *Bifidobacterium* sp., and *Akkermansia muciniphila* (a Gram-negative anaerobe belonging to the *Verrucomicrobia*) – are able to completely degrade mucins on their own. Many other organisms can contribute to mucin degradation by producing sialidases (*Bacteroides* spp., *Bifidobacterium* spp., *Clostridium* spp., *Prevotella* spp., *Bacteroides* spp., *Escherichia coli*, and *Enterococcus faecalis*) and/or glycosidases (*Bacteroides* spp., *Bifidobacterium* spp., *Clostridium* spp., *Lactobacillus* spp., *Prevotella* spp., *Bacteroides* spp., *E. coli*, and *Ent. faecalis*).

The monosaccharides and amino acids generated by mucin degradation are then used by many colonic bacteria as carbon and energy sources. The main end products of bacterial metabolism are SCFAs, CO_2, and H_2. In fact, SCFAs (mainly acetate, propionate, and butyrate) are the principal products of fermentation in the colon – between 300 and 400 mmol is produced per day. As well as acting as nutrients for other colonic bacteria, SCFAs are absorbed by the colonic mucosa and are metabolized by the host to provide between 3 and 9% of his/her energy requirements. Because of their utilization by resident bacteria, the carbohydrate concentration of the colonic contents decreases with distance from the cecum, and so becomes less significant as a carbon and energy source for resident microbes.

Undigested fats and unabsorbed digestion products of fats that reach the colon appear not to be used by colonic microbes and are excreted in feces.

The concentrations of urea, ammonia, and free amino acids are low in the effluent from the ileum, which means that the main sources of nitrogen for colonic bacteria are proteins and peptides. However, large quantities of ammonia are produced during the microbial fermentation of amino acids released from peptides and proteins, and this constitutes an important nitrogen source for many resident microbes. Proteins found in the colon include those from the diet, tissue proteins (e.g. collagen), serum albumin, antibodies, pancreatic enzymes, and those derived from exfoliated mucosal cells. Bacterial proteases, rather than host proteases, are responsible for most of the protein degradation that occurs in the colon. Proteolytic organisms present in the colon include *Clostridium* spp. (e.g. *Cl. perfringens* and *Cl. bifermentans*), *Bacteroides* spp. (e.g. *B. fragilis* group and *B. splanchnicus*), *Fusobacterium* spp., *Prevotella* spp., *Propionibacterium* spp., enterococci, staphylococci, and lactobacilli. The resulting amino acids can serve as sources of nitrogen, carbon, and energy; they become increasingly important nutrient sources, as carbohydrate levels in the lumenal contents become depleted as they pass along the colon. Many colonic bacteria are able to carry out amino acid fermentation, and these include *Clostridium* spp., *Eggerthella lenta*, *Fusobacterium* spp., *Peptostreptococcus* spp. and other Gram-positive anaerobic cocci (GPAC), *Prevotella melaninogenica*, *Acidaminococcus* spp., and *Peptococcus* spp. The main products are SCFAs, ammonia, amines, indoles, organic acids, alcohols, and hydrogen. The main SCFAs produced include acetate, propionate, and butyrate as well as branched-chain fatty acids (BCFAs) – isobutyrate, 2-methylbutyrate, and isovalerate.

A number of vitamins are present in the colon and are derived both from the diet and from the colonic microbiota – particularly *Bacteroides* spp., *Bifidobacteria* spp., *Clostridium* spp., and enterobacteria. Vitamins produced by these organisms include vitamin K, nicotinic acid, folate, pyridoxine, vitamin B_{12}, and thiamine.

Although the environment within the large intestine gradually changes from the cecum to the anus, it is possible to recognize three main environmental regions, corresponding to distinct anatomical sites, and these are summarized in Fig. 9.15.

9.4 THE INDIGENOUS MICROBIOTA OF THE GASTROINTESTINAL TRACT

A major problem in determining the composition of the microbiota of the GIT is obtaining samples for analysis. Access to most regions of the GIT is difficult and is either uncomfortable for the individual or requires some form of anesthesia, and this can affect the motility and secretory activity of the GIT. Ideally, samples should be taken endoscopically, as patients are not usually anesthetized during the procedure, and are not either fasting or being given antibiotics.

Analysis of the microbial communities present in the human GIT has shown them to be extremely complex, with estimates of the number of different species present ranging from 500 to 1000. Furthermore, 16S rRNA gene-based analysis of the colonic microbiota has shown that as many as 80% of phylotypes defined by 16S rRNA gene sequences do not correspond to known cultured bacterial species, and more than half are entirely novel species. The total number of microbial genes in these communities (known as the microbiome or metagenome) is between 2 and 4 million (i.e. 70–140 times greater than the number of genes involved in the human genome), which represents an enormous metabolic potential that is far greater than that of their host.

The relationship between *Homo sapiens* and the indigenous microbiota of the GIT is an excellent example of a symbiosis. In the upper regions of the GIT (the stomach, duodenum, and jejunum), the host provides an environment that is not conducive to extensive colonization by microbes, and it is in these regions that the host extracts many of the nutrients from the constituents of his/her diet. In the terminal ileum and the colon, however, the environment is suitable for the establishment of large and diverse microbiotas. Within these regions, the host takes advantage of the enormous metabolic potential of its indigenous microbiota

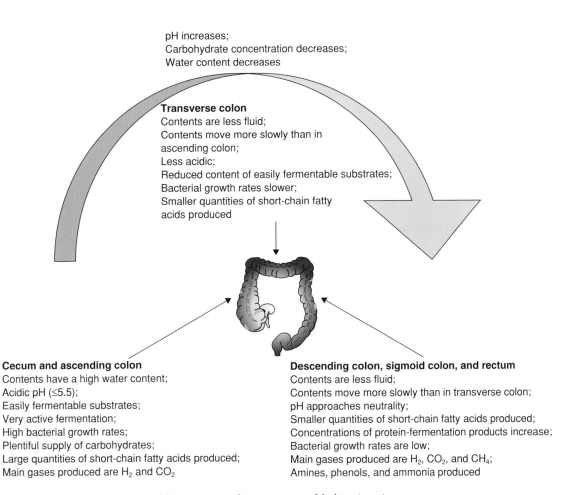

pH increases;
Carbohydrate concentration decreases;
Water content decreases

Transverse colon
Contents are less fluid;
Contents move more slowly than in
ascending colon;
Less acidic;
Reduced content of easily fermentable substrates;
Bacterial growth rates slower;
Smaller quantities of short-chain fatty
acids produced

Cecum and ascending colon
Contents have a high water content;
Acidic pH (≤5.5);
Easily fermentable substrates;
Very active fermentation;
High bacterial growth rates;
Plentiful supply of carbohydrates;
Large quantities of short-chain fatty acids produced;
Main gases produced are H_2 and CO_2

Descending colon, sigmoid colon, and rectum
Contents are less fluid;
Contents move more slowly than in transverse colon;
pH approaches neutrality;
Smaller quantities of short-chain fatty acids produced;
Concentrations of protein-fermentation products increase;
Bacterial growth rates are low;
Main gases produced are H_2, CO_2, and CH_4;
Amines, phenols, and ammonia produced

Fig. 9.15 Important environmental determinants in the main regions of the large intestine.

to degrade dietary constituents that it is itself unable to digest. Many of the resulting products are then absorbed by the colon, along with vitamins synthesized by some microbes. The microbes, in turn, are supplied with an environment that is suitable for their growth, and a constant supply of nutrients. As well as supplying nutrients, the indigenous microbiota plays an important role in the development of the intestinal mucosa and immune system of its host.

The environment of the GIT varies considerably along its length, resulting in distinct microbial communities in each of the main regions. In addition, there is also a horizontal stratification at any point within the GIT. Hence, the lumen, the mucous layer, and the epithelial surface all offer different environments within which different microbial communities could develop. In some regions of the GIT, the crypts at the bases of villi, as well as various glands, would also provide different sets of environmental conditions and so may harbor distinct microbiotas. Furthermore, particulate matter within the lumen provides a substratum for biofilm formation, thereby enabling the establishment of communities whose composition differs from that present in the fluid phase of the lumenal contents. However, because of the difficulties in obtaining from these microhabitats samples free of contamination by microbes from other sites, few studies have been able to provide information regarding the composition of these communities.

9.4.1 Members of the intestinal microbiota

The organisms that are most frequently cultivated from the GIT belong to the genera *Bacteroides*, *Eubacterium*, *Clostridium*, *Bifidobacterium*, *Streptococcus*, *Lactobacillus*, *Peptostreptococcus*, *Peptococcus*, *Ruminococcus*, *Fusobacterium*, *Veillonella*, *Enterococcus*, *Propionibacterium*, *Actinomyces*, *Desulfovibrio*, *Helicobacter*, *Porphyromonas*, *Prevotella*, *Escherichia*, *Enterobacter*, *Citrobacter*, *Serratia*, *Candida*, *Gemella*, and *Proteus*. Many of these genera have been described in previous chapters, and those that have not been discussed earlier are described below.

As mentioned above, an important feature of the microbiota of the GIT is that most of the organisms present have not yet been grown in the laboratory. Fortunately, an increasing number of investigations of the intestinal microbiota have used culture-independent approaches. The application of such techniques in analyses of the most densely populated region of the GIT, the large intestine, has revealed that the phylotypes present belong to only nine of the 70 known bacterial divisions and to only one of the 13 archaeal divisions. In fact, the vast majority of phylotypes belong to only two of the bacterial divisions – *Firmicutes* and *Bacteroidetes* (the latter also known as *Cytophaga–Flavobacterium–Bacteroides*). The other main divisions represented include *Actinobacteria*, *Verrucomicrobia*, *Proteobacteria*, and *Fusobacteria*. Three classes are recognized within the *Firmicutes* – *Bacilli*, *Clostridia*, and *Mollicutes*, and most inhabitants of the large intestine belong to the *Clostridia* class, which is further divided into a number of phylogenetic clusters. The main phylogenetic clusters to which intestinal phylotypes belong are clusters XIVa, IV, and XVI, each of which contains a number of genera (Table 9.5). Many of the organisms belonging to these clusters have not yet been grown in the laboratory.

9.4.1.1 *Bacteroides*

The genus *Bacteroides* consists of Gram-negative pleomorphic rods, and their main characteristics are as follows:
- Gram-negative bacilli;
- Nonmotile;

Table 9.5 Major phylogenetic groups present in the microbiota of the large intestine.

Phylogenetic group	Examples of intestinal organisms within the group
Bacteroidetes (*Cytophaga–Flavobacterium–Bacteroides*)	*Bacteroides* (at least 13 species) including *B. distasonis*, *B. fragilis*, *B. thetaiotaomicron*; *Prevotella* (at least eight species); *Porphyromonas* (at least five species); *Rikenella microfusus*; *Cytophaga fermentans*
Cl. coccoides–Eub. rectale group (*Clostridium* cluster XIVa)	*Clostridium* (at least 12 species) including *Cl. coccoides*; *Eubacterium* (at least nine species) including *Eub. rectale*; *Ruminococcus* (at least five species); *Butyrivibrio*; *Streptococcus hansenii*; *Coprococcus eutactus*; *Roseburia*; *Lachnospira*
Cl. leptum group (*Clostridium* cluster IV)	*Clostridium* (*Cl. leptum*, *Cl. sporosphaeroides*, *Cl. methylpentosum*); *Eubacterium siraeum*; *Ruminococcus* (*Rum. bromii*, *Rum. albus*, *Rum. flavifaciens*, *Rum. callidus*); *Faecalibacterium prausnitzii*; *Anaerofilum*
Eubacterium cylindroides group (*Clostridium* cluster XVI)	*Eub. cylindroides*; *Eub. biforme*; *Eub. tortuosum*; *Eub. dolichum*; *Clostridium innocuum*
Gammaproteobacteria	*Haemophilus*; *Klebsiella*; *Escherichia*;
Lactobacillus–Enterococcus group	*Lactobacillus*; *Enterococcus*; *Pediococcus*; *Leuconostoc*
Actinobacteria	*Actinomyces*; *Bifidobacterium*; *Propionibacterium*; *Atopobium*; *Collinsella*; *Coriobacterium*; *Eggerthella*
Verrucomicrobia	*Verrucomicrobium*
Fusobacteria	*Fusobacterium*

- Nonsporing;
- Anaerobic;
- Aerotolerant;
- G+C content of their DNA is 40−48 mol%;
- Simple nutritional requirements;
- Most species require hemin and vitamin B_{12} for growth;
- NH_4^+ is their primary source of nitrogen.

Most species can utilize a variety of carbohydrates as an energy and carbon source and can degrade a wide range of polysaccharides including starch, cellulose, xylan, pectin, dextran, arabinogalactan, guar gum, laminarin, chondroitin sulfate, glycosaminoglycans, malto-oligosaccharides, gum arabic, hyaluronic acid, and heparin. Mucins can also be degraded, and some species are proteolytic and can hydrolyze fibrinogen, casein, trypsin, chymotrypsin, transferrin, and ovalbumin. The major end products of carbohydrate metabolism are succinate, propionate, and acetate.

One of the most abundant organisms in the lower regions of the GIT is *Bacteroides thetaiotaomicron*, and its 6.26-Mbp genome has been sequenced (http://cmr.tigr.org/tigr-scripts/CMR/GenomePage.cgi?org=ntbt01). A large proportion of the proteome is devoted to the acquisition, degradation, and utilization of complex carbohydrates. Hence, many of the outer-membrane proteins are likely to be involved in the acquisition of a variety of carbohydrates, and 20 putative sugar-specific transporters and 20 permease subunits of ATP-binding cassette transporters have been identified. A large number (172) and variety (23 different activities) of glycosylhydrolases are present in *B. thetaiotaomicron* – more than that of any other bacterium sequenced. This shows an adaptation to its role in the degradation of carbohydrates that the host cannot digest. Most (61%) of these glycosylhydrolases are predicted to be in either the periplasm or the outer membrane, or are extracellular, which implies that their degradation products could also be used either by the host or by other bacteria. Many of the predicted proteins are enzymes that enable the degradation of host-derived polymers such as mucin, chondroitin sulfate, hyaluronate, and heparin.

Another remarkable feature of the organism's proteome is the large number of one- and two-component signal-transduction systems, which suggests that the organism is well adapted to sense and respond to changes in its environment. A number of genetic elements that can be mobilized are also present, including a plasmid, 63 transposases, and four conjugative transposons. This suggests that the organism is able to engage in the horizontal transfer of genes encoding antibiotic resistance, virulence factors, etc.

Another frequently detected species is *B. vulgatus*. This organism is capsulated and ferments a range of monosaccharides and disaccharides. It can grow on amylase and amylopectin, and some strains can also ferment pectin, polygalacturonate, xylan, and arabinogalactan. Recently, its 5.16-Mbp genome has been sequenced (http://www.pubmedcentral.nih.gov/articlerender.fcgi?tool=pubmed&pubmedid=17579514) and was found to encode a predicted 4088-member proteome. Its proteome is enriched (compared with that of *Bacteroidetes* which do not inhabit the gut) for genes related to polysaccharide metabolism, environmental sensing, gene regulation, and membrane transport. It has 159 predicted glycoside hydrolases and seven polysaccharide lyases, together with a large complement of enzymes for the degradation of pectin. It is the only sequenced gut *Bacteroidetes* with a gene encoding a xylanase.

Bacteroides spp. have a high pathogenic potential and account for approximately two thirds of all anaerobes isolated from clinical specimens – the most frequently isolated species being *B. fragilis*.

9.4.1.2 *Eubacterium*

Until recently, *Eubacterium* was a heterogeneous group of organisms which included anaerobic Gram-positive rods whose taxonomic position was uncertain. Two main groups were recognized – saccharolytic and asaccharolytic. Most of the latter are currently retained within the genus *Eubacterium*, while many of the former have been assigned to new genera. Hence, two species regularly present in the GIT of humans, *Eub. lentum* and *Eub. aerofaciens*, have been re-named *Eggerthella lenta* and *Collinsella aerofaciens*, respectively and are often referred to as "*Eubacterium*-like organisms". Species currently recognized as belonging to the genus *Eubacterium* that are present in the GIT include: *Eub. biforme, Eub. contortum, Eub. rectale, Eub. cylindroides, Eub. hadrum, Eub. ventriosum, Eub. barkeri, Eub. limosum*, and *Eub. moniliforme*. All of these ferment sugars to produce a mixture of fatty acids. *Eubacterium* spp. and *Eubacterium*-like organisms are anaerobic, nonmotile, nonsporing Gram-positive bacilli. They are not very aerotolerant. Some intestinal species can hydrolyze starch and pectin. Most of these organisms produce one or more fatty acids from glucose, but *Eg. lenta* does not. *Eg. lenta* can use amino acids as an energy source and produces a mixture of lactate, acetate, formate, and succinate. *Col. aerofaciens* ferments sugars and produces formate, lactate, and hydrogen.

9.4.1.3 *Roseburia*

Roseburia are anaerobic, non-spore-forming, motile, Gram-positive, slightly curved rods. The G+C content of their DNA is 29–31 mol%. They can ferment a range of carbohydrates and produce mainly butyrate, lactate, and hydrogen. Growth is improved by the presence of acetate. Species found in the GIT include *Ros. intestinalis*, *Ros. hominis*, *Ros. faecis*, and *Ros. inulinivorans*. Together with the closely related *Eub. rectale*, these organisms have been reported to comprise up to 7% of the microbiota of the human colon.

9.4.1.4 *Clostridium*

Clostridium are obligately anaerobic, spore-forming, Gram-positive rods. Most species are motile – important exceptions include *Cl. perfringens*, *Cl. ramosum*, and *Cl. innocuum*. The genus is one of the largest among the prokaryotes and contains 146 species, most of which are fermentative and/or proteolytic. Many produce a number of SCFAs (e.g. acetate and butyrate) when grown in carbohydrate-containing media, as well as a variety of other fermentation products such as acetone and butanol. Species regularly isolated from the human GIT include *Cl. perfringens*, *Cl. ramosum*, *Cl. innocuum*, *Cl. paraputrificum*, *Cl. sporogenes*, *Cl. tertium*, *Cl. bifermentans*, and *Cl. butyricum*. Of these, *Cl. perfringens* is the most frequent cause of gas gangrene – a life-threatening infection. *Cl. bifermentans* (and possibly *Cl. tertium* and *Cl. sporogenes*) is also able to cause gas gangrene. A number of the clostridia listed above may also be involved in polymicrobial infections such as peritonitis, intra-abdominal abscesses, and septicemia.

Cl. perfringens is proteolytic, is able to hydrolyze various polysaccharides and glycoproteins (including starch, mucin, and hyaluronic acid), and can ferment a range of sugars, producing mainly acetate and butyrate. Its 3.25-Mbp genome has been sequenced (http://cmr.tigr.org/tigr-scripts/CMR/GenomePage.cgi?org=bcl).

9.4.1.5 *Bifidobacterium*

Bifidobacterium spp. are pleomorphic Gram-positive rods which occur singly, in chains, or in clumps.

The main characteristics of *Bifidobacterium* spp. are as follows:

- Gram-positive bacilli;
- Nonsporing;
- Nonmotile;
- Obligate anaerobes (some can grow in CO_2-enriched air);
- G+C content of their DNA is 45–67 mol%;
- Growth over the temperature range 20–49.5°C;
- Optimum growth temperature is 37–41°C;
- Growth over pH range 4.0–8.5;
- Optimum pH range is 6.5–7.0;
- Acid-tolerant but not acidophilic;
- Produce acetate and lactate from sugars;
- Hydrolyze a wide range of polysaccharides;
- Proteolytic;
- Produce several vitamins.

Thirty-two species are recognized, most of which have been detected in the human GIT, and several also occur in the vagina and oral cavity. They all produce acid from glucose, the end products being acetate and lactate. They ferment a wide range of sugars and can hydrolyze a variety of polysaccharides, including starch, xylan, pectin, inulin, gum arabic, and dextran. They can also hydrolyze proteins and peptides and produce a number of enzymes that are important in the degradation of mucins, including sialidases, α- and β-glycosidases, α- and β-D-glucosidases, α- and β-D-galactosidases, and β-D-fucosidase. All of the species inhabiting humans can utilize NH_4^+ as the sole source of nitrogen, and many excrete large quantities of a variety of amino acids. They also produce a number of vitamins including thiamine, folic acid, nicotinic acid, pyridoxine, cyanocobalamin, and biotin. Growth is stimulated by a number of oligosaccharides, protein hydrolysates, and glycoproteins, and these are often termed "bifidogenic factors". The low pH generated by the acetate and lactate that they produce inhibits the growth of a number of microbes. Some species also produce a bacteriocin, bifidocin B, as well as other unidentified antimicrobial compounds.

On the basis of culture-dependent analyses, bifidobacteria have been shown to comprise approximately 10% of the human adult intestinal microbiota, but evaluations based on culture-independent methods suggest that they constitute less than 3% of the microbiota. The most commonly occurring species in the GIT are *Bif. pseudocatenulatum*, *Bif. longum*, *Bif. angulatum*, *Bif. adolescentis*, *Bif. catenulatum*, *Bif. bifidum*, *Bif. gallicum*, *Bif. infantis*, and *Bif. breve*. Several studies have shown that an individual is colonized by a limited number of bifidobacterial strains, and that these strains persist in that person for considerable periods of time.

Species usually present in the vagina include *Bif. breve* and *Bif. adolescentis* as well as, to a lesser extent, *Bif. longum* and *Bif. bifidum*. The most frequently isolated species from the oral cavity are *Bif. denticolens*, *Bif. adolescentis*, *Bif. inopinatum*, and *Bif. dentium*.

There is considerable interest in the use of bifidobacteria as probiotics, and a number of studies have shown that they extert antagonistic effects against a variety of other organisms. The antagonistic properties of *Bifidobacterium* spp. are as follows:
• They prevent the attachment of enteropathogens to intestinal epithelial cells.
• They inhibit the growth of intestinal organisms by producing acidic metabolic end products.
• They produce a broad-spectrum antimicrobial compound that can inhibit the growth of *B. fragilis*, *Cl. perfringens*, and several enteropathogens.
• They produce a low molecular mass (3500 Da) lipophilic compound able to kill *Salmonella typhimurium*, *Listeria monocytogenes*, *Yersinia pseudotuberculosis*, *Staphylococcus aureus*, *Pseudomonas aeruginosa*, *Klebsiella pneumoniae*, and *Escherichia coli*.
• They produce a bacteriocin, bifidocin B, which has a broad antimicrobial spectrum, inhibiting the growth of species of *Listeria*, *Enterococcus*, *Bacillus*, *Lactobacillus*, *Leuconostoc*, and *Pediococcus*.

The 2.26-Mbp genome of *Bif. longum* has been sequenced (http://cmr.tigr.org/tigr-scripts/CMR/GenomePage.cgi?org=ntbl01), and analysis has revealed many features that demonstrate its adaptation to the environment of the colon:
• Homologues of enzymes needed to ferment a wide range of sugars;
• Homologues of enzymes for the fermentation of amino acids;
• More than 20 predicted peptidases;
• Numerous predicted proteins for carbohydrate transport and metabolism, including more than 40 glycosyl hydrolases;
• Enzymes with predicted activities for carbohydrate hydrolysis, including xylanases, arabinosidases, galactosidases, neopullulanase, isomaltase, maltase, and inulinase;
• Three α-mannosidases and an endo-NAc glucosaminidase.

The findings of this genome analysis show that the organism has the potential to hydrolyze, and to utilize the degradation products from, a range of complex plant polymers that survive digestion in the small intestine and also host glycoproteins. *Bifidobacterium* spp. are very rarely implicated in human infections.

9.4.1.6 *Enterococcus*

The genus *Enterococcus* consists of Gram-positive cocci that occur singly, in pairs, or in chains. The main characteristics of *Enterococcus* spp. are as follows:
• Gram-positive cocci;
• Facultative anaerobes;
• Nonsporing;
• Catalase-negative;
• Growth over temperature range 10–45°C;
• Optimum growth is between 35 and 37°C;
• Complex nutritional requirements;
• Ferment sugars to produce mainly lactate;
• Can tolerate adverse environmental conditions;
• Halotolerant (can grow in 6.5% NaCl);
• Bile-resistant (can tolerate 40% bile salts).

The genus consists of 17 species, but only two of these, *Ent. faecalis* and *Ent. faecium*, are regularly found in the human GIT – the former is more frequently found and in greater proportions. Some species (including *Ent. faecalis*) can hydrolyze proteins, including collagen, casein, insulin, hemoglobin, fibrinogen, and gelatin, together with a number of peptides, etc. They can also hydrolyze lipids and hyaluronic acid. Enterococci are very hardy organisms and can tolerate a wide variety of growth conditions, including temperatures of 10–45°C, as well as hypotonic, hypertonic, acidic, or alkaline environments. They are also resistant to bile salts, desiccation, detergents, and many antimicrobial agents.

Ent. faecalis and, to a lesser extent, *Ent. faecium* can cause infections in humans, particularly in hospitalized individuals, and many of these are difficult to treat because of the resistance of many strains to a variety of antibiotics. The 3.35-Mbp genome of a vancomycin-resistant strain of *Ent. faecalis* has been sequenced (http://cmr.tigr.org/tigr-scripts/CMR/GenomePage.cgi?org=gef). The chromosome of this organism, together with its three plasmids, has a total of 3337 protein-encoding genes. A remarkable feature of the organism's genome is that more than one quarter of it consists of mobile and/or exogenously acquired DNA – this is the highest proportion of mobile elements detected in any bacterium sequenced so far. These include 38 insertion elements, seven regions probably derived from integrated phages, multiple conjugative and composite transposons, a putative pathogenicity island, and integrated plasmid genes. The genome of *Ent. faecalis*, therefore, is highly malleable and is likely to have undergone multiple rearrangement events. Many putative sugar-uptake systems are present, as are pathways for the metabolism of more than 15 different sugars, thereby demonstrating its adaptation to the utilization of carbohydrates in the GIT. It has a variety of cation homeostasis mechanisms, which probably contribute to its ability to survive extreme pHs, desiccation, and high salt and metal concentrations. A remarkable number (134) of putative surface proteins that may function as adhesins are present.

9.4.1.7 *Helicobacter pylori*

Hel. pylori is a Gram-negative curved rod, and its main characteristics are as follows:
- Gram-negative curved bacillus;
- Microaerophile;
- Motile;
- Non sporing;
- G+C content of its DNA is 35–40 mol%;
- Optimum temperature for growth is 37°C;
- Growth over the pH range 6.0–8.0;
- Optimum pH for growth is 7.0;
- Catalase-positive;
- Oxidase-positive;
- Urease-positive;
- Nutritionally fastidious.

The organism grows best in an atmosphere with a reduced oxygen content (5–10%), together with an elevated concentration of CO_2 (5–12%) and hydrogen (5–10%). During adverse conditions, it can revert to a coccoid form – this can occur as a result of nutritional deprivation, increased oxygen tension, alkaline pH, increased temperature, exposure to antibiotics, and aging. These coccoid structures remain viable for up to 4 weeks and can then revert to the normal bacillary form. It is possible that the coccoid form may enable survival of the organism in the environment and so facilitate its transmission in water or via fomites. Its 1.66-Mbp genome has been sequenced (http://cmr.tigr.org/tigr-scripts/CMR/GenomePage.cgi?org=ghp).

The main habitat of *Hel. pylori* is the mucous layer overlying the gastric epithelium as well as the epithelial surface itself. It reaches these sites by burrowing into and through the mucous gel, powered by its monopolar flagella – its curved morphology and ability to move in a helical manner facilitate gel penetration. Within

these habitats, it creates a microenvironment with a neutral pH (the organism is not acidophilic or acid-tolerant) by using its surface-associated urease to produce ammonia from the urea in gastric juice. *Hel. pylori* expresses urease at a level higher than any other known microbe, and the ammonia produced is also thought to provide a protective "shell", enabling it to move through the acidic gastric fluid until it reaches the mucous layer. The antrum is the preferred site of colonization because little acid production occurs in this region – it cannot produce sufficient ammonia to neutralize the acid produced in the corpus, i.e. the main acid-secreting region of the stomach. *Hel. pylori* cannot survive at a pH lower than 4.0, and so is vulnerable within the gastric lumen, although it may be able to survive by transformation to its coccoid form.

The organism possesses a number of adhesins that mediate its binding to the gastric epithelium, and these include BabA and lipopolysaccharide (LPS).

Hel. pylori is present in the stomach of 20–50% of adults in Europe and North America, but is found in far higher proportions of the adult populations of developing countries (section 9.4.2.2). DNA fingerprinting of strains isolated from different individuals has shown that each strain is unique unless the patients are linked epidemiologically. This population structure is similar to that found in organisms such as *Neisseria meningitidis* and is characteristic of organisms that are naturally competent for genetic transformation.

9.4.1.8 *Enterobacteriaceae*

Enterobacteriaceae are nonsporing, facultatively anaerobic Gram-negative bacilli that ferment glucose to produce acid and reduce nitrate to nitrite – most species are motile. They contain an antigenic polysaccharide known as the "enterobacterial common antigen". The G+C content of their DNA is 38–60 mol%. They grow well at between 25 and 37°C. The *Enterobacteriaceae* consist of approximately 30 genera and more than 150 species – many of these are normal inhabitants of the human GIT. The genera most frequently present in the human GIT, and in the greatest proportions, are *Escherichia*, *Proteus*, *Citrobacter*, and *Enterobacter*. Space limitations preclude a description of all of the many species encountered in the human GIT, and so only one of these, *E. coli*, will be described further.

E. coli is usually motile and ferments glucose and many other sugars to produce mainly acetic, lactic, and succinic acids together with CO_2 and H_2. The classification system used for distinguishing between different strains of the organism is based on the type of O antigen (i.e. LPS), K antigen (i.e. capsule), and H antigen (i.e. flagellum) possessed by the strain. Currently, more than 170 O antigens, 80 K antigens, and 56 H antigens have been recognized.

E. coli is responsible for urinary tract infections and is also an important cause of meningitis in neonates. A variety of enteropathogenic strains of *E. coli* also exist, but they are not members of the indigenous microbiota of the GIT.

9.4.1.9 *Ruminococcus*

The genus *Ruminococcus* consists of nonsporing, non-motile, anaerobic Gram-positive cocci. The G+C content of their DNA is 37–48 mol%. Thirteen species are recognized, and all use carbohydrates as an energy source and produce acetate – some also produce other fatty acids such as formate, lactate, and succinate. They utilize NH_4^+ as a nitrogen source. Species frequently isolated from the human GIT include *Rum. obeum*, *Rum. torques*, *Rum. flavefaciens*, *Rum. gnavus*, and *Rum. bromii*. Many ruminococci produce sialidases and/or glycosidases and/or proteases, and so are able to partially degrade mucins and other glycoproteins. Few bacterial species are able to produce the full complement of enzymes necessary to degrade mucins entirely,

but *Rum. torques* and *Rum. gnavus* are able to do so. Ruminococci rarely cause infections in humans.

9.4.1.10 Methanogenic organisms

Methanogenic organisms produce large quantities of methane as a byproduct of their energy-generating reactions. Two main genera are found in the human colon – *Methanobrevibacter* and *Methanosphaera*. Both of these belong to the domain *Archaea* and so are only distantly related to the domain *Bacteria*. The main characteristics of those species most frequently detected in the human GIT are listed in Table 9.6.

H_2 is produced by colonic bacteria during the fermentation of carbohydrates, and its utilization by methanogens is an important means by which the gas is disposed of in this environment.

The 1.77-Mbp genome of *Methanosphaera stadtmanae* has been sequenced (http://archaea.ucsc.edu/cgi-bin/hgGateway?org=Methanosphaera+stadtmanae&db=methStad1).

9.4.1.11 *Desulfovibrio*

The genus *Desulfovibrio* consists of motile, anaerobic Gram-negative curved rods with a G+C content of 46–61 mol%. They grow best at a temperature of 30–38°C. Members of the genus utilize sulfate as an

Table 9.6 Main characteristics of methanogenic organisms frequently present in the human intestinal tract.

Characteristic	*Methanobrevibacter smithii*	*Methanosphaera stadtmanae*
Morphology	Gram-positive short rods	Gram-positive cocci that occur singly, in tetrads or in clusters.
Atmospheric requirements	Anaerobic	Anaerobic, needs CO_2
Motility	Non-motile	Nonmotile
G+C content of DNA	29–31 mol%	26 mol%
Optimum pH for growth	7.0	6.5–6.9
Optimum temperature for growth	37–39°C	36–40°C
Nutritional requirements	Nutritionally fastidious; requires acetate, amino acids, and B vitamins for growth	Nutritionally fastidious; requires acetate, CO_2, several amino acids, NH_4^+, thiamin, and biotin for growth
Energy generation	From the oxidation of H_2, using CO_2 as the electron acceptor, which is converted to CH_4	From the oxidation of H_2, using methanol as the electron acceptor, which is converted to CH_4

energy source, and hydrogen, lactate, and ethanol as electron donors, and this results in the generation of hydrogen sulfide. The main source of sulfate in the colon is mucin, from which it is liberated by the sulfatases produced by organisms such as *Bacteroides* spp. *Desulfovibrio* spp. constitute the major sulfate-reducing organisms in the human colon, and frequently encountered species include *Des. desulfuricans*, *Des. fairfieldensis*, and *Des. vulgaris*.

9.4.1.12 *Acidaminococcus*

The genus *Acidaminococcus* consists of non-motile, non-sporing, anaerobic Gram-negative cocci which belong to the family *Veillonellaceae*. The G+C content of their DNA is 56.6 mol%, and they are similar to *Veillonella* spp. (section 8.4.1.5), apart from the fact that they degrade amino acids to acetate and butyrate.

9.4.1.13 *Faecalibacterium prausnitzii*

Faecalibacterium prausnitzii (formerly known as *Fusobacterium prausnitzii*) is a nonmotile, nonsporing Gram-negative anaerobic bacillus. The G+C content of its DNA is 47–57 mol%. It can utilize glucose or fructose as a carbon and energy source and produces butyrate, lactate, and formate. Growth is stimulated by acetate. The organism can hydrolyze fructo-oligosaccharide, starch, and inulin. Its genome is currently being sequenced at the Washington University School of Medicine, St. Louis, Missouri, USA, as part of the Human Gut Microbiome Initiative.

9.4.2 Community composition in different regions of the intestinal tract

As can be appreciated from section 9.3, there are profound differences in the environmental selection factors operating at the various regions of the GIT. Consequently, different microbiotas are found within each region. In addition to this horizontal variation along the GIT, there may also be a cross-sectional stratification at any point along the tract. Hence, potential colonization sites include the epithelial surface, the mucous layer, the lumen, glands, and crypts.

In the upper regions of the GIT (esophagus, stomach, duodenum, and jejunum), material rarely completely fills the lumen and is usually present for relatively short periods of time. The organisms found in the luminal

material are invariably transients (derived from the oral cavity and the upper respiratory tract as well as from food) and are not usually regarded as being members of the indigenous microbiota of these regions. The main sites of colonization, therefore, are the mucosal surface (including glands and crypts) and its associated layer of mucus. In contrast, the lumen of the large intestine is generally completely filled with material, and transit times through it are much longer – organisms present in luminal material, therefore, are considered to be members of the indigenous microbiota. Whether or not microbes colonize the epithelial surface of the colon (as opposed to being attached to, residing within, or underlying the mucous layer) in healthy individuals is uncertain at present, and is a cause of controversy.

9.4.2.1 Esophagus

The lumen of the esophagus differs from that of other regions of the GIT in that it is a passageway rather than a receptacle and, therefore, contains material for only short periods of time. The esophageal wall and its associated mucous layer, therefore, are the only possible sites for microbial colonization. Closely adherent bacteria have been observed attached to the mucosa at a density of approximately 10^4 bacteria per mm^2 (Fig. 9.16).

Because of sampling difficulties, there have been few investigations of the microbiota of the esophagus and, in the two most recently published studies, bacteria could be cultivated from only approximately half of the individuals sampled. While streptococci were found to be the most frequently isolated organisms, lactobacilli were present in the greatest proportion (Fig. 9.17).

In a culture-independent study, 833 unique sequences belonging to 95 taxa from 41 genera were identified in the esophagus of four adults, and most (82%) of the taxa identified corresponded to species that could be cultivated in the laboratory. Fourteen taxa were detected in all four individuals: *Streptococcus mitis*, *Streptococcus thermophilus*, *Streptococcus parasanguinis*, *Veillonella atypica*, *Veillonella dispar*, *Rothia mucilaginosas*, *Megasphaera micronuciformis*, *Granulicatella adiacens*, *Prevotella pallens*, *Bacteroides* AF385513, TM7 AF385520, *Clostridium* AY278618, *Bulleidia moorei*, and *Actinomyces odontolyticus*. *Streptococcus* spp., *Prevotella* spp., and *Veillonella* spp. dominated the microbiota and together comprised 70% of the clones (Fig. 9.18). The absence of spirochaetes and many of the not-yet-cultivated organisms that are indigenous to the oral cavity suggests that the

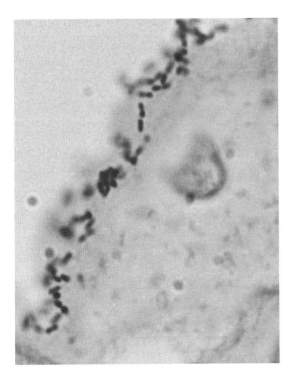

Fig. 9.16 Photomicrograph showing Gram-stained bacteria adhering closely to the esophageal wall in a biopsy taken from an adult. Reprinted with permission from Pei, Z., Yang, L., Peek, R. Jr., Levine, S., Pride, D. and Blaser, M. (2005) Bacterial biota in reflux esophagitis and Barrett's esophagus. *World J Gastroenterol* 11, 7277–83.

esophageal microbiota is distinctly different from that of the oral cavity.

9.4.2.2 Stomach

A number of problems are associated with defining the indigenous microbiota of the stomach. First of all, each day approximately 10^{10} bacteria from the oral cavity and upper respiratory tract are transferred to the stomach by swallowing. In addition, the stomach receives microbes that are present in the food and beverages consumed by the host. One of the main factors limiting microbial colonization of the stomach is its low pH, which kills many, but not necessarily all, incoming microbes. Organisms likely to survive will be aciduric species such as some streptococci and lactobacilli. However, as pointed out in section 9.3.2, gastric pH not only varies between individuals but also fluctuates widely in an individual throughout the day as a result of food and beverage intake (Fig. 9.19). Consequently, for short periods of time, the pH of the stomach may be conducive to the survival and growth of a range of bacterial species originating from the oral cavity and/or the diet. It is often difficult, therefore, to ascertain whether an organism detected in the stomach is an autochthonous species. This is particularly problematic when detection is based on the analysis of DNA taken from gastric samples, because DNA from swallowed, or food-borne, organisms may persist in the stomach for long periods of time. A further, separate complication stems from the fact that samples for analysis are usually obtained by endoscopy, which requires that the individual fasts for 12 h. The results obtained can hardly be regarded as reflecting the "normal" situation.

The number of viable microbes that can be recovered from the stomach of individuals with a gastric pH of less than 4 (i.e. most healthy adults) is usually less than 10^3 cfu/ml. However, viable counts 100–1000 times greater than this can be found after a meal because of the transient increase in pH resulting from the buffering action of food (Fig. 9.19), which enables the survival of organisms present in food and in the saliva swallowed with it. Individuals with a gastric pH of more than 4 tend to have viable counts of the order of 10^5–10^6 cfu/ml.

Microbes frequently isolated from gastric juice (i.e. from the gastric lumen) are mainly acid-tolerant species of streptococci and lactobacilli, although these are considered to be transients from the oral and/or nasal cavity. Organisms that have been cultivated from stomach contents include viridans streptococci (*Strep. sanguinis* and *Strep. salivarius*), lactobacilli (*L. plantarum*, *L. salivarius*, *L. fermentum*, and *L. gasseri*), *Staph. aureus*, *Staphylococcus epidermidis*, *Neisseria* spp., *Haemophilus* spp., *Bacteroides* spp., *Micrococcus* spp., *Bifidobacterium* spp., coryneforms, and *Veillonella* spp.

The range of organisms associated with the gastric mucosa is similar to that present in the gastric lumen except that an additional organism, *Hel. pylori*, may be present in large proportions of the population (Fig. 9.20).

Unlike many of the organisms found in the gastric lumen, *Hel. pylori* can be considered to be a true member of the indigenous gastric microbiota, i.e. it is an autochthonous species (Fig. 9.21). Initial colonization by the organism occurs during childhood when the gastric pH may be lower than in adults and the immune

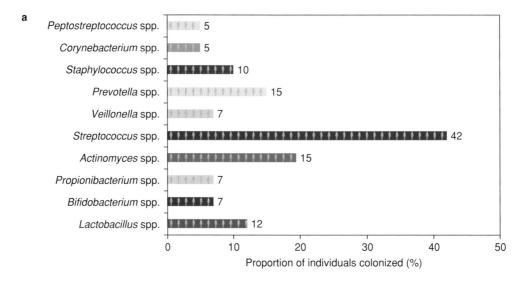

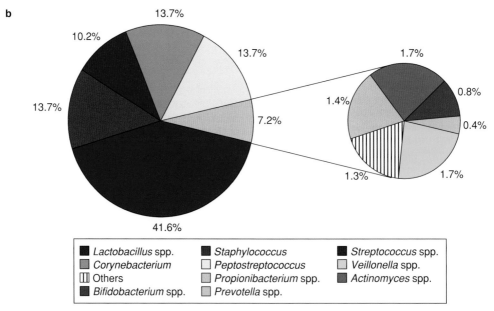

Fig. 9.17 Culture-dependent study of the esophageal microbiota. (a) Frequency of isolation of the various bacterial genera detected. (b) Relative proportions of the organisms present. Data are mean values derived from the results of two studies involving 17 healthy adults (1, 70).

system may not be fully developed. Another contributing factor may be that children frequently sample their environment through their mouths. Once acquired, the organism generally persists in the individual for life. It has been suggested that, prior to the 20th century, all humans were colonized by *Hel. pylori*, but improved living conditions have resulted in a decrease in the prevalence of the organism – this is particularly evident in developed countries in which prevalence rates continue to fall.

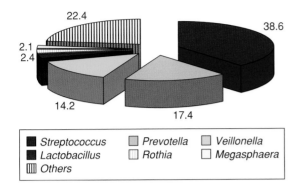

Fig. 9.18 Culture-independent analysis of the esophageal microbiota. DNA was extracted from esophageal biopsies obtained from four adults, the bacterial 16S rRNA genes present were amplified by PCR, and the resulting products were cloned and sequenced. A total of 900 clones were analyzed, and the figures represent the proportion (%) of clones corresponding to the most frequently detected genera (2).

Currently, approximately half of the world's population is colonized by *Hel. pylori*, but there are marked regional variations in colonization rates (Fig. 9.22), which also vary with age, ethnicity, and socioeconomic status. In general, colonization rates are higher in developing countries than in developed countries and are declining in the latter. The frequency of colonization increases with age among children and adolescents, levels off during middle age, and then often decreases in the elderly.

Other organisms detected on the gastric mucosa include streptococci, micrococci, *Peptostreptococcus* spp., Gram-negative anaerobic rods, *Veillonella* spp., *Neisseria* spp., and lactobacilli. Few studies have reported the actual proportions of the various species comprising the cultivable microbiota of the gastric mucosa. The results of one such study are shown in Fig. 9.23, from which it can be seen that the microbiota is dominated by streptococci, *Peptostreptococcus* spp., micrococci, and Gram-negative anaerobic bacilli. Unfortunately, although *Hel. pylori* was present in each of the subjects,

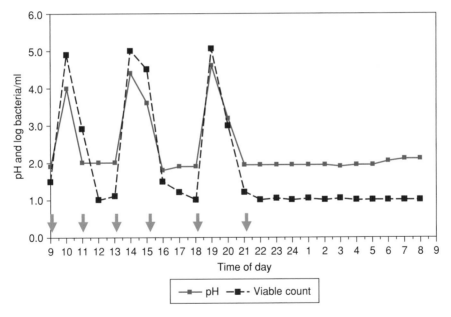

Fig. 9.19 Effect of food and drink on the pH and bacterial content of gastric juice over a 24-h period. The bacterial concentration is expressed as $\log_{10}$ of total bacteria per ml of gastric juice. Arrows denote when food or a beverage (the latter together with a biscuit) were taken. Note that the concentration of bacteria in the gastric juice correlates closely with the pH – a rise in pH results in an increase in the bacterial concentration. From: Hill, M.J. (1989) Factors controlling the microflora of the healthy upper gastrointestinal tract. In: Hill, M.J. and Marsh, P.D. (eds) *Human Microbial Ecology*. Copyright 1989 by CRC Press. Reprinted by permission of CRC Press via the Copyright Clearance Center.

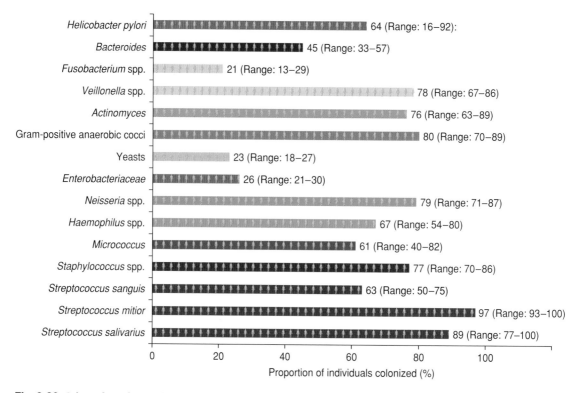

Fig. 9.20 Culture-dependent analysis of the microbiota associated with the gastric mucosa. Figures represent mean values of two studies involving 58 adults, except for *Hel. pylori*, for which a median value is given of nine studies involving 95 666 adults in nine countries (3–13).

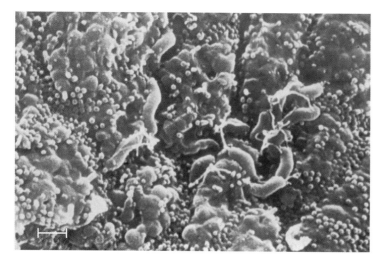

Fig. 9.21 Electron micrograph showing *Hel.* pylori in the mucus layer overlaying the gastric epithelium. Scale bar, 1 μm. Image kindly supplied by Martin J. Blaser, New York University School of Medicine, New York, New York, USA.

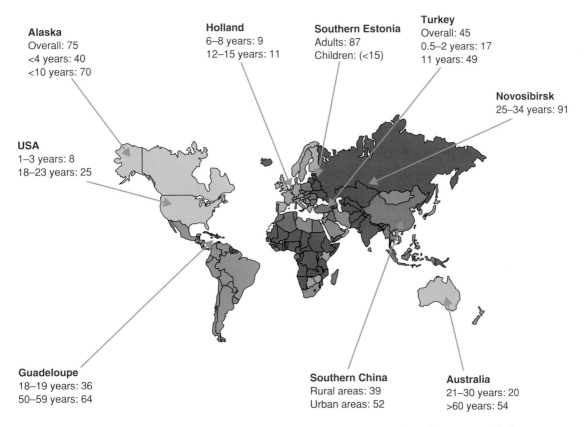

Alaska
Overall: 75
<4 years: 40
<10 years: 70

Holland
6–8 years: 9
12–15 years: 11

Southern Estonia
Adults: 87
Children: (<15)

Turkey
Overall: 45
0.5–2 years: 17
11 years: 49

Novosibirsk
25–34 years: 91

USA
1–3 years: 8
18–23 years: 25

Guadeloupe
18–19 years: 36
50–59 years: 64

Southern China
Rural areas: 39
Urban areas: 52

Australia
21–30 years: 20
>60 years: 54

Fig. 9.22 Prevalence of gastric colonization by *Helicobacter pylori* in various countries and population groups. The figures denote the proportions (%) of the population group colonized by the organism.

viable counts of the organism were not reported. Several studies have found that the number of viable *Hel. pylori* on the gastric mucosa is approximately 10^5 to 10^6 cfu/g of mucosa, which means that, if such levels were present in the study summarized in Fig. 9.23, the organism would comprise approximately 90% of the cultivable gastric microbiota.

At the time of writing (mid-2007), few culture-independent studies of the indigenous gastric microbiota of healthy individuals have been published. In one such study (Fig. 9.24), the most frequently detected bacteria were *Pseudomonas* spp. which are generally regarded as being environmental organisms. As studies of this type will detect DNA from non-viable as well as from viable organisms, the pseudomonal DNA detected may have originated from organisms present in food (e.g. plant material) or beverages consumed by the subjects.

Staphylococci and enterococci also accounted for appreciable proportions of the microbiota.

In a study of the gastric mucosal microbiota of 23 individuals suffering from a variety of diseases, 128 phylotypes from eight bacterial phyla were detected in 16S rRNA gene clone libraries prepared from DNA extracted from mucosal biopsies (Fig. 9.25). *Hel. pylori* dominated the microbiota, which also contained a variety of phylotypes that have not been reported in culture-dependent analyses of the gastric microbiota, e.g. *Caulobacter, Actinobacillus, Corynebacterium, Rothia, Gemella, Leptotrichia, Porphyromonas, Capnocytophaga,* TM7, *Flexistipes,* and *Deinococcus.* However, many of these organisms are frequently detected in the oral cavity, and the detection of their DNA is not surprising. Although interesting, these findings cannot be regarded as representing the indigenous gastric microbiota of

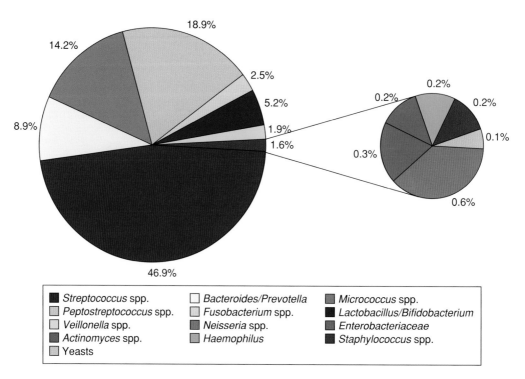

Fig. 9.23 Culture-dependent analysis of the gastric mucosa of 30 healthy adults. Microbes were cultured from biopsies taken from the antrum and corpus in each individual, and the mean values from these two sites were used to determine the relative proportions of each organism in the cultivable microbiota. Unfortunately, although *Hel. pylori* was present in the subjects, viable counts of the organism were not reported; the actual proportion of each of the organisms in the study will therefore have been much lower than the percentage shown (3).

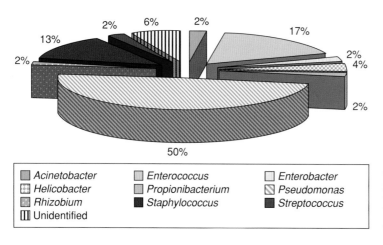

Fig. 9.24 Culture-independent analysis of organisms present on the gastric mucosa of five healthy adults. DNA was extracted from biopsies and subjected to temporal temperature gradient gel electrophoresis of PCR-amplified 16S rRNA gene fragments. The figures denote the relative proportions (%) of species belonging to the indicated genera detected in the five individuals (14).

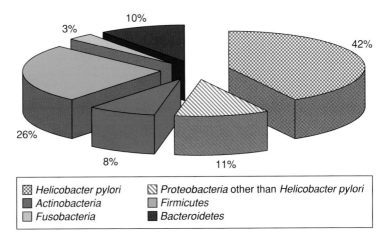

Fig. 9.25 Culture-independent analysis of the microbiota of the gastric mucosa of 23 individuals suffering from various diseases of the upper gastrointestinal tract. DNA was extracted from mucosal biopsies, the bacterial 16S rRNA genes present were amplified by PCR, and the resulting products were cloned and sequenced. The figures denote mean values for the proportions (%) of clones with sequences corresponding to each of the phylogenetic groups detected (15).

⬚ *Helicobacter pylori*   ▨ *Proteobacteria* other than *Helicobacter pylori*
■ *Actinobacteria*   ▨ *Firmicutes*
⬚ *Fusobacteria*   ■ *Bacteroidetes*

healthy adults, as all of the individuals involved had some gastric or other intestinal abnormality. The study has been included because, at the time of writing, it is one of the few published culture-independent studies of the microbiota of the gastric mucosa.

9.4.2.3 Small intestine

Both the lumen and mucosa of the upper regions of the small intestine (duodenum and jejunum) are generally sparsely populated. This is because: (1) the pH is low; (2) rapid peristalsis results in a short transit time for the lumenal contents; (3) chyme and intestinal secretions exert a flushing action; and (4) a range of antimicrobial compounds (bile, proteolytic enzymes, antimicrobial peptides and proteins, etc.) are present. Both culture-based and culture-independent studies have shown that the numbers of organisms in the lumen and on the mucosa, as well as the complexity of the resident communities, gradually increase along the small intestine (Fig. 9.26).

9.4.2.3.1 Duodenum

Most culture-based studies of samples from the duodenal lumen or mucosa have either failed to detect viable microbes or have found them to be present only in low concentrations – approximately 10^2–10^4 cfu/ml of luminal contents. The organisms detected are mainly acid-tolerant, facultative anaerobes such as *Streptococcus* spp. and *Lactobacillus* spp. (derived from the oral and nasal cavities) and *Enterococcus* spp. However, *Bacteroides* spp., *Veillonella* spp., *Bifidobacterium* spp., enterobacteria, *Staphylococcus* spp., and yeasts may also be isolated in

smaller numbers. The results of one culture-dependent study of the microbiota of the duodenal mucosa are shown in Fig. 9.27. The number of strains isolated from each of 26 healthy individuals ranged from zero to seven, with a median value of three.

The time of sampling in relation to food intake has a profound effect on the recovery of viable microbes. Hence, bacteria are more frequently recovered from samples taken after chyme from the stomach has entered the duodenum.

9.4.2.3.2 Jejunum

The jejunum also appears to have a sparse microbiota, although viable organisms tend to be recovered more frequently, and in greater numbers, from this region than from the duodenum. Organisms present in the contents of the lumen are similar to those found in the duodenum and include members of the oral and respiratory microbiotas, e.g. streptococci, lactobacilli, *Fusobacterium nucleatum*, *Neisseria* spp., *Actinomyces* spp., and *Veillonella* spp. (Fig. 9.28). As in the duodenum, viridans streptococci are among the dominant organisms, but *Neisseria* spp. and *F. nucleatum* are also present in high proportions.

The range of microbes cultured from the jejunal mucosa is similar to that found in the lumen (Fig. 9.29), but viridans streptococci dominate the microbiota (Fig. 9.30).

In a culture-independent analysis (using PCR-amplified 16S rRNA gene clone libraries) of the mucosal microbiota of the jejunum in one healthy adult, 88 clones were analyzed, and these corresponded to 22 different phylotypes, most of which are

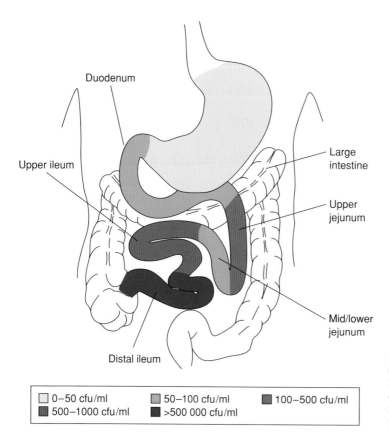

Fig. 9.26 Extent of colonization of the stomach and various regions of the small intestine. Figures denote the concentration of viable microbes present in aspirates from each region. The study involved 18 healthy adults (16).

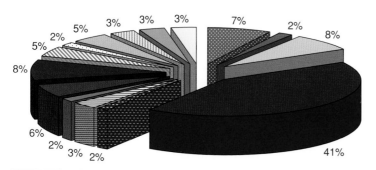

Fig. 9.27 Culture-based analysis of the microbiota of the duodenal mucosa. Samples of mucosa were obtained from 26 healthy adults by endoscopy, and a total of 65 isolates were obtained. Figures denote the relative proportions (%) of the various species detected (17).

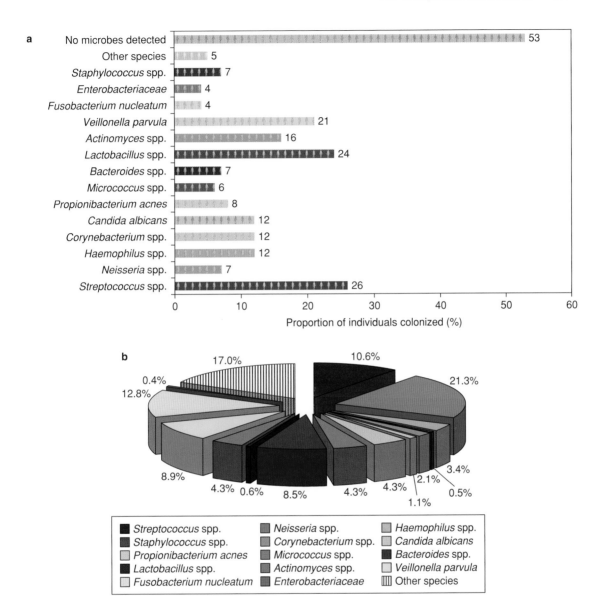

Fig. 9.28 Culture-dependent analysis of the microbiota of the jejunal contents of 85 healthy, fasting adults. (a) Proportions of individuals colonized by the various organisms detected. (b) Relative proportions of the predominant microbes present (18).

members of the oral or respiratory microbiotas (Fig. 9.31). The results are similar to those obtained from culture-based studies, in that the microbiota is dominated by streptococci. Of the sequences obtained, 68% corresponded to streptococci – among which *Strep. mitis* predominated.

In contrast, another culture-independent study (using 16S rRNA gene libraries) of the jejunal contents of three healthy adults produced results that were very different from those generated in culture-based studies. Streptococci were found not to be the dominant organisms in any of the individuals – one was dominated by

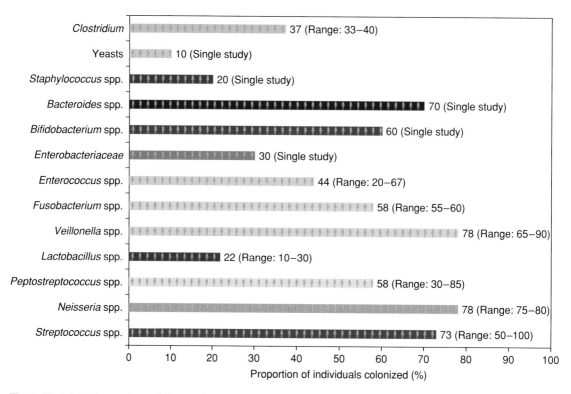

Fig. 9.29 Cultivable microbiota of the jejunal mucosa. Figures are mean values derived from the results of three studies involving 42 healthy adults in different countries (19–21).

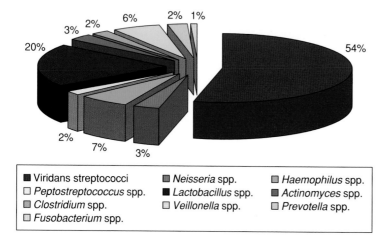

- ■ Viridans streptococci
- □ Peptostreptococcus spp.
- □ Clostridium spp.
- □ Fusobacterium spp.
- ■ Neisseria spp.
- ■ Lactobacillus spp.
- □ Veillonella spp.
- □ Haemophilus spp.
- ■ Actinomyces spp.
- □ Prevotella spp.

Fig. 9.30 Relative proportions of organisms comprising the cultivable microbiota of the jejunal mucosa of 20 healthy adults (19).

lactobacilli, another by the *Enterobacter asburiae* subgroup, and the third by the *Klebsiella planticola* subgroup. The composition of the microbial community in each individual was relatively simple. The overall

relative proportions of the various organisms detected in the three individuals are shown in Fig. 9.32.

In contrast to the results obtained from populations in developed countries, the jejunum of individuals from

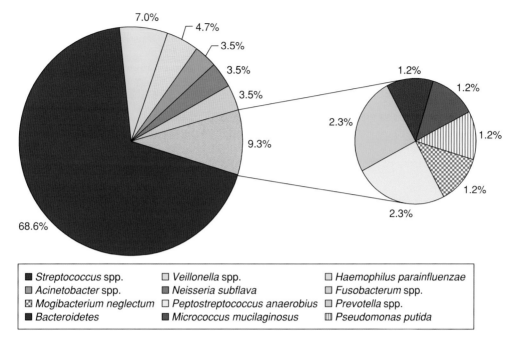

Fig. 9.31 Culture-independent analysis of the microbiota of the jejunal mucosa. DNA was extracted from biopsies of the jejunal mucosa obtained from a healthy adult female, the bacterial 16S rRNA genes present were amplified by PCR, and the resulting products were cloned and sequenced. A total of 88 clones were analyzed, and the figures denote the proportion (%) of clones of each phylotype detected (22).

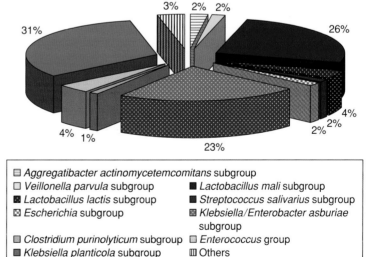

Fig. 9.32 Culture-independent analysis of the jejunal contents of three adults. DNA was extracted from the jejunal contents, the bacterial 16S rRNA genes present were amplified by PCR, and the resulting products were cloned and sequenced. A total of 273 clones were analyzed, and the figures denote mean values for the proportions (%) of clones with sequences corresponding to each of the phylogenetic groups detected (23).

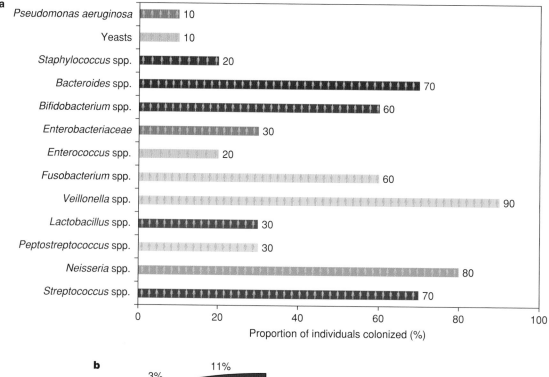

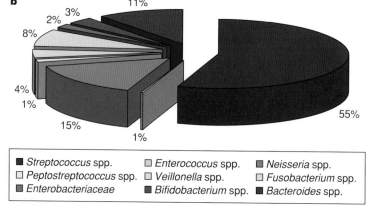

Fig. 9.33 Culture-dependent analysis of the microbiota of the jejunal mucosa in ten Indian adults. (a) Frequency of isolation of the various organisms. (b) Relative proportions of the predominant species (21).

developing countries appears to have a more substantial and varied microbiota (Fig. 9.33). The viable counts of the jejunal contents are considerably higher than those encountered in individuals from developed countries. The organisms present include enterobacteria, enterococci, streptococci, *Neisseria* spp., lactobacilli,

Bifidobacterium spp., *Fusobacterium* spp., *Bacteroides* spp., *Veillonella* spp., *Ps. aeruginosa*, *Peptostreptococcus* spp., coagulase-negative staphylococci (CNS), and yeasts.

Interestingly, the mucosa of the small intestine of Indians differs considerably from that of individuals in developed countries, in being relatively flat and having leaf-like villi. Whether these differences contribute to the establishment of a qualitatively and quantitatively different microbiota has not been established.

9.4.2.3.3 Ileum

In this region, conditions are more conducive to the establishment of a resident microbiota for a number of reasons: (1) peristalsis is slower; (2) intestinal juice dilutes the antimicrobial effects of pancreatic enzymes and bile; (3) the content of bile acids is severely reduced because most are absorbed in the jejunum; and (4) the pH is neutral or slightly alkaline. The ileal microbiota is derived from organisms arriving in material from the jejunum and by the reflux of cecal contents through the ileocecal sphincter, which separates the ileum from the large intestine. In terms of its cultivable microbiota, its composition can be regarded as being intermediate between that of the upper small intestine (sparse, mainly Gram-positive species, with many facultative anaerobes) and that of the large intestine (large numbers of microbes, mainly Gram-negative species, dominated by obligate anaerobes).

Most of what is known concerning the ileal microbiota has been derived from studies of patients with an ileostomy, i.e. individuals who have had their colon removed and an opening made from the ileum to the abdomen to serve as an anus. The lumenal contents of the ileum contain between 10^6 and 10^8 cfu/ml, and the cultivable microbiota is dominated by facultative bacteria (Fig. 9.34), which are between 20 and 50 times more numerous than obligate anaerobes (mainly *Veillonella* spp., *Clostridium* spp., and *Bacteroides* spp.). Facultative organisms (mainly streptococci, enterococci, and coliforms) dominate the lumenal microbiota for up to 6 h following a meal.

The host's diet has a dramatic effect on the relative proportions of microbes in the ileal contents. An increase in dietary protein results in increased proportions of streptococci, enterococci, and coliforms, whereas a high-fat diet favors increased proportions of *Bacteroides* spp. and *Clostridium* spp (Fig. 9.35).

In contrast to the situation in the lumen, obligately anaerobic species comprise an appreciable proportion (approximately 50%) of the cultivable microbiota associated with the mucosal surface of the ileum – findings that are supported by the results of culture-independent studies. Organisms isolated from the mucosa include *Bacteroides* spp., *Col. aerofaciens*, clostridia, bifidobacteria, peptostreptococci, eubacteria, and propionibacteria.

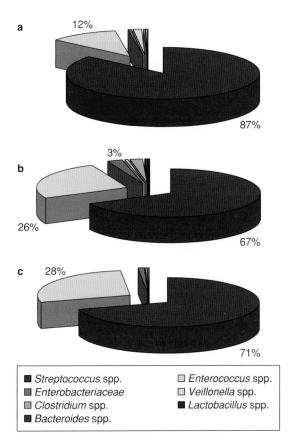

Fig. 9.34 Culture-dependent analysis of the bacterial composition of the ileal contents of ten adults (a) 0–1 h, (b) 3–4 h, and (c) 5–6 h after a meal rich in sucrose and low in dietary fiber, i.e. characteristic of a Western diet (24).

Few culture-independent studies of the ileal microbiota have been carried out, but the results obtained are broadly in agreement with those from culture-dependent studies. In one such study (based on an analysis of 16S rRNA gene clone libraries), the ileal contents were found to be dominated by facultative anaerobes – mainly *Enterobacteriaceae*, streptococci, and enterococci (Fig. 9.36).

In a study using 16S rRNA gene clone libraries, obligate anaerobes were found to be important constituents of the microbiota of the ileal mucosa – indeed, they were the dominant organisms present (Fig. 9.37). This is in keeping with the results of culture-dependent studies. Most (88%) of the *Bacteroidetes* were identified

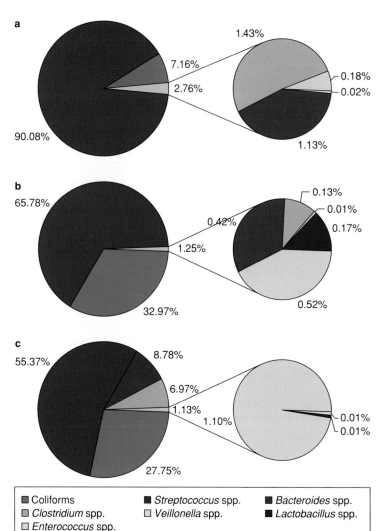

Fig. 9.35 Effect of diet on the cultivable microbiota of the ileal contents of nine adults with ileostomies fed on (a) a normal diet, (b) a high-protein diet, and (c) a high-fat diet (25).

as *B. vulgatus*, while all of the sequences in *Clostridium* clusters XIVa and XI corresponded to those of uncultured bacteria published in GenBank, i.e. mpn-group 24, AF54, adhufec 66.25, and swine feces F17. The remaining clones had sequences corresponding to *Clostridium ramosum* (*Clostridium* cluster XVIII), *Haemophilus influenzae* (Gammaproteobacteria), and *Streptococcus salivarius* (*Bacillus–Lactobacillus–Streptococcus*). Obligate anaerobes were also shown to comprise an appreciable proportion of those phylotypes identified by fluorescence in situ hybridization (FISH) on the ileal mucosa of healthy adults (Figure 9.38). However, bacteria were

detected on the ileal mucosa of only 50% of the individuals studied (Fig. 9.38).

9.4.2.4 Large intestine

Transit times through the large intestine are much longer than in the small intestine, thereby enabling the establishment of substantial microbial communities in the lumen and making it one of the most densely populated microbial ecosystems on Earth. The total number of microbial cells present is approximately 10^{13}–10^{14}, and recent studies have shown that approximately half

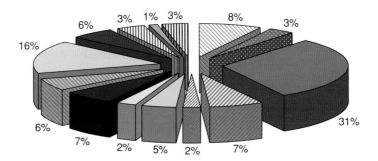

Fig. 9.36 Culture-independent analysis of the ileal contents. DNA was extracted from the ileal contents of three adults, the bacterial 16S rRNA genes present were amplified by PCR, and the resulting products were cloned and sequenced. A total of 272 clones were analyzed, and the figures denote mean values for the proportions (%) of clones with sequences corresponding to each of the phylogenetic groups detected (23).

- *Aggregatibacter actinomycetemcomitans* subgroup
- *Escherichia* subgroup
- *Klebsiella* subgroup/*Enterococcus asburiae* subgroup
- *Klebsiella planticola* subgroup
- *Xenorhabdus* subgroup/*Proteus vulgaris* subgroup
- *Veillonella parvula* subgroup
- *Lactobacillus mali* subgroup
- *Enterococcus* group
- *Streptococcus pneumoniae* subgroup
- *Clostridium leptum*
- *Peptostreptococcus anerobius*
- *Lactococcus lactis* subgroup
- *Streptococcus salivarius* subgroup
- Others

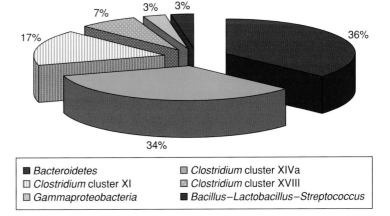

Fig. 9.37 Culture-independent analysis of the microbiota of the ileal mucosa. 16S rRNA gene clone libraries were prepared from the DNA extracted from a biopsy of the terminal ileal mucosa of a healthy adult female. The figures denote the proportions of clones with sequences corresponding to each phylogenetic group (26).

- *Bacteroidetes*
- *Clostridium* cluster XI
- *Gammaproteobacteria*
- *Clostridium* cluster XIVa
- *Clostridium* cluster XVIII
- *Bacillus–Lactobacillus–Streptococcus*

of these cells are viable and one third are dead, while the remaining cells are injured.

It has been recognized for many years that the large intestine is colonized by a wide variety of microbes, but we are only now beginning to appreciate the extent of this diversity. For example, a recent culture-independent study (based on sequencing of cloned small-subunit rRNA genes) of the communities inhabiting the mucosa and lumen of the large intestine in three healthy adults has revealed the presence

of a total of 395 bacterial phylotypes and one archaeal phylotype – *Methanobrevibacter smithii*. Most of the inferred organisms were members of the *Firmicutes* (95% of which were *Clostridia*) and *Bacteroidetes* phyla (Fig. 9.39). As well as revealing the diversity of the microbial communities within the large intestine, the results of this study have shown that a large majority of the organisms present (80%) have not yet been cultivated and, furthermore, that most of them are novel (62%).

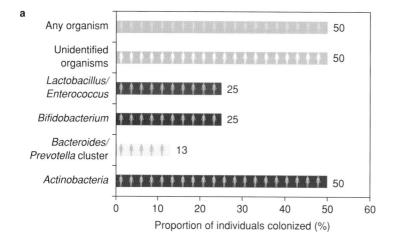

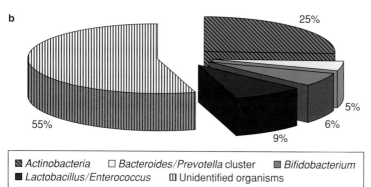

Fig. 9.38 Culture-independent analysis of the microbiota of the ileal mucosa. Biopsies from the terminal ileum of eight healthy adults were examined by FISH using a panel of 14 different 16S/23S rRNA gene-targeted oligonucleotide probes. (a) Frequency of detection of the various phylotypes. (b) Relative proportions of the phylotypes detected (27).

A number of studies have revealed that the microbiota of the lumen and that associated with the mucosa differ in composition. With regard to the mucosal microbiotas of the various regions of the large intestine, culture-dependent as well as culture-independent studies (using denaturing gradient gel electrophoresis [DGGE] and temperature gradient gel electrophoresis [TGGE] profiling, as well as sequencing of clones from 16S rRNA gene libraries) have shown that they are broadly similar.

The communities present in the main regions of the large intestine, the cecum, colon, and rectum, will now be described.

9.4.2.4.1 Cecum

The environment in the cecum differs substantially from that of other regions of the large intestine in that it has a more acidic pH, its contents have a higher water content, and easily fermentable compounds are present at higher concentrations. Because of sampling difficulties, there are very few reports on the composition of the microbial communities inhabiting the cecum. Culture-based studies have shown that facultative anaerobes, mainly *Enterobacteriaceae*, comprise a greater proportion of the lumenal microbiota than they do in the colon, where obligate anaerobes predominate (Fig. 9.40). Other numerically dominant groups include *Lactobacillus* spp. and *Bacteroides* spp.

The results of a culture-independent study are broadly in agreement with these findings as *Enterobacteriaceae* (specifically, *E. coli*) and the *Lactobacillus–Enterococcus* group were also found to comprise substantial proportions of the lumenal microbiota (Fig. 9.41). However, in contrast to the results of the culture-dependent study, members of the *Clostridia*, rather than *Bacteroidetes*, were found to comprise a greater proportion of the anaerobes present.

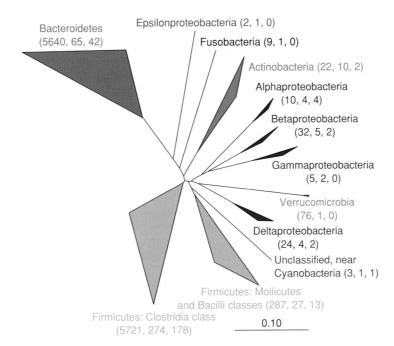

Bacteroidetes
(5640, 65, 42)

Epsilonproteobacteria (2, 1, 0)

Fusobacteria (9, 1, 0)

Actinobacteria (22, 10, 2)

Alphaproteobacteria
(10, 4, 4)

Betaproteobacteria
(32, 5, 2)

Gammaproteobacteria
(5, 2, 0)

Verrucomicrobia
(76, 1, 0)

Deltaproteobacteria
(24, 4, 2)

Unclassified, near
Cyanobacteria (3, 1, 1)

Firmicutes: Mollicutes
and Bacilli classes (287, 27, 13)

Firmicutes: Clostridia class
(5721, 274, 178)

0.10

Fig. 9.39 Phylogenetic tree constructed from sequences of bacterial 16S rRNA genes present in microbial DNA obtained from the large intestine of three healthy adults (28). The DNA was extracted from samples taken from the mucosa of the caecum, ascending colon, transverse colon, descending colon, sigmoid colon, and rectum as well as from feces. In parentheses after the name of each clade is given, in order, the total number of recovered sequences, phylotypes, and novel phylotypes. The angle where each triangle joins the tree represents the relative abundance of sequences, and the lengths of the two adjacent sides indicate the range of branching depths within that clade. A total of 1524 archaeal sequences were also obtained, and these all corresponded to a single phylotype – *Methanobrevibacter smithii*. Reprinted with permission from AAAS, Washington, DC, USA, from: Eckburg, P.B., Bik, E.M., Bernstein, C.N., Purdom, E., Dethlefsen, L., Sargent, M., Gill, S.R., Nelson, K.E. and Relman, D.A. (2005) *Science* 308, 1635–38.

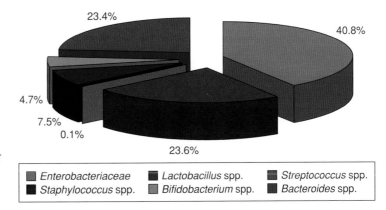

23.4%

40.8%

4.7%

7.5%

0.1%

23.6%

| Enterobacteriaceae | Lactobacillus spp. | Streptococcus spp. |
| Staphylococcus spp. | Bifidobacterium spp. | Bacteroides spp. |

Fig. 9.40 Cultivable microbiota of the contents of the cecum. Figures denote mean values of the relative proportions of the various genera and are derived from three studies involving 21 adults (16, 29, 30).

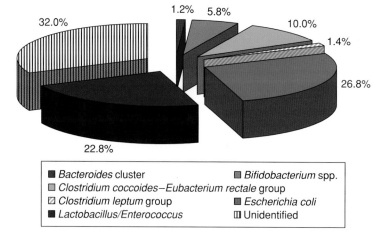

Fig. 9.41 Culture-independent analysis of the microbial community in the lumen of the cecum. RNA was extracted from the cecal contents of eight healthy adults, and the predominant phylogenetic groups were quantified by hybridization using a set of six 16S rRNA-targeted probes. Figures denote the relative proportions of each phylogenetic group (29).

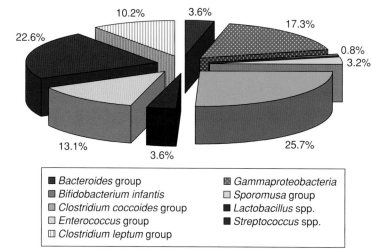

Fig. 9.42 Culture-independent analysis of the contents of the cecum. 16S rRNA gene clone libraries were prepared from DNA extracted from the cecal contents of three adults. The figures denote the proportions of clones with sequences corresponding to each phylogenetic group (23).

In another culture-independent study (which used PCR-amplified 16S rRNA gene clone libraries), the mean number of phylotypes detected in each of three individuals was 25. *Strep. salivarius*, *Enterococcus* spp., the *Ent. asburiae* subgroup, and *Clostridium leptum* were the predominant phylotypes, accounting for 15%, 13%, 13%, and 10% of the clones, respectively (Fig. 9.42). As in culture-dependent studies, facultative anaerobes were found to comprise a substantial proportion of the microbiota – more than half in this study.

Little is known with regard to the composition of the microbiota associated with the cecal mucosa. The limited number of studies published have shown that *Bacteroides* spp. are among the dominant organisms (Fig. 9.43).

9.4.2.4.2 Colon

The human colon harbors a large and diverse microbial community and is the most densely populated region of the body. Most of the resident microbiota is present in the lumenal contents – each gram of which contains 10^{11}–10^{12} microbes. Because of the difficulties associated with obtaining representative samples of the lumen contents (particularly from the upper regions), most studies of the lumenal microbiota have been carried out on feces. Studies have verified that the microbial composition of feces is similar to that found in the lumen of the various regions of the colon, although the absolute numbers may differ. Another important habitat for colonic microbes is the mucosal surface (Fig. 9.44), and evidence suggests that the composition

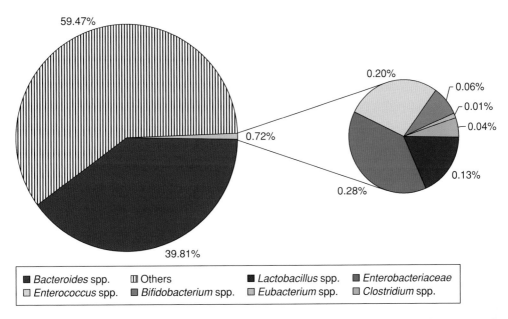

Fig. 9.43 Cultivable microbiota of the cecal mucosa. Figures denote the proportions (%) of each group of organisms and are mean values derived from the results of two studies involving a total of 19 adults (31, 32).

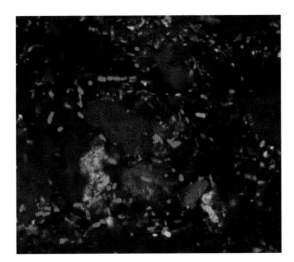

Fig. 9.44 Light micrograph of the mucosal surface of the colon stained with 16S rRNA oligonucleotide probes targeted against *Bacteroides* (red, Cy 3), *Bifidobacterium* (blue, Cy 5), and *Escherichia* (green, fluorescein isothiocyanate). Reproduced with the permission of Cambridge University Press, Cambridge, UK, from: Macfarlane, S. and Macfarlane, G.T. (2003) Bacterial growth on mucosal surfaces and biofilms in the large bowel. In: Wilson, M. and Devine, D. (eds), *Medical Implications of Biofilms*.

of the community associated with the mucosa differs from that of the lumenal microbiota. Exactly where such "mucosa-associated" organisms are located is a matter of controversy at present. Possible sites are (1) attached directly to epithelial cells; (2) positioned between the epithelial cells (but not attached to them) and the mucous layer; (3) positioned within the mucous layer; (4) attached to the lumenal surface of the mucous layer. The results of studies aimed at shedding light on this issue are very much influenced by the sample-preparation methods employed. Feces can be considered to be comprised of microbes that have been shed from the mucosal surfaces, as well as those organisms that have never been mucosa-associated.

The microbiota of the colon is extremely complex, and more than 800 species or phylotypes from more than 190 genera have been detected using a combination of culture-based and culture-independent approaches. However, many species are present in small numbers, and it has been estimated that between 30 and 40 species belonging to six genera account for 99% of the cultivable microbiota. The number of species that can be cultivated from the colon of an adult is usually between 13 and 30. However, studies have shown that the proportion of the colonic microbiota

Table 9.7 Most frequently isolated species of common genera present in human feces.

Genus/group	Most frequently isolated species
Bacteroides	*B. thetaiotaomicron, B. vulgatus, B. distasonis, B. eggerthii, B. fragilis, B. ovatus, B. uniformis, B. caccae*
Prevotella	*Prev. tannerae, Prev. loeschei*
Eubacterium and *Eubacterium*-like organisms	*Col. aerofaciens, Eg. lenta, Eub. contortum, Eub. cylindroides, Eub. rectale, Eub. biforme, Eub. ventriosum, Eub. alactolyticum, Eub. timidum*
Bifidobacterium	*Bif. adolescentis, Bif. infantis, Bif. angulatum, Bif. catenulatum, Bif. pseudocatenulatum, Bif. breve, Bif. longum*
Clostridium	*Cl. ramosum, Cl. bifermentans, Cl. butyricum, Cl. perfringens, Cl. difficile, Cl. indolis, Cl. sordellii, Cl. septicum, Cl. sporogenes, Cl. innocuum*
Enterococcus	*Ent. faecalis, Ent. faecium, Ent. durans*
Fusobacterium	*F. mortiferum, F. necrophorum, F. nucleatum, F. varium*
Gram-positive anaerobic cocci	*Peptostreptococcus productus, Micromonas micros (Peptostreptococcus micros), Anaerococcus prevotii (Peptostreptococcus prevotii), Schleiferella asaccharolytica (Peptostreptococcus asaccharolyticus), Finegoldia magna (Peptostreptococcus magnus)*
Ruminococcus	*Rum. albus, Rum. obeum, Rum. torques, Rum. flavefaciens, Rum. gnavus, Rum. bromii*
Enterobacteria	*E. coli, Enterococcus aerogenes, Proteus mirabilis, K. pneumoniae, Moraxella morganii*
Lactobacillus	*L. acidophilus, L. brevis, L. casei, L. salivarius, L. plantarum, L. gasseri, L. ruminis, L. crispatus, L. paracasei, L. rhamnosus, L. delbrueckii*
Actinomyces	*A. naeslundii, A. odontolyticus*
Propionibacterium	*P. acnes, P. avidum*
Streptococcus	*Strep. salivarius, Strep. bovis, Strep. equinus*

that can be cultured in the laboratory may be as low as 20%. The organisms most frequently isolated from feces include *Bacteroides* spp., *Bifidobacterium* spp., Enterobacteriaceae, *Eubacterium* spp. (including former members of this genus), *Clostridium* spp., and *Enterococcus* spp. (Fig. 9.45 and Table 9.7).

With regard to the relative proportions of the various organisms present, the cultivable microbiota of the colon is dominated by obligate anaerobes, which are approximately 1000-fold more abundant than facultative anaerobes. The main anaerobic genera are *Bacteroides*, *Eubacterium* (including *Eubacterium*-like organisms), and *Bifidobacterium*, while *Streptococcus* spp. and enterobacteria comprise the main facultative organisms (Fig. 9.46). Species belonging to the genus *Bacteroides* are the most numerous organisms and constitute 20–30% of the cultivable microbiota – *B. thetaiotaomicron* and *B. vulgatus* are generally the

predominant species present. *Eubacterium* spp. (including *Eubacterium*-like organisms) and *Bifidobacterium* spp. are the next most abundant organisms.

Although the relative proportions of these genera vary considerably between individuals, the composition (at the genus level) of the fecal microbiota of an individual remains relatively stable over time. However, within some genera (e.g. *Lactobacillus*) the particular species present in the feces of an individual may vary markedly from day to day. Other cultivable genera that are regularly present in human feces, but in smaller proportions than those shown in Fig. 9.46, include *Staphylococcus* spp., *Ruminococcus* spp., *Peptostreptococcus* spp. and other GPAC, *Fusobacterium* spp., *Prevotella* spp., *Acidaminococcus* spp., *Actinomyces* spp., *Propionibacterium* spp., *Desulfovibrio* spp., *Candida* spp., and other yeasts.

Recently, it has been shown that approximately 5% of the bacterial cell mass in the lumen of the colon is

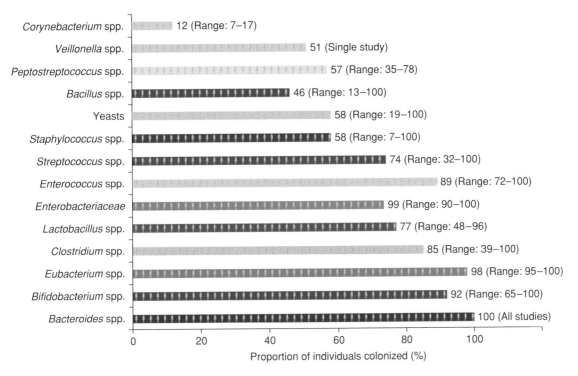

Fig. 9.45 Cultivable microbiota of feces. Figures represent mean values of the proportions (%) of individuals colonized and are derived from nine studies involving 263 healthy adults in several countries (33–41).

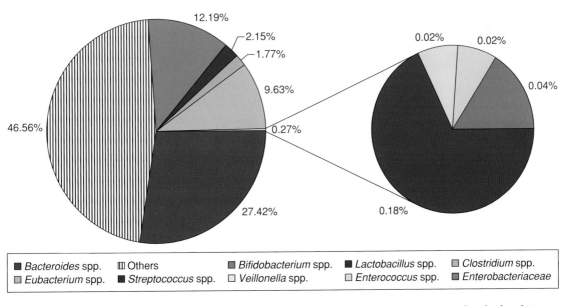

Fig. 9.46 Culture-based analysis of the composition of the fecal microbiota. The figures represent mean values for the relative proportions of the various genera – these have been derived from the results of ten studies involving 212 healthy adults from several countries (33–35, 37, 42–47).

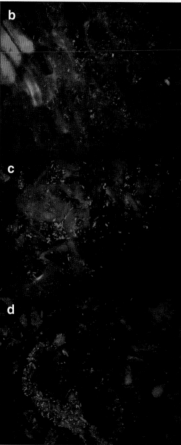

present as a biofilm on the surface of food particles, while an even higher proportion is more loosely associated with such particles (Fig. 9.47). The composition of the microbial community on the food particles does not appear to differ from that of the feces as a whole – with the cultivable microbiota of the biofilms being dominated by *Bacteroides* spp. (mainly *B. thetaiotaomicron* and *B. vulgatus*) and *Bifidobacterium* spp. (mainly *Bif. adolescentis* and *Bif. angulatum*).

As mentioned previously, the application of culture-independent techniques to the study of the fecal microbiota has shown that large proportions of the microbes present (up to 80% in some studies) have not yet been grown in the laboratory. The application of such techniques has resulted in a very different view of the composition of the fecal microbiota. The use of oligonucleotide probes has shown that while Gram-negative anaerobes (e.g. *Bacteroides* spp.) comprise a large proportion of the fecal microbiota, Gram-positive anaerobes (particularly the *Clostridium coccoides–Eub. rectale* and *Cl. leptum* groups) are far more abundant than is implied by culture-based analyses (Fig. 9.48). Furthermore, bifidobacteria appear to be present in much lower proportions than those reported in culture-based studies.

Detailed analysis (involving the sequencing of clones from 16S rRNA gene libraries) of the phylotypes present in feces has also shown that the microbiota is dominated by members of the *Cl. coccoides–Eub. rectale*, *Cl. leptum*, and *Bacteroides–Prevotella* groups (Fig. 9.49). However, approximately three-quarters of the sequences detected in these studies do not correspond to known species. Of those that do, species frequently detected include: *Faecalibacterium prausnitzii* and *Rum. bromii* (from the *Cl. leptum* cluster); *Eub. rectale*, *Eubacterium gnavus*, and *Eubacterium hallii* (from the *Cl. coccoides–Eub. rectale* cluster); *B. thetaiotaomicron*, *Bacteroides uniformis*, *Bacteroides caccae*, and *Bacteroides ovatus*; *Bif. infantis* and *Bif. bifidum*; *Streptococcus sanguis* and *Strep. salivarius*; *Phascolarctobacterium faecium*; and *Coll. aerofaciens*.

Fig. 9.47 Bacterial biofilms on food particles in the human large intestine. (a) Scanning electron micrograph of a bacterial biofilm on the surface of a food particle. Scale bar, 5 μm. (b–d) Bacteria on food particles labeled with 16S rRNA gene oligonucleotide probes. (b) Members of the *Eubacterium rectale–Cl. coccoides* group are stained with Cy 3 (red), and bifidobacteria are stained with Cy 5 (blue). (c) Total eubacteria are stained with Cy 5 (blue), and bifidobacteria are stained with Cy 3 (red). (d) Enterobacteria are stained with Cy 5 (blue), and members of the *Eubacterium rectale–Cl. coccoides* group are stained with Cy 3 (red). Images kindly supplied by Dr Sandra Macfarlane, Gut Group, University of Dundee, Dundee, UK.

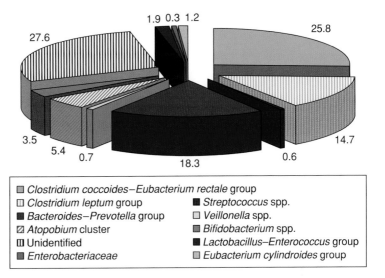

Fig. 9.48 Culture-independent analysis of the composition of the fecal microbiota. Labeled 16S rRNA-targeted oligonucleotide probes were used to detect and quantify 16S rRNA in either RNA extracted from fecal samples (dot-blot hybridization) or in bacterial cells (FISH) present in the samples. The probes used were designed to detect major phylogenetic groups or genera rather than species. The figures are mean values for the relative proportions of the various bacterial groups, and are derived from ten studies involving 228 subjects from several countries (29, 48–56).

- *Clostridium coccoides–Eubacterium rectale* group
- *Clostridium leptum* group
- *Bacteroides–Prevotella* group
- *Atopobium* cluster
- Unidentified
- *Enterobacteriaceae*
- *Streptococcus* spp.
- *Veillonella* spp.
- *Bifidobacterium* spp.
- *Lactobacillus–Enterococcus* group
- *Eubacterium cylindroides* group

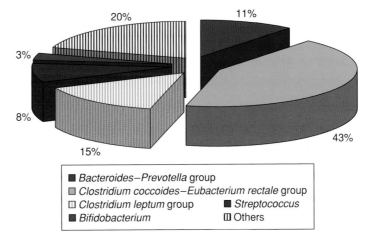

Fig. 9.49 Culture-independent analysis of the fecal microbiota. DNA was extracted from feces, the bacterial 16S rRNA genes present were amplified by PCR, and the resulting products were cloned and sequenced. The figures denote mean values for the proportions (%) of clones with sequences corresponding to each of the phylogenetic groups detected, and are derived from four studies involving a total of nine adults (23, 57–59).

- *Bacteroides–Prevotella* group
- *Clostridium coccoides–Eubacterium rectale* group
- *Clostridium leptum* group
- *Bifidobacterium*
- *Streptococcus*
- Others

In addition to the lumen and particulate matter within the lumen, the colon offers a number of other habitats for microbial colonization, i.e. the mucous layer, the mucosal surface, and intestinal crypts. Within these regions, the environment is likely to differ from that found in the lumen in a number of ways: (1) the oxygen concentration will be considerably higher (due to diffusion from the underlying oxygen-rich tissues); (2) the range of nutrients is likely to differ due to the abundance of mucins and host cell secretory and excretory products; and (3) the concentration of host antimicrobials is likely to be much greater. Furthermore, only those organisms with appropriate adhesins will be able to adhere to the epithelium or to the mucus. These factors would be expected to enable the establishment of microbial communities different from those existing in the lumen. A number of studies (many based on DGGE and TGGE profiling) comparing the two communities have shown statistically significant differences between them, although the exact nature of such differences (e.g. which phylotypes are present at one site but absent from the other) remains to be established. Differences between the communities may also be attributable to differences in the relative proportions of

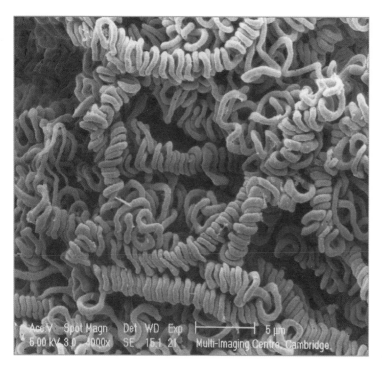

Fig. 9.50 Scanning electron micrograph of large spiral bacterial forms associated with the colonic epithelium. Reproduced with the permission of Cambridge University Press, Cambridge, UK, from: Macfarlane, S. and Macfarlane, G.T. (2003) Bacterial growth on mucosal surfaces and biofilms in the large bowel. In: Wilson, M. and Devine, D. (eds), *Medical Implications of Biofilms*.

the organisms present. Microscopic examination of the colonic mucosa has revealed the presence of organisms that have an unusual morphology and that are absent from feces (Fig. 9.50).

Although most studies have found differences between the lumenal and mucosal microbiotas, others have failed to do so. A major complication is the difficulty in obtaining samples of mucosa and/or mucus that are uncontaminated by lumenal contents, and this may obscure differences in composition between the various communities. Sample processing, particularly the extent of washing needed to remove microbes considered to be part of the lumenal, rather than the mucosal, microbiota also differs between studies and can, therefore, make inter-study comparisons difficult. Because of technical difficulties and ethical considerations, it is also often difficult to distinguish between bacterial populations that adhere to the epithelium and those that are present within the mucous layer itself. A number of investigators have concluded that in disease-free individuals, all "mucosa-associated" microbes are, in fact, present within the mucous layer itself or otherwise are associated with the lumenal

surface of the mucous layer overlying the epithelium (Fig. 9.51). Other investigators, on the other hand, assert that bacteria do actually adhere to the epithelial cells.

Culture-independent analysis (involving the sequencing of clones from 16S rRNA gene libraries) of the microbiota associated with the colonic mucosa has shown that it is dominated by members of the *Cl. coccoides–Eub. rectale* group, *Cl. leptum*, and the *Bacteroides–Prevotella* group (Fig. 9.52). In terms of the relative proportions of the major phylogenetic groups, the overall composition is similar to that of the fecal microbiota determined using a similar culture-independent technique (Fig. 9.49). As DGGE and other molecular-profiling approaches have found significant differences between the fecal and mucosal microbiotas, these distinctions must be attributable to differences in the phylotypes present.

The cultivable microbiota of the colonic mucosa appears to be dominated by *Bacteroides* spp., although few studies have identified all of the organisms present (Fig. 9.53).

Apart from bacteria, archaea, and yeasts, the colon also harbors approximately 1200 different viruses,

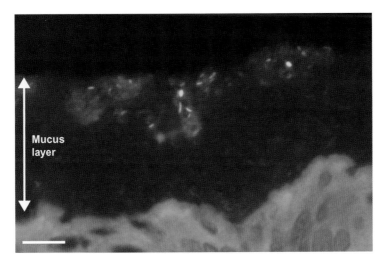

Fig. 9.51 Colon from a healthy adult hybridized with rhodamine-BACT338 (which hybridizes with all bacteria) and fluorescein-EREC482 (which hybridizes with members of the *Cl. coccoides–Eub. rectale* group). The mucus and epithelial cells are visible due to autofluorescence. Bacteria can be seen in the outer region of the mucous layer. EREC482-hybridizing bacteria stain green and yellow (bright rhodamine and bright fluorescein results in yellow staining), and are randomly dispersed among the other bacteria (stained orange-red). Scale bar, 10 μm. Reproduced with the permission of Lippincott, Williams & Wilkins, Philadelphia, Pennsylvania, USA, from: van der Waaij, L.A., Harmsen, H.J., Madjipour, M., Kroese, F.G., Zwiers, M., van Dullemen, H.M., de Boer, N.K., Welling, G.W. and Jansen, P.L. (2005) Bacterial population analysis of human colon and terminal ileum biopsies with 16S rRNA-based fluorescent probes: Commensal bacteria live in suspension and have no direct contact with epithelial cells. *Inflamm Bowel Dis* 11, 865–71.

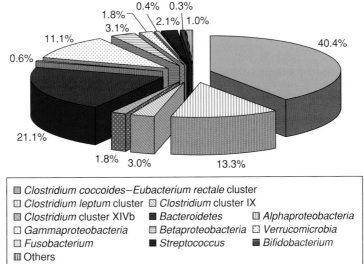

Fig. 9.52 Culture-independent analysis of the mucosa-associated microbiota of the colon. DNA was extracted from biopsies taken from the colonic mucosa, the bacterial 16S rRNA genes present were amplified by PCR, and the resulting products were cloned and sequenced. The figures denote mean values for the proportions (%) of clones with sequences corresponding to each of the phylogenetic groups detected, and are derived from five studies involving a total of 17 adults (22, 57, 60–62).

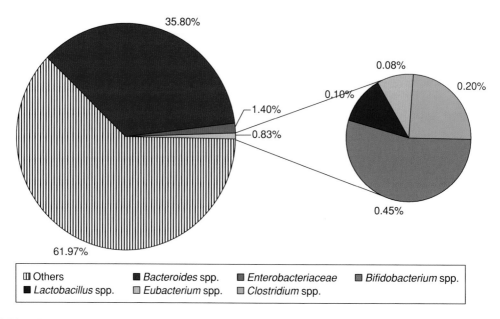

Fig. 9.53 Cultivable microbiota of the colonic mucosa. Figures denote mean values for the relative proportions of the various organisms isolated, and are derived from the results of three studies involving 61 healthy adults (32, 63, 64).

the majority of which are bacteriophages. Phages can affect the composition of bacterial communities as a result of specific predation on their hosts, but nothing is known about such predator–prey interactions in vivo and the effects that they may have on the microbiota of the colon.

It is well established that diet can affect the composition of the colonic microbiota in infants, where characteristic differences are observed between those who are breast-fed and those who are formula-fed. Hence, *Bifidobacterium* spp. dominate the fecal microbiota of 1-week-old breast-fed infants, while in formula-fed infants, the microbiota is more complex with high proportions of *Bifidobacterium* spp., *Bacteroides* spp., enterobacteria, and *Streptococcus* spp. Formula-fed infants also have much higher counts of *Clostridium* spp. in their feces than breast-fed infants.

Although the effect of diet on the colonic microbiota of adults has been studied extensively, the complexity of the microbiota and the large variations in its composition between individuals have made the results of such studies difficult to interpret. Nevertheless, individuals in the UK and the USA have been found to have a higher ratio of obligate anaerobes to facultative

anaerobes than individuals in Japan, India, or Uganda, and this has been attributed to dietary influences. A diet that is rich in sulfate has also been shown to favor sulfate-reducing bacteria (e.g. *Desulfovibrio* spp.) over methanogenic archaea. In a recent study of the effect of diet on the colonic microbiota, 12 obese adults were placed on either a fat-restricted or carbohydrate-restricted diet, and the microbial content of their feces was monitored over a 12-month period. The relative proportions of *Firmicutes* and *Bacteroidetes* gradually changed with time and approached a value similar to that found in two lean individuals (Fig. 9.54). The increase in the relative abundance of *Bacteroidetes* correlated with a decrease in body mass in the obese individuals.

The most likely constituents to evade digestion in the upper GIT and, consequently, to exert an effect on the colonic microbiota are complex carbohydrates. There is, therefore, considerable interest in using such compounds as "prebiotics" to alter the composition of the colonic microbiota in order to improve health.

As has been mentioned previously, the microbiota of the large intestine is able to degrade complex polysaccharides and other materials that the host cannot itself

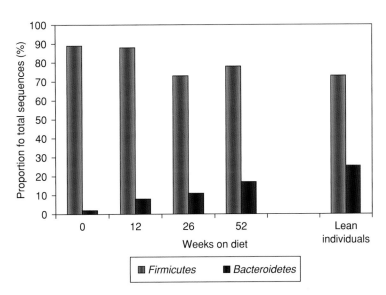

Fig. 9.54 Culture-independent analysis of the relative proportions of *Firmicutes* and *Bacteroidetes* in the feces of 12 obese individuals on a fat-restricted or carbohydrate-restricted diet. Fecal samples were also obtained from two lean, control individuals. DNA was extracted from fecal samples, the bacterial 16S rRNA genes present were amplified by PCR, and the resulting products were cloned and sequenced. A total of 18 348 sequences were obtained, and the figures denote the proportion of sequences belonging to the *Firmicutes* and *Bacteroidetes* divisions (65).

Table 9.8 Comparative functional analysis of the human gut microbiome. Metabolic functions that are enriched in the human gut microbiome are shown compared to the human genome and compared to an average content of all sequenced bacterial genomes (71).

Metabolic functions enriched in the human gut microbiome compared to the human genome	Metabolic functions enriched in the human gut microbiome compared to the average content of all sequenced bacterial genomes
Energy production and conversion	Energy production and conversion
Carbohydrate transport and metabolism	Carbohydrate transport and metabolism
Amino acid transport and metabolism	Amino acid transport and metabolism
Nucleotide transport and metabolism	Nucleotide transport and metabolism
Glycan biosynthesis and metabolism	Coenzyme transport and metabolism
Lipid metabolism	—
Metabolism of cofactors and vitamins	—
Biosynthesis of secondary metabolites	—
Biodegradation of xenobiotics	—

utilize, and can convert these materials into their low molecular mass constituents that can then be metabolized by both the host and by the resident microbiota. In a recent study of the collective genome ("microbiome") of the gut microbiota of two individuals, the metabolic functions of the identified genes were compared with those of the human genome and with those of all sequenced bacterial genomes. The gut microbiome was found to be significantly enriched, with genes encoding a number of activities (Table 9.8).

The microbiota of the colon is extremely complex and is a stable, climax community, which is maintained in this state because of the stable environment and nutrient supply provided by the host, together with a plethora of positive and negative interactions between its constituent members (see section 9.4.3).

The host, in return, benefits in a number of ways from the presence of this microbiota. The beneficial effects of the colonic microbiota can be summarized as follows:

- Helps to exclude exogenous pathogens;
- Involved in the development of host immune functions;
- Involved in the differentiation and development of host tissues;
- Involved in the development of host nutritional capabilities;
- Provides nutrients to the host;
- Provides vitamins to the host;
- Provides butyrate, propionate, and acetate, which are important energy sources for intestinal epithelial cells;
- Detoxifies harmful dietary constituents;
- Helps to prevent bowel cancer.

9.4.2.4.3 Rectum

Microscopic examination of biopsies from the rectal mucosa of healthy individuals has revealed the presence of bacteria in microcolonies and also spread throughout the mucous layer (Fig. 9.55). In contrast, the crypts appear to be devoid of bacteria. Details of the structure of one of the adherent microcolonies are shown in Fig. 9.56. This mushroom-shaped microcolony was approximately 16.5 μm in height and consisted of mainly live cells at its base, with increasing proportions of dead cells as the distance from the mucosal surface increased.

Culture-based studies have shown that the mucosal microbiota of the rectum, like that of the colonic mucosa, is dominated by *Bacteroides* spp. (Fig. 9.57). *Bacteroides* spp. are also among the most frequently detected organisms (Fig. 9.58). Few studies have identified the isolated organisms to the species level, but in one such study, a wide range of *Bacteroides* spp. (12 species), *Bifidobacterium* spp. (five species), streptococci (five species), *Prevotella* spp. (four species), and *Clostridium* spp. (four species) were isolated from ten healthy individuals (Fig. 9.58). The most frequently detected species included *Bif. adolescentis*, *Bif. angulatum*, *E. coli*, *B. thetaiotaomicron*, *B. vulgatus*, and *Strep. parasanguinis*.

Although culture-independent analysis of the mucosa-associated microbiota has also shown the predominance of *Bacteroides* spp., much higher proportions of organisms belonging to clostridial clusters XIVa, IV, and IX are also found – many of these phylotypes have not yet been cultured in the laboratory (Fig. 9.59).

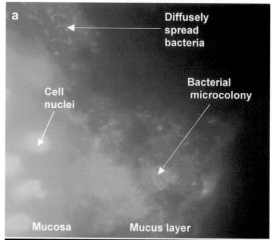

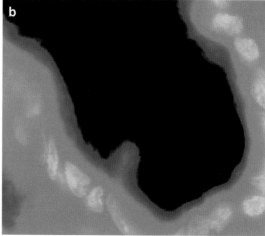

Fig. 9.55 Light micrographs of transverse sections of healthy rectal mucosa stained with 4′,6′-diamidino-2-phenylindole, showing (a) the presence of bacterial microcolonies and individual bacterial cells in the mucous layer and (b) a crypt uncontaminated by bacteria. Original magnification ×60. Reproduced with the permission of the University of Chicago Press, Chicago, Illinois, USA, from: Macfarlane, S., Furrie, E., Cummings, J.H. and Macfarlane, G.T. (2004) Chemotaxonomic analysis of bacterial populations colonizing the rectal mucosa in patients with ulcerative colitis. *Clin Infect Dis* 38, 1690–99. © 2004 by the Infectious Diseases Society of America, Arlington, Virginia, USA.

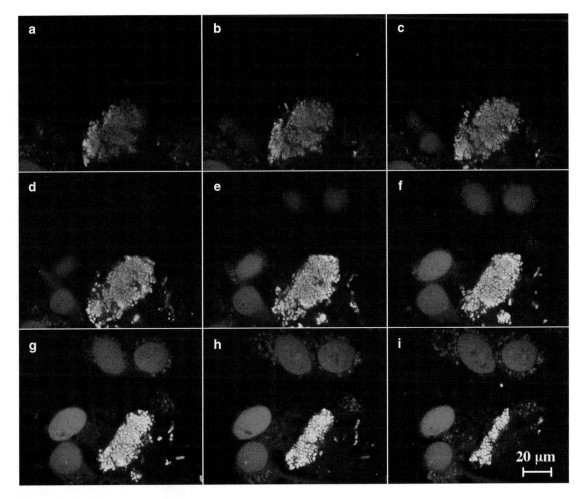

Fig. 9.56 Confocal laser scanning microscopy of a bacterial microcolony on healthy rectal mucosa stained with a live/dead stain – live cells stain yellow, while dead cells stain red. The microcolony has been sliced in 1.5-μm sections from the top (a) to the mucosal surface (i). The proportion of live cells was far greater at the base of the microcolony than at its tip. Original magnification ×60. Reproduced with the permission of the University of Chicago Press, Chicago, Illinois, USA, from: Macfarlane, S., Furrie, E., Cummings, J.H. and Macfarlane, G.T. (2004) Chemotaxonomic analysis of bacterial populations colonizing the rectal mucosa in patients with ulcerative colitis. *Clin Infect Dis* 38, 1690–9. © 2004 by the Infectious Diseases Society of America, Arlington, Virginia, USA.

9.4.3 Microbial interactions in the gastrointestinal tract

Within the extremely complex microbiotas of the GIT, particularly in the terminal ileum, cecum, colon, and rectum, there is potential for a range of microbial inter-actions to occur. Many of these interactions, however, remain theoretical as few have been demonstrated to take place in vitro, and there is little evidence to support their occurrence in vivo. Examples of possible positive and negative interactions are given in Tables 9.9 and 9.10, respectively.

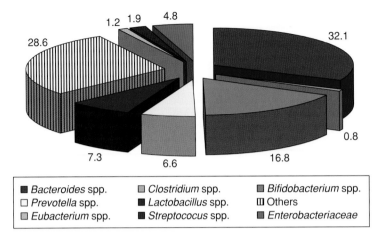

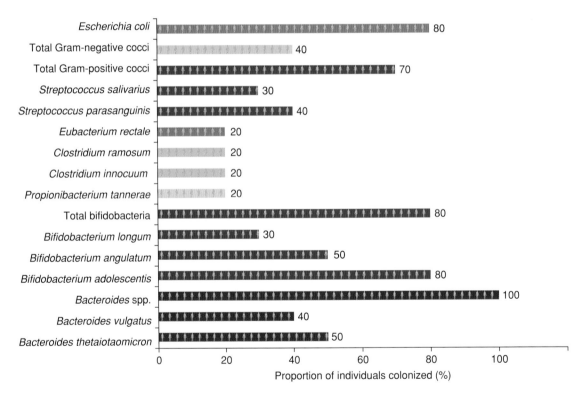

Fig. 9.57 Cultivable microbiota of the rectal mucosa. Figures denote mean values of the relative proportions (%) of the various genera comprising the cultivable microbiota, and are derived from the results of five studies involving 42 adults (20, 66–69).

Fig. 9.58 Prevalence of cultivable bacteria colonizing the rectal mucosa of ten healthy individuals. Figures denote the proportion (%) of individuals colonized by each organism (66).

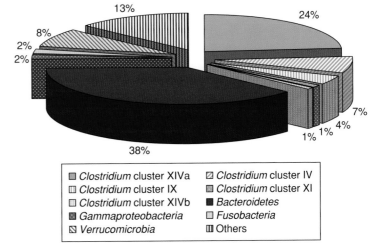

Fig. 9.59 Culture-independent analysis of the microbiota associated with the rectal mucosa of one adult female. DNA was extracted from biopsies of the rectal mucosa, the bacterial 16S rRNA genes present were amplified by PCR, and the resulting products were cloned and sequenced. A total of 88 clones were sequenced, and the figures denote the proportions (%) of clones with sequences corresponding to each of the phylogenetic groups detected (22).

Table 9.9 Positive interactions that may occur between members of the intestinal microbiota.

Process	Benefit to effector organism or to other organisms
Oxygen utilization by aerobes and facultative anaerobes	Creates an environment that is suitable for growth of microaerophiles and anaerobes
Degradation of polysaccharides	Provides monosaccharides to be used as carbon and/or energy sources
Degradation of proteins	Provides amino acids to be used as carbon or nitrogen or as energy sources
Degradation of mucins and other glycoproteins	Provides sugars, amino acids, and sulfate
Excretion of metabolic end products (e.g. lactate, acetate, butyrate, ethanol, hydrogen, and ammonia)	Products serve as nutrients for other organisms
Quorum sensing	Production of bacteriocins and virulence factors

Table 9.10 Negative interactions that may occur between members of the intestinal microbiota.

Process	Adverse effects
Production of acidic end products of metabolism	Generates a pH that can kill or inhibit the growth of other species
Production of toxic metabolic end products, e.g. H_2S, organic acids, H_2O_2	Can result in death of susceptible organisms, or inhibition of their growth
Production of bacteriocins	Can result in death of susceptible organisms, or inhibition of their growth
Binding to a particular receptor site	Prevents adhesion of other organisms if they utilize adhesins that recognize the same particular receptor site
Competition for a specific nutrient	Can result in death of competing organisms, or inhibition of their growth

9.5 OVERVIEW OF THE INDIGENOUS MICROBIOTA OF THE GASTROINTESTINAL TRACT

The GIT has a number of distinct regions, each housing a characteristic microbial community or communities. In the upper GIT (oral cavity, pharynx, and esophagus), the resident microbiota is associated with surfaces, and because material (food, secretions, etc.) passes rapidly through these regions, microbial communities cannot become established in their lumens. As the passage of material becomes slower in the lower regions of the GIT, there is an opportunity for communities to develop within the lumen as well as on the mucosal surface – in the distal ileum, cecum, colon and rectum such communities are substantial. Because of its complex anatomy, the oral cavity provides a large variety of habitats for microbial colonization and harbors a number of complex microbial communities – these are discussed in Chapter 8.

Very few studies have investigated the esophageal microbiota. Culture-dependent studies of this region have shown that it is dominated by staphylococci, lactobacilli, and *Corynebacterium* spp., while a culture-independent study found streptococci and Gram-negative anaerobes (*Prevotella* spp. and *Veillonella* spp.) to be dominant. Because of its low pH, the stomach is a hostile environment for a wide range of organisms. The vast numbers of bacteria (in saliva and food) that are continually entering the stomach make it difficult to distinguish between autochthonous species and transients. Organisms detected in the lumen are mainly acid-tolerant species of streptococci and lactobacilli together with staphylococci, *Neisseria* spp., and various anaerobes. These organisms are also present in the mucosa-associated community which, in addition, often contains the important pathogen, *Hel. pylori*.

The environments within the duodenum and jejunum are also largely inimical to many microbes because of the low pH, the presence of bile (and other antimicrobial compounds), and the rapid transit of material. Consequently, the mucosa and the lumen of both of these regions have sparse microbiotas consisting mainly of acid-tolerant streptococci and lactobacilli. In the ileum, especially the terminal region, conditions are less hostile to microbes, and the microbiotas within the lumen and on the mucosa are more substantial. Streptococci, enterococci, and coliforms are the dominant organisms in the lumen, but the microbiota of the mucosa is very different and consists of high proportions of anaerobes including *Bacteroides* spp., *Clostridium* spp., GPAC, and *Bifidobacterium* spp.

The cecum has a lower pH and a higher content of easily fermentable compounds than the more distal regions of the GIT and, consequently, it harbors microbial communities very different from those in the rest of the large intestine. The lumen is dominated by facultative organisms (mainly *Enterobacteriaceae* and lactobacilli), although substantial proportions of anaerobes (*Bacteroides* spp. and *Clostridium* spp.) are also present. Little is known about the mucosal microbiota, although it appears to be dominated by *Bacteroides* spp.

The colon is colonized by a very large and diverse microbial population. Up to 80% of the organisms present have not yet been grown in the laboratory, and many of these are novel phylotypes. Most of our knowledge of the colonic microbiota has come from fecal studies in which a substantial proportion of the organisms present are in the form of biofilms on the surface of particulate matter. Molecular profiling has shown that the communities associated with the colonic mucosa differ from those in feces, although the exact nature of such differences remains to be established. Culture-dependent studies of the fecal microbiota have revealed that obligate anaerobes are 1000-fold greater in number than facultative organisms – the predominant genera being *Bacteroides*, *Eubacterium* (and related genera), *Bifidobacterium*, and *Clostridium*. However, the use of culture-independent approaches has given us a different picture and has revolutionized our understanding of the microbial communities that are present in the colon. Studies using these techniques have revealed that the dominant organisms in feces are members of the *Firmicutes* (mainly the *Cl. coccoides–Eub. rectale* group, and *Cl. leptum*) and, to a lesser extent, the *Bacteroidetes*. Other groups present in substantial proportions include *Bifidobacterium* spp. and members of the *Atopobium* cluster. Both culture-dependent and culture-independent studies of the microbiota of the rectal mucosa have shown that members of the *Bacteroidetes* are the dominant organisms.

9.6 SOURCES OF DATA USED TO COMPILE FIGURES

1 Pajecki, D., Zilberstein, B., dos Santos, M.A., Ubriaco, J.A., Quintanilha, A.G., Cecconello, I. and Gama-Rodrigues, J. (2002) *J Gastrointest Surg* 6, 723–9.
2 Pei, Z., Bini, E.J., Yang, L., Zhou, M., Francois, F. and Blaser, M.J. (2004) *Proc Natl Acad Sci U S A* 101, 4250–5.

3 Adamsson, I., Nord, C.E., Lundquist, P., Sjostedt, S. and Edlund, C. (1999) *J Antimicrob Chemother* 44, 629–40.

4 Stark, C.A., Adamsson, I., Edlund, C., Sjosted, S., Seensalu, R., Wikstrom, B. and Nord, C.E. (1996) *J Antimicrob Chemother* 38, 927–39.

5 Santos, I.S., Boccio, J., Santos, A.S., Valle, N.C., Halal, C.S., Bachilli, M.C. and Lopes, R.D. (2005) *BMC Public Health* 5, 118.

6 Shimatani, T., Inoue, M., Iwamoto, K., Hyogo, H., Yokozaki, M., Saeki, T., Tazuma, S. and Horikawa, Y. (2005) *J Gastroenterol Hepatol* 20, 1352–7.

7 Sung, K.C., Rhee, E.J., Ryu, S.H. and Beck, S.H. (2005) *Int J Cardiol* 102, 411–17.

8 Robertson, M.S., Cade, J.F., Savoia, H.F. and Clancy, R.L. (2003) *Intern Med J* 33, 163–7.

9 Sporea, I., Popescu, A., van Blankenstein, M., Stirli, R., Focsea, M. and Danila M. (2003) *Rom J Gastroenterol* 12, 15–18.

10 Robinson, L.G., Black, F.L., Lee, F.K., Sousa, A.O., Owens, M., Danielsson, D., Nahmias, A.J. and Gold, B.D. (2002) *J Infect Dis* 186, 1131–7.

11 Abasiyanik, M.F., Sander, E. and Salih, B.A. (2002) *Can J Gastroenterol* 16, 527–32.

12 Bakka, A.S. and Salih, B.A. (2002) *Diagn Microbiol Infect Dis* 43, 265–8.

13 Moayyedi, P., Axon, A.T., Feltbower, R., Duffett, S., Crocombe, W., Braunholtz, D., Richards, I.D., Dowell, A.C. and Forman, D. (2002) *Int J Epidemiol* 31, 624–31.

14 Monstein, H.J., Tiveljung, A., Kraft, C.H., Borch, K. and Jonasson, J. (2000) *J Med Microbiol* 49, 817–22.

15 Bik, E.M., Eckburg, P.B., Gill, S.R., Nelson, K.E., Purdom, E.A., Francois, F., Perez-Perez, G., Blaser, M.J. and Relman, D.A. (2006) *Proc Natl Acad Sci U S A* 103, 732–7.

16 Gorbach, S.L., Plaut, A.G., Nahas, L., Weinstein, L., Spanknebel, G. and Levitan, R. (1967) *Gastroenterology* 53, 856–67.

17 Skar, V., Skar, A.G. and Osnes, M. (1989) *Scand J Gastroenterol* 24, 649–56.

18 Justesen, T., Nielsen, O.H., Jacobsen, I.E., Lave, J. and Rasmussen, S.N. (1984) *Scand J Gastroenterol* 19, 279–82.

19 Sullivan, A., Tornblom, H., Lindberg, G., Hammarlund, B., Palmgren, A.C., Einarsson, C. and Nord, C.E. (2003) *Anaerobe* 9, 11–14.

20 Johansson, M.L., Molin, G., Jeppsson, B., Nobaek, S., Ahrne, S. and Bengmark, S. (1993) *Appl Environ Microbiol* 59, 15–20.

21 Bhat, P., Albert, M.J., Rajan, D., Ponniah, J., Mathan, V.I. and Baker, S.J. (1980) *J Med Microbiol* 13, 247–56.

22 Wang, M., Ahrne, S., Jeppsson, B. and Molin, G. (2005) *FEMS Microbiol Ecol* 54, 219–31.

23 Hayashi, H., Takahashi, R., Nishi, T., Sakamoto, M. and Benno, Y. (2005) *J Med Microbiol* 54, 1093–101.

24 Berghouse, L., Hori, S., Hill, M., Hudson, M., Lennard-Jones, J.E. and Rogers, E. (1984) *Gut* 25, 1071–7.

25 Fernandez, F., Kennedy, H., Hill, M. and Truelove, S. (1985) *Microbiologie Aliments Nutrition* 47, 3.

26 Wang, X., Heazlewood, S.P., Krause, D.O. and Florin, T.H. (2003) *J Appl Microbiol* 95, 508–20.

27 Kleessen, B., Kroesen, A.J., Buhr, H.J. and Blaut, M. (2002) *Scand J Gastroenterol* 37, 1034–41.

28 Eckburg, P.B., Bik, E.M., Bernstein, C.N., Purdom, E., Dethlefsen, L., Sargent, M., Gill, S.R., Nelson, K.E. and Relman, D.A. (2005) *Science* 308, 1635–8.

29 Marteau, P., Pochart, P., Dore, J., Bera-Maillet, C., Bernalier, A. and Corthier, G. (2001) *Appl Environ Microbiol* 67, 4939–42.

30 Bentley, D.W., Nichols, R.L., Condon, R.E. and Gorbach, S.L. (1972) *J Lab Clin Med* 79, 421–9.

31 Croucher, S.C., Houston, A.P., Bayliss, C.E. and Turner, R.J. (1983) *Appl Environ Microbiol* 45, 1025–33.

32 Langlands, S.J., Hopkins, M.J., Coleman, N. and Cummings, J.H. (2004) *Gut* 53, 1610–16.

33 Matto, J., Maunuksela, L., Kajander, K., Palva, A., Korpela, R., Kassinen, A. and Saarela, M. (2005) *FEMS Immunol Med Microbiol* 43, 213–22.

34 Yamano, T., Iino, H., Takada, M., Blum, S., Rochat, F. and Fukushima, Y. (2006) *Br J Nutr* 95, 303–12.

35 Watanabe, S., Narisawa, Y., Arase, S., Okamatsu, H., Ikenaga, T., Tajiri, Y. and Kumemura, M.J. (2003) *Allergy Clin Immunol* 111, 587–91.

36 Delgado, S., Ruas-Madiedo, P., Suarez, A. and Mayo, B. (2006) *Dig Dis Sci* 51, 737–43.

37 Tannock, G.W., Munro, K., Harmsen, H.J., Welling, G.W., Smart, J. and Gopal, P.K. (2000) *Appl Environ Microbiol* 66, 2578–88.

38 Si, J.M., Yu, Y.C., Fan, Y.J. and Chen, S.J. (2004) *World J Gastroenterol* 10, 1802–5.

39 Finegold, S.M., Ingram-Drake, L., Gee, R., Reinhardt, J., Edelstein, M.A.C., Macdonald, K. and Wexler, H. (1987) *Antimicrob Agents Chemother* 31, 443–6.

40 Wilson, M. (2005) The gastrointestinal tract and its indigenous microbiota. In: *Microbial Inhabitants of Humans: Their Ecology and Role in Health and Disease.* Cambridge University Press, Cambridge, UK, pp. 665–836.

41 Finegold, S.M., Sutter, V.L., Boyle, J.D. and Shimada, K. (1970) *J Infect Dis* 122, 376–81.

42 Woodmansey, E.J., McMurdo, M.E., Macfarlane, G.T. and Macfarlane, S. (2004) *Appl Environ Microbiol* 70, 6113–22.

43 Hopkins, M.J. and Macfarlane, G.T. (2002) *J Med Microbiol* 51, 448–54.

44 Chen, H.L., Cheng, H.C., Liu, Y.J., Liu, S.Y. and Wu, W.T. (2006) *Nutrition* 22, 1112–19.

45 Brown, A.C., Shovic, A., Ibrahim, S.A., Holck, P. and Huang, A. (2005) *Altern Ther Health Med* 11, 58–64.

46 Robinson, R.R., Feirtag, J. and Slavin, J.L. (2001) *J Am Coll Nutr* 20, 279–85.

47 Goossens, D., Jonkers, D., Russel, M., Stobberingh, E., Van Den Bogaard, A. and Stockbrugger, R. (2003) *Aliment Pharmacol Ther* 18, 495–505.

48 Lay, C., Rigottier-Gois, L., Holmstrom, K., Rajilic, M., Vaughan, E.E., de Vos, W.M., Collins, M.D., Thiel, R., Namsolleck, P., Blaut, M. and Dore, J. (2005) *Appl Environ Microbiol* 71, 4153–5.

49 Stebbings, S., Munro, K., Simon, M.A., Tannock, G., Highton, J., Harmsen, H., Welling, G., Seksik, P., Dore, J., Grame, G. and Tilsala-Timisjarvi, A. (2002) *Rheumatology (Oxford)* 41, 1395–401.

50 Maukonen, J., Satokari, R., Matto, J., Soderlund, H., Mattila-Sandholm, T. and Saarela, M. (2006) *J Med Microbiol* 55, 625–33.

51 Sghir, A., Gramet, G., Suau, A., Rochet, V., Pochart, P. and Dore, J. (2000) *Appl Environ Microbiol* 66, 2263–6.

52 Clavel, T., Henderson, G., Alpert, C.A., Philippe, C., Rigottier-Gois, L., Dore, J. and Blaut, M. (2005) *Appl Environ Microbiol* 71, 6077–85.

53 Gostner, A., Blaut, M., Schaffer, V., Kozianowski, G., Theis, S., Klingeberg, M., Dombrowski, Y., Martin, D., Ehrhardt, S., Taras, D., Schwiertz, A., Kleessen, B., Luhrs, H., Schauber, J., Dorbath, D., Menzel, T. and Scheppach, W. (2006) *Br J Nutr* 95, 40–50.

54 Franks, A.H., Harmsen, H.J., Raangs, G.C., Jansen, G.J., Schut, F. and Welling, G.W. (1998) *Appl Environ Microbiol* 64, 3336–45.

55 Harmsen, H.J., Raangs, G.C., He, T., Degener, J.E. and Welling, G.W. (2002) *Appl Environ Microbiol* 68, 2982–90.

56 Garrido, D., Suau, A., Pochart, P., Cruchet, S. and Gotteland, M. (2005) *FEMS Microbiol Lett* 248, 249–56.

57 Delgado, S., Suarez, A. and Mayo, B. (2006) *Dig Dis Sci* 51, 744–51.

58 Hayashi, H., Sakamoto, M. and Benno, Y. (2002) *Microbiol Immunol* 46, 535.

59 Suau, A., Bonnet, R., Sutren, M., Godon, J.J., Gibson, G.R., Collins, M.D. and Dore J. (1999) *Appl Environ Microbiol* 65, 4799–807.

60 Lucke, K., Miehlke, S., Jacobs, E. and Schuppler, M. (2006) *J Med Microbiol* 55, 617–24.

61 Prindiville, T., Cantrell, M. and Wilson, K.H. (2004) *Inflamm Bowel Dis* 10, 824–33.

62 Hold, G.L., Pryde, S.E., Russell, V.J., Furrie, E. and Flint, H.J. (2002) *FEMS Microbiol Ecol* 39, 33–9.

63 Swidsinski, A., Ladhoff, A., Pernthaler, A., Swidsinski, S., Loening-Baucke, V., Ortner, M., Weber, J., Hoffmann, U., Schreiber, S., Dietel, M. and Lochs, H. (2002) *Gastroenterology* 122, 44–54.

64 Poxton, I.R., Brown, R., Sawyerr, A. and Ferguson, A. (1997) *Clin Infect Dis* 25 (Suppl 2), S111–13.

65 Turnbaugh, P.J., Ley, R.E., Mahowald, M.A., Magrini, V., Mardis, E.R. and Gordon, J.I. (2006) *Nature* 444, 1027–31.

66 Macfarlane, S., Furrie, E., Cummings, J.H. and Macfarlane, G.T. (2004) *Clin Infect Dis* 38, 1690–9.

67 Poxton, I.R., Brown, R., Sawyerr, A. and Ferguson, A. (1997) *J Med Microbiol* 46, 85–91.

68 Parthmakanthan, S., Thornley, J.P. and Hawkey, C.J. (1999) *Microb Ecol Health Dis* 11, 169–74.

69 Matsuda, H., Fujiyama, Y., Andoh, A., Ushijima, T., Kajinami, T. and Bamba, T. (2000) *J Gastroenterol Hepatol* 15, 61–68.

70 Macfarlane, S., Furrie, E., Macfarlane, G.T. and Dillon, J.F. (2007) *Clin Infect Dis* 45, 29–38.

71 Gill, S.R., Pop, M., DeBoy, R.T., Eckburg, P.B., Turnbaugh, P.J., Samuel, B.S., Gordon, J.I., Relman, D.A., Fraser-Liggett, C.M. and Nelson, K.E. (2006) *Science* 312, 1355–9.

9.7 FURTHER READING

9.7.1 Books

Fuller, R. and Perdigon, G. (eds) (2003) *Gut Flora, Nutrition, Immunity and Health.* Blackwell Publishing, Oxford, UK.

Gibson, G.R. and Macfarlane, G.T. (eds) (1995) *Human Colonic Bacteria: Role in Nutrition, Physiology and Pathology.* CRC Press, Boca Raton, Florida, USA.

Gibson, G.R. and Roberfroid, M.B. (eds) (1999) *Colonic Microbiota, Nutrition and Health.* Kluwer Academic Publishers, Dordrecht, The Netherlands.

Mackie, R.I., White, B.A. and Isaacson, R.E. (eds) (1997) *Gastrointestinal Microbiology; Volume 2. Gastrointestinal Microbes and Host Interactions.* Chapman and Hall, New York, New York, USA.

Ouwehand, A.C. and Vaughan, E.E. (eds) (2006) *Gastrointestinal Microbiology.* Taylor & Francis, New York, New York, USA.

Rowland, I.R. (ed.) (1997) *Role of the Gut Flora in Toxicity and Cancer.* Academic Press, London, UK.

9.7.2 Reviews and papers

Abell, G.C.J. and McOrist, A.L. (2007) Assessment of the diversity and stability of faecal bacteria from healthy adults using molecular methods. *Microb Ecol Health Dis* 19, 229–40.

Abreu, M.T., Fukata, M. and Arditi, M. (2005) TLR signaling in the gut in health and disease. *J Immunol* 174, 4453–60.

Alvaro, E., Andrieux, C., Rochet, V., Rigottier-Gois, L., Lepercq, P., Sutren, M., Galan, P., Duval, Y., Juste, C. and Dore, J. (2007) Composition and metabolism of the intestinal microbiota in consumers and non-consumers of yogurt. *Br J Nutr* 97, 126–33.

Andoh, A., Sakata, S., Koizumi, Y., Mitsuyama, K., Fujiyama, Y. and Benno, Y. (2007) Terminal restriction fragment length polymorphism analysis of the diversity of fecal microbiota in patients with ulcerative colitis. *Inflamm Bowel Dis* 13, 955–62.

Andrieux, C., Membre, J.M., Cayuela, C. and Antoine, J.M. (2002) Metabolic characteristics of the faecal microflora in humans from three age groups. *Scandinavian J Gastroenterol* 37, 792–8.

Belenguer, A., Duncan, S.H., Calder, A.G., Holtrop, G., Louis, P., Lobley, G.E. and Flint, H.J. (2006) Two routes of metabolic cross-feeding between *Bifidobacterium adolescentis* and butyrate-producing anaerobes from the human gut. *Appl Environ Microbiol* 72, 3593–9.

Ben-Amor, K., Heilig, H., Smidt, H., Vaughan, E.E., Abee, T. and de Vos, W.M. (2005) Genetic diversity of viable, injured, and dead fecal bacteria assessed by fluorescence-activated cell sorting and 16S rRNA gene analysis. *Appl Environ Microbiol* 71, 4679–89.

Ben-Neriah, Y. and Schmidt-Supprian, M. (2007) Epithelial NF-kappaB maintains host gut microflora homeostasis. *Nat Immunol* 8, 479–81.

Bik, E.M., Eckburg, P.B., Gill, S.R., Nelson, K.E., Purdom, E.A., Francois, F., Perez-Perez, G., Blaser, M.J. and Relman, D.A. (2006) Molecular analysis of the bacterial microbiota in the human stomach. *Proc Natl Acad Sci U S A* 103, 732–7.

Blaser, M.J. and Atherton, J.C. (2004) *Helicobacter pylori* persistence: Biology and disease. *J Clin Invest* 113, 321–33.

Blaut, M. and Clavel, T. (2007) Metabolic diversity of the intestinal microbiota: Implications for health and disease. *J Nutr* 137, 751S–5S.

Blaut, M., Collins, M.D., Welling, G.W., Dore, J., van Loo, J. and de Vos, W. (2002) Molecular biological methods for studying the gut microbiota: The EU human gut flora project. *Br J Nutr* 87 (Suppl 2), S203–11.

Booijink, C.C.G.M., Zoetendal, E.G., Kleerebezem, M. and de Vos, W.M. (2007) Microbial communities in the human small intestine: Coupling diversity to metagenomics. *Fut Microbiol* 2, 285–95.

Carey, C.M., Kirk, J.L., Ojha, S. and Kostrzynska, M. (2007) Current and future uses of real-time polymerase chain reaction and microarrays in the study of intestinal microbiota, and probiotic use and effectiveness. *Can J Microbiol* 53, 537–50.

Cash, H.L., Whitham, C.V., Behrendt, C.L. and Hooper, L.V. (2006) Symbiotic bacteria direct expression of an intestinal bactericidal lectin. *Science* 313, 1126–30.

Chen, H.L., Cheng, H.C., Liu, Y.J., Liu, S.Y. and Wu, W.T. (2006) Konjac acts as a natural laxative by increasing stool bulk and improving colonic ecology in healthy adults. *Nutrition* 22, 1112–19.

Clavel, T. and Haller, D. (2007) Molecular interactions between bacteria, the epithelium, and the mucosal immune system in the intestinal tract: implications for chronic inflammation. *Curr Issues Intest Microbiol* 8, 25–43.

Collier-Hyams, L.S. and Neish, A.S. (2005) Innate immune relationship between commensal flora and the mammalian intestinal epithelium. *Cell Mol Life Sci* 62, 1339–48.

Comstock, L.E. and Coyne, M.J. (2003) *Bacteroides thetaio-taomicron*: A dynamic, niche-adapted human symbiont. *Bioessays* 25, 926–9.

Cummings, J.H. (1998) Dietary carbohydrates and the colonic microflora. *Curr Opin Clin Nutr Metab Care* 1, 409–14.

Dann, S.M. and Eckmann, L. (2007) Innate immune defenses in the intestinal tract. *Curr Opin Gastroenterol* 23, 115–20.

de Graaf, A.A. and Venema, K. (2007) Gaining insight into microbial physiology in the large intestine: A special role for stable isotopes. *Adv Microb Physiol* 53, 73–314.

Delgado, S., Suarez, A. and Mayo, B. (2006) Identification of dominant bacteria in feces and colonic mucosa from healthy Spanish adults by culturing and by 16S rDNA sequence analysis. *Dig Dis Sci* 51, 744–51.

Donskey, C.J., Hujer, A.M., Das, S.M., Pultz, N.J., Bonomo, R.A. and Rice, L.B. (2003) Use of denaturing gradient gel electrophoresis for analysis of the stool microbiota of hospitalized patients. *J Microbiol Methods* 54, 249–56.

Duncan, S.H., Louis, P. and Flint, H.J. (2007) Cultivable bacterial diversity from the human colon. *Lett Appl Microbiol* 44, 343–50.

Eckburg, P.B., Bik, E.M., Bernstein, C.N., Purdom, E., Dethlefsen, L., Sargent, M., Gill, S.R., Nelson, K.E. and Relman, D.A. (2005) *Science* 308, 1635–8.

Edwards, C.A. and Parrett, A.M. (2002) Intestinal flora during the first months of life: New perspectives. *Br J Nutr* 88 (Suppl 1), S11–18.

Egert, M., de Graaf, A.A., Smidt, H., de Vos, W.M. and Venema, K. (2006) Beyond diversity: Functional microbiomics of the human colon. *Trends Microbiol* 14, 86–91.

Flint, H.J., Duncan, S.H., Scott K.P. and Louis, P. (2007) Interactions and competition within the microbial community of the human colon: Links between diet and health. *Environ Microbiol* 9, 1101–11.

Fukata, M. and Abreu, M.T. (2007) TLR4 signalling in the intestine in health and disease. *Biochem Soc Trans* 35, 1473–8.

Furrie, E. (2006) A molecular revolution in the study of intestinal microflora. *Gut* 55, 141–3.

Furrie, E., Macfarlane, S., Thomson, G. and Macfarlane, G.T. (2005) Toll-like receptors-2, -3 and -4 expression patterns on human colon and their regulation by mucosal-associated bacteria. *Immunology* 115, 565–74.

Gill, S.R., Pop, M., DeBoy, R.T., Eckburg, P.B., Turnbaugh, P.J., Samuel, B.S., Gordon, J.I., Relman, D.A., Fraser-Liggett, C.M. and Nelson, K.E. (2006) Metagenomic analysis of the human distal gut microbiome. *Science* 312, 1355–9.

Gilmore, M.S. and Ferretti, J.F. (2003) The thin line between gut commensal and pathogen. *Science* 299, 1999–2002.

Go, M.G. (2002) Natural history and epidemiology of *Helicobacter pylori* infection. *Aliment Pharmacol Ther* 16 (Suppl 1), 3–15.

Gostner, A., Blaut, M., Schaffer, V., Kozianowski, G., Theis, S., Klingeberg, M., Dombrowski, Y., Martin, D., Ehrhardt, S., Taras, D., Schwiertz, A., Kleessen, B., Luhrs, H., Schauber, J., Dorbath, D., Menzel, T. and Scheppach, W. (2006) Effect of isomalt consumption on faecal microflora and colonic metabolism in healthy volunteers. *Br J Nutr* 95, 40–50.

Guarner, F. (2006) Enteric flora in health and disease. *Digestion* 73 (Suppl 1), 1–8.

Guarner, F. and Malagelada, J.R. (2003) Gut flora in health and disease. *Lancet* 361, 512–19.

Gunn, J.S. (2000) Mechanisms of bacterial resistance and response to bile. *Microbes Infect* 2, 907–13.

Hayashi, H., Sakamoto, M. and Benno, Y. (2002) Phylogenetic analysis of the human gut microbiota using 16S rDNA clone libraries and strictly anaerobic culture-based methods. *Microbiol Immunol* 46, 535–48.

Hayashi, H., Takahashi, R., Nishi, T., Sakamoto, M. and Benno, Y. (2005) Molecular analysis of jejunal, ileal, caecal and recto-sigmoidal human colonic microbiota using 16S rRNA gene libraries and terminal restriction fragment length polymorphism. *J Med Microbiol* 54, 1093–101.

Hebuterne, X. (2003) Gut changes attributed to ageing: Effects on intestinal microflora. *Curr Opin Clin Nutr Metab Care* 6, 49–54.

Heller, F. and Duchmann, R. (2003) Intestinal flora and mucosal immune responses. *Int J Med Microbiol* 293, 77–86.

Hill, M.J. (1998) Composition and control of ileal contents. *Eur J Cancer Prev* 7 (Suppl 2), S75–8.

Hopkins, M.J., Sharp, R. and Macfarlane, G.T. (2002) Variation in human intestinal microbiota with age. *Dig Liv Dis* 34(Suppl 2), S12–18.

Huijsdens, X.W., Linskens, R.K., Mak, M., Meuwissen, S.G.M., Vandenbroucke-Grauls, C.M.J.E. and Savelkoul, P.H.M. (2002) Quantification of bacteria adherent to gastrointestinal mucosa by real-time PCR. *J Clin Microbiol* 40, 4423–7.

Kassinen, A., Krogius-Kurikka, L., Makivuokko, H., Rinttila, T., Paulin, L., Corander, J., Malinen, E., Apajalahti, J. and Palva, A. (2007) The fecal microbiota of irritable bowel syndrome patients differs significantly from that of healthy subjects. *Gastroenterology* 133, 24–33.

Kato, S., Fujimura, S., Kimura, K., Nishio, T., Hamada, S., Minoura, T. and Oda, M. (2006) Non-*Helicobacter* bacterial flora rarely develops in the gastric mucosal layer of children. *Dig Dis Sci* 51, 641–6.

Kawakubo, M., Ito, Y., Okimura, Y., Kobayashi, M., Sakura, K., Kasama, S., Fukuda, M.N., Fukuda, M., Katsuyama, T. and Nakayama, J. (2004) Natural antibiotic function of a human gastric mucin against *Helicobacter pylori* infection. *Science* 305, 1003–6.

Kelly, D., Conway, S. and Aminov, R. (2005) Commensal gut bacteria: Mechanisms of immune modulation. *Trends Immunol* 26, 326–33.

Kivi, M. and Tindberg, Y. (2006) *Helicobacter pylori* occurrence and transmission: A family affair? *Scand J Infect Dis* 38, 407–17.

Klijn, A., Mercenier, A. and Arigoni, F. (2005) Lessons from the genomes of bifidobacteria. *FEMS Microbiol Rev* 29, 491–9.

Kusters, J.G., van Vliet, A.H. and Kuipers, E.J. (2006) Pathogenesis of *Helicobacter pylori* infection. *Clin Microbiol Rev* 19, 449–90.

Langlands, S.J., Hopkins, M.J., Coleman, N. and Cummings, J.H. (2004) Prebiotic carbohydrates modify the mucosa associated microflora of the human large bowel. *Gut* 53, 1610–16.

Lay, C., Rigottier-Gois, L., Holmstrom, K., Rajilic, M., Vaughan, E.E., de Vos, W.M., Collins, M.D., Thiel, R., Namsolleck, P., Blaut, M. and Dore, J. (2005) Colonic microbiota signatures across five northern European countries. *Appl Environ Microbiol* 71, 4153–5.

Lepage, P., Seksik, P., Sutren, M., de la Cochetière, M.F., Jian, R., Marteau, P. and Doré, J. (2005) Biodiversity of the mucosa-associated microbiota is stable along the distal digestive tract in healthy individuals and patients with IBD. *Inflamm Bowel Dis* 11, 473–80.

Li, F., Hullar, M.A. and Lampe, J.W. (2007) Optimization of terminal restriction fragment polymorphism (TRFLP) analysis of human gut microbiota. *J Microbiol Methods* 68, 303–11.

Lievin-Le Moal, V. and Servin, A.L. (2006) The front line of enteric host defense against unwelcome intrusion of harmful microorganisms: Mucins, antimicrobial peptides, and microbiota. *Clin Microbiol Rev* 19, 315–37.

Lotz, M., Menard, S. and Hornef, M. (2007) Innate immune recognition on the intestinal mucosa. *Int J Med Microbiol* 297, 379–92.

Louis, P., Scott, K.P., Duncan, S.H. and Flint, H.J. (2007) Understanding the effects of diet on bacterial metabolism in the large intestine. *J Appl Microbiol* 102, 1197–208.

Lu, L. and Walker, W.A. (2001) Pathologic and physiologic interactions of bacteria with the gastrointestinal epithelium. *Am J Clin Nutr* 73, 1124S–30S.

Lucke, K., Miehlke, S., Jacobs, E. and Schuppler, M. (2006) Prevalence of *Bacteroides* and *Prevotella* spp. in ulcerative colitis. *J Med Microbiol* 55, 617–24.

Macfarlane, G.T. and Macfarlane, S. (1997) Human colonic microbiota: Ecology, physiology and metabolic potential of intestinal bacteria. *Scand J Gastroenterol Supplement* 222, 3–9.

Macfarlane, S., Furrie, E., Cummings, J.H. and Macfarlane, G.T. (2004) Chemotaxonomic analysis of bacterial populations colonizing the rectal mucosa in patients with ulcerative colitis. *Clin Infect Dis* 38, 1690–9.

Macfarlane, S., Furrie, E., Macfarlane, G.T. and Dillon, J.F. (2007) Microbial colonization of the upper gastrointestinal tract in patients with Barrett's esophagus. *Clin Infect Dis* 45, 29–38.

Macpherson, A.J. and Slack, E. (2007) The functional interactions of commensal bacteria with intestinal secretory IgA. *Curr Opin Gastroenterol* 23, 673–8.

Macpherson, A.J., Hapfelmeier, S. and McCoy, K.D. (2007) The armed truce between the intestinal microflora and host mucosal immunity. *Semin Immunol* 19, 57–8.

Magalhaes, J.G., Tattoli, I. and Girardin, S.E. (2007) The intestinal epithelial barrier: How to distinguish between the microbial flora and pathogens. *Semin Immunol* 19, 106–15.

Manichanh, C., Rigottier-Gois, L., Bonnaud, E., Gloux, K., Pelletier, E., Frangeul, L., Nalin, R., Jarrin, C., Chardon, P., Marteau, P., Roca, J. and Dore, J. (2006) Reduced diversity of faecal microbiota in Crohn's disease revealed by a metagenomic approach. *Gut* 55, 205–11.

Marchesi, J. and Shanahan, F. (2007) The normal intestinal microbiota. *Curr Opin Infect Dis* 20, 508–13.

Marteau, P., Pochart, P., Dore, J., Bera-Maillet, C., Bernalier, A. and Corthier, G. (2001) Comparative study of bacterial groups within the human cecal and fecal microbiota. *Appl Environ Microbiol* 67, 4939–42.

Maukonen, J., Satokari, R., Matto, J., Soderlund, H., Mattila-Sandholm, T. and Saarela, M. (2006) Prevalence and temporal stability of selected clostridial groups in irritable bowel syndrome in relation to predominant faecal bacteria. *J Med Microbiol* 55, 625–33.

McCole, D.F. and Barrett, K.E. (2007) Varied role of the gut epithelium in mucosal homeostasis. *Curr Opin Gastroenterol* 23, 647–54.

Michelsen, K.S. and Arditi, M. (2007) Toll-like receptors and innate immunity in gut homeostasis and pathology. *Curr Opin Hematol* 14, 48–54.

Moal, V.L.L. and Servin, A.L. (2006) The front line of enteric host defense against unwelcome intrusion of harmful microorganisms: Mucins, antimicrobial peptides, and microbiota. *Clin Microbiol Rev* 19, 315–37.

Monstein, H-J., Tiveljung, A., Kraft, C.H., Borch, K. and Jonasson, J. (2000) Profiling of bacterial flora in gastric biopsies from patients with *Helicobacter pylori*-associated gastritis and histologically normal control individuals by temperature gradient get electrophoresis and 16S rDNA sequence analysis. *J Med Microbiol* 49, 817–22.

Mountzouris, K.C., McCartney, A.L. and Gibson, G.R. (2002) Intestinal microflora of human infants and current trends for its nutritional modulation. *Br J Nutr* 87, 405–20.

Mueller, S., Saunier, K., Hanisch, C., Norin, E., Alm, L., Midtvedt, T., Cresci, A., Silvi, S., Orpianesi, C., Verdenelli, M.C., Clavel, T., Koebnick, C., Zunft, H.J-F., Doré, J. and Blaut, M. (2006) Differences in fecal microbiota in different European study populations in relation to age, gender, and country: A cross-sectional study. *Appl Environ Microbiol* 72, 1027–33.

Niess, J.H. and Reinecker, H.C. (2006) Dendritic cells in the recognition of intestinal microbiota. *Cell Microbiol* 4, 558–64.

O'Hara, A.M. and Shanahan, F. (2006) The gut flora as a forgotten organ. *EMBO Reports* 7, 688–93.

O'Hara, A.M. and Shanahan, F. (2007) Gut microbiota: Mining for therapeutic potential. *Clin Gastroenterol Hepatol* 5, 274–84.

Pei, Z., Bini, E.J., Yang, L., Zhou, M., Francois, F. and Blaser, M.J. (2004) Bacterial biota in the human distal esophagus. *Proc Natl Acad Sci U S A*. 101, 4250–5.

Peterson, D.A., McNulty, N.P., Guruge, J.L. and Gordon, J.I. (2007) IgA response to symbiotic bacteria as a mediator of gut homeostasis. *Cell Host Microbe* 2, 328–39.

Probert, H.M. and Gibson, G.R. (2002) Bacterial biofilms in the human gastrointestinal tract. *Curr Issues Intest Microbiol* 3, 23–27.

Pryde, S.E., Duncan, S.H., Hold, G.L., Stewart, C.S. and Flint, H.J. (2002) The microbiology of butyrate formation in the human colon. *FEMS Microbiol Letters* 217, 133–9.

Ramakrishna, B.S. (2007) The normal bacterial flora of the human intestine and its regulation. *J Clin Gastroenterol* 41 (Suppl 1), S2–6.

Rath, H.C. (2003) The role of endogenous bacterial flora: Bystander or the necessary prerequisite? *Eur J Gastroenterol Hepatol* 15, 615–20.

Salzman, N.H., Underwood, M.A. and Bevins, C.L. (2007) Paneth cells, defensins, and the commensal microbiota: A hypothesis on intimate interplay at the intestinal mucosa. *Semin Immunol* 19, 70–83.

Samuel, B.S., Hansen, E.E., Manchester, J.K., Coutinho, P.M., Henrissat, B., Fulton, R., Latreille, P., Kim, K., Wilson, R.K. and Gordon, J.I. (2007) Genomic and metabolic adaptations of *Methanobrevibacter smithii* to the human gut. *Proc Natl Acad Sci U S A* 104, 10643–8.

Savage, D.C. (1977) Microbial ecology of the gastrointestinal tract. *Ann Rev Med* 31, 107–33.

Sears, C.L. (2005) A dynamic partnership: Celebrating our gut flora. *Anaerobe* 11, 247–51.

Sghir, A., Gramet, G., Suau, A., Rochet, V., Pochart, P. and Dore, J. (2000) Quantification of bacterial groups within human fecal flora by oligonucleotide probe hybridization. *Appl Environ Microbiol* 66, 2263–6.

Stewart, J.A., Chadwick, V.S. and Murray, A. (2005) Investigations into the influence of host genetics on the predominant eubacteria in the faecal microflora of children. *J Med Microbiol* 54, 1239–42.

Swidsinski, A., Sydora, B.C., Doerffel, Y., Loening-Baucke, V., Vaneechoutte, M., Lupicki, M., Scholze, J., Lochs, H. and Dieleman, L.A. (2007) Viscosity gradient within the mucus layer determines the mucosal barrier function and the spatial organization of the intestinal microbiota. *Inflamm Bowel Dis* 13, 963–70.

Swift, S., Vaughan, E.E. and de Vos, W.M. (2000) Quorum sensing within the gut ecosystem. *Microb Ecol Health Dis* 12 (Suppl 2), 81–92.

Tannock, G.W. (2000) The intestinal microflora: Potentially fertile ground for microbial physiologists. *Adv Microb Physiol* 42, 25–46.

Tannock, G.W. (2002) The bifidobacterial and *Lactobacillus* microflora of humans. *Clin Rev Allergy Immunol* 22, 231–53.

Tannock, G.W. (2007) What immunologists should know about bacterial communities of the human bowel. *Semin Immunol* 19, 94–105.

Thompson-Chagoyan, O.C., Maldonado, J. and Gil, A. (2007) Colonization and impact of disease and other factors on intestinal microbiota. *Dig Dis Sci* 52, 2069–77.

Tuohy, K.M., Pinart-Gilberga, M., Jones, M., Hoyles, L., McCartney, A.L. and Gibson, G.R. (2007) Survivability of a probiotic *Lactobacillus casei* in the gastrointestinal tract of healthy human volunteers and its impact on the faecal microflora. *J Appl Microbiol* 102, 1026–32.

Turnbaugh, P.J., Ley, R.E., Mahowald, M.A., Magrini, V., Mardis, E.R. and Gordon, J.I. (2006) An obesity-associated gut microbiome with increased capacity for energy harvest. *Nature* 444, 1027–31.

Vaughan, E.E., Heilig, H.G.H.J., Ben-Amor, K. and de Vos, W.M. (2005) Diversity, vitality and activities of intestinal lactic acid bacteria and bifidobacteria assessed by molecular approaches. *FEMS Microbiol Rev* 29, 477–90.

Wang, M., Ahrne, S., Jeppsson, B. and Molin, G. (2005) Comparison of bacterial diversity along the human intestinal tract by direct cloning and sequencing of 16S rRNA genes. *FEMS Microbiol Ecol* 54, 219–31.

Wang, X., Heazlewood, S.P., Krause, D.O. and Florin, T.H. (2003) Molecular characterization of the microbial species that colonize human ileal and colonic mucosa by using 16S rDNA sequence analysis. *J Appl Microbiol* 95, 508–20.

Wehkamp, J., Chu, H., Shen, B., Feathers, R.W., Kays, R.J., Lee, S.K. and Bevins, C.L. (2006) Paneth cell antimicrobial peptides: Topographical distribution and quantification in human gastrointestinal tissues. *FEBS Lett* 580, 5344–50.

Winkler, P., Ghadimi, D., Schrezenmeir, J. and Kraehenbuhl, J.P. (2007) Molecular and cellular basis of microflora-host interactions. *J Nutr* 137, 756S–72S.

Woodmansey, E.J. (2007) Intestinal bacteria and ageing. *J Appl Microbiol* 102, 1178–86.

Xu, J., Chiang, H.C., Bjursell, M.K. and Gordon, J.I. (2004) Message from a human gut symbiont: Sensitivity is a prerequisite for sharing. *Trends Microbiol* 12, 21–8.

Xu, J., Mahowald, M.A., Ley, R.E., Lozupone, C.A., Hamady, M., Martens, E.C. et al. (2007) Evolution of symbiotic bacteria in the distal human intestine. *PLoS Biol* 5, e156

Zilberstein, B., Quintanilha, A.G., Santos, M.A., Pajecki, D., Moura, E.G., Alves, P.R., Maluf Filho, F., de Souza, J.A. and Gama-Rodrigues, J. (2007) Digestive tract microbiota in healthy volunteers. *Clinics* 62, 47–54.

Zocco, M.A., Ainora, M.E., Gasbarrini, G. and Gasbarrini, A. (2007) *Bacteroides thetaiotaomicron* in the gut: Molecular aspects of their interaction. *Dig Liver Dis* 39, 707–12.

Zoetendal, E.G., Collier, C.T., Koike, S., Mackie, R.I. and Gaskins, H.R. (2004) Molecular ecological analysis of the gastrointestinal microbiota: A review. *J Nutr* 134, 465–72.

Chapter 10

THE FUTURE

The main aim of this book was to provide a comprehensive and detailed description of the microbial communities colonizing the body surfaces of healthy humans. However, while searching for, collecting, and analyzing the data to be used in the book, it soon became apparent that a number of issues would make this aim difficult to achieve. The main problems encountered were the following:

1 There is tremendous variation regarding the amount of research that has been carried out on the resident microbial communities of different body sites. Hence, the colon represents one extreme with innumerable studies of its microbiota, while the conjunctiva and the female urethra are sites that have received relatively little attention and, consequently, few studies of their resident microbial communities have been published.

2 Interest in a particular body site has waxed and waned over the years. Hence, relatively few major studies of the microbiota of the skin, urethra, and conjunctiva have been carried out during the past decade, whereas the intestinal, oral, and vaginal microbiotas have been extensively studied during this period.

3 While molecular approaches have frequently been used to analyze the microbiota of the intestinal tract, oral cavity, and vagina, such techniques have been applied only rarely, if at all, to the resident microbial communities of the conjunctiva, skin, urethra, or respiratory tract.

4 Few longitudinal studies of the microbiota of any body site have been carried out, which means that, in general, what we have is a series of "snap-shots" of the microbial community inhabiting a body site at a particular time. Consequently, we know little of the development of microbial communities from the initial colonization stage through to the formation of a climax community. The main exception to this is the oral cavity, where detailed studies have been carried out on the development of dental plaque on the tooth surface.

5 While many studies have been directed at establishing the presence or absence of a particular organism at a body site, few have tried to determine the whole range of microbes present at a site, together with their relative proportions. Hence, in general, we know little of the overall composition of the communities inhabiting particular sites. This is particularly true of most regions of the respiratory tract, the skin, and the genito-urinary tract, whereas the composition of some of the communities inhabiting the oral cavity and intestinal tract have been investigated to some extent.

6 There are few reports of how the age, gender, genotype, or lifestyle of the host affects the microbiota of any body site.

7 Virtually nothing is known of the viral constituents of the microbiota of any body site in healthy individuals.

8 With the exception of *Candida* spp. (mainly in the oral cavity, vagina, and intestinal tract) and *Mallasezia* spp. (mainly on the skin), very little is known of the fungal constituents of the indigenous microbiota.

9 Little is known of the protoctistal members of the indigenous microbiota.

All of these factors, together with other problems discussed in Chapter 1 (see section 1.1.1), have combined to render it difficult to provide a description of the various microbial communities that is balanced with respect to the various body sites. Hence, in the case of the female urethra, what is currently known of its microbiota is based mainly on the results of culture-dependent studies carried out more than 20 years ago. In contrast, during the past 5 years, there have been a large number of culture-independent studies of the colonic microbiota, although the results of such studies have also provided us with an intimidating indication of its enormous complexity.

However, there are indications that this rather bleak situation is not going to last much longer. As a research area, the indigenous microbiota of humans is undergoing a renaissance, and researchers in this field certainly live in exciting times. Having been for many years a relatively dormant field, there has been a resurgence of interest in trying to ascertain the composition of the microbial communities that live on the surfaces of humans. The reasons for this are many and include the following:

1 There is a growing realization that these microbial communities are not passive bystanders, but are involved in the development of the host's immune system and nutritional capabilities, the induction of angiogenesis, the supply of energy to host epithelial cells, the detoxification of potentially toxic ingested chemicals, the supply of vitamins to the host, the harvesting of nutrients from polymers that the host cannot digest, the regulation of host fat storage, and the protection of the host against endogenous and exogenous pathogens. Many of these remarkable roles attributed to the indigenous microbiota have been discovered only recently and, as we learn more about our "intimate strangers", it is very likely that there will be even more startling revelations of the ways in which they affect our growth, development, and physiology.

2 The development of a range of molecular tools has provided us with new ways of analyzing what are, in many cases, extremely complex communities. Such approaches have enabled us to identify, for the first time, those not-yet-cultivated taxa which are known to account for appreciable proportions of the indigenous microbiota of many sites – particularly the intestinal tract, female reproductive system, and skin.

3 There is evidence that an imbalance in the composition of some indigenous communities may be responsible for a number of chronic diseases and conditions (e.g. inflammatory bowel diseases, periodontal diseases, carcinomas, obesity), and this has provided the impetus to fully characterize these communities.

4 There is increasing interest in manipulating the composition of the indigenous microbiota of some body sites (particularly the colon) using prebiotics, probiotics, and synbiotics in order to confer some benefit on human health. The development of such approaches, as well as the monitoring of their performance, requires a much improved knowledge of the composition of the indigenous microbiota.

5 Because of the spread of antibiotic-resistant organisms and the increasing limitations that this places on our ability to treat some infectious diseases, there is growing interest in the use of probiotic and other microbe-based approaches for preventing and/or treating infectious diseases. The success of such approaches, again, requires a much greater knowledge of our indigenous microbiota.

6 The enormous diversity of the microbiota of many sites (particularly the intestinal tract and oral cavity) with their large proportions of not-yet-cultivated taxa constitute a potential treasure-trove of sources of yet-to-be-discovered pharmacologically active compounds.

7 Last, but not least, it is a fascinating subject – who could fail to be interested in knowing more about the microbes that have chosen us as their home?

Such renewed interest in the indigenous microbiota of humans has resulted in the launch of a number of initiatives, some of which are listed below.

1 The European Commission has already financed large-scale studies of the intestinal microbiota and, as part of the Seventh Research Framework Programme, has issued calls for large-scale research projects into the nature of the indigenous microbiota of all body sites.

2 In the USA, the Human Gut Microbiome Initiative (HGMI) involves determining whole-genome sequences of 100 species representing the bacterial divisions known to reside in the colon. The organisms to be sequenced have been chosen from the 11 831-member 16S rRNA-sequence dataset generated from the human colonic microbiota of three healthy adults. This is being undertaken at the Genome Sequencing Center, Washington University School of Medicine, St. Louis, Missouri, USA (http://genome.wustl.edu/sub_genome_group.cgi?GROUP=3&SUB_GROUP=4).

3 An international effort involving a metagenomic analysis of the microbial communities residing on the skin and in the nose, mouth, respiratory tract, stomach, vagina, and colon of healthy individuals, the Human Microbiome Project, will lay the foundation for further studies of human-associated microbial communities. With a 2008 kick-off and a 5-year timescale, funding is the responsibility of the Roadmap 1.5 Program of the US National Institutes of Health, Bethesda, Maryland, USA (http://nihroadmap.nih.gov/hmp).

The results of such studies, and the development of the technologies associated with them, will ultimately contribute to a greater understanding of the composition of the microbial communities that we carry around with us. However, knowing which microbes are present is only one facet of the human–microbe symbiosis; we also need to know why these particular microbes are present and exactly what they are doing. In order to address these aspects of our relationship with our microbiota, we will require a greater knowledge of the micro-environments existing at epithelial surfaces, the genes that are being expressed by members of the microbial communities colonizing these surfaces, and the roles played by the innate and acquired immune systems in recognizing and controlling these microbial populations. We remain remarkably ignorant of many of the important environmental selection factors (e.g. oxygen and carbon dioxide levels, pH, redox potential, range, and concentration of nutrients, etc.) operating at most epithelial surfaces, and know even less about in situ gene expression in the microbial colonizers of these surfaces. On the other hand, tremendous strides are being made in our understanding of the immune responses to our indigenous microbiota, particularly with regard to innate immune mechanisms – and this is a very rapidly advancing field.

Recent technological advances in analyzing complex microbial communities combined with a renewed interest in the indigenous microbiota and a need to understand these complex communities will, hopefully, enable us to address a number of fundamental questions about our microbial symbionts and the relationship that we have with them. Such questions include:

1 Is there a core microbiota at each body site that is shared by all humans?

2 If such a core microbiota does exist, what are the major host–microbe and microbe–microbe interactions that are responsible for its maintenance?

3 What dictates the composition of the "peripheral" communities surrounding such a core?

4 How robust are these indigenous communities?

5 How do diet and other lifestyle factors affect the composition of the indigenous microbiota?

6 What effects do external environmental factors, such as climate, have on the composition of the microbiota of each body site?

7 How do age and gender affect the composition of the microbiota at a body site?

8 What effect does the host genotype have on the composition of the microbiota of a particular body site?

9 How redundant or how modular are the contributions of individual microbial constituents to community function and to host biology?

10 What effect does the indigenous microbiota have on host development and physiology?

11 How does the composition of the indigenous microbiota affect the health of an individual, and vice versa?

10.1 FURTHER READING

Bengmark, S. and Gil, A. (2006) Bioecological and nutritional control of disease: Prebiotics, probiotics and synbiotics. *Nutr Hosp* 21 (Suppl 2), 72–86.

Blaut, M. and Clavel, T. (2007) Metabolic diversity of the intestinal microbiota: Implications for health and disease. *J Nutr* 137, 751S–5S.

Booijink, C.C., Zoetendal, E.G., Kleerebezem, M. and de Vos, W.M. (2007) Microbial communities in the human small intestine: Coupling diversity to metagenomics. *Future Microbiol* 2, 285–95.

Carey, C.M., Kirk, J.L., Ojha, S. and Kostrzynska, M. (2007) Current and future uses of real-time polymerase chain reaction and microarrays in the study of intestinal microbiota, and probiotic use and effectiveness. *Can J Microbiol* 53, 537–50.

Clavel, T. and Haller, D. (2007) Molecular interactions between bacteria, the epithelium, and the mucosal immune system in the intestinal tract: Implications for chronic inflammation. *Curr Issues Intest Microbiol* 8, 25–43.

Committee on Metagenomics: Challenges and Functional Applications, National Research Council. (2007) *The New Science of Metagenomics*: *Revealing the Secrets of Our Microbial Planet*. Available: http://www.nap.edu/catalog/11902.html. Accessed 9 October 2007.

de Graaf, A.A. and Venema, K. (2007) Gaining insight into microbial physiology in the large intestine: a special role for stable isotopes. *Adv Microb Physiol* 53, 73–314.

de Vrese, M. and Marteau, P.R. (2007) Probiotics and prebiotics: Effects on diarrhea. *J Nutr* 137 (3 Suppl 2), 803S–11S.

Dethlefsen, L., Eckburg, P.B., Bik, E.M. and Relman, D.A. (2006) Assembly of the human intestinal microbiota. *Trends Ecol Evol* 21, 517–23.

Eckburg, P.B. and Relman, D.A. (2007) The role of microbes in Crohn's disease. *Clin Infect Dis* 44, 256–62.

Falagas, M.E., Betsi, G.I. and Athanasiou, S. (2006) Probiotics for prevention of recurrent vulvovaginal candidiasis: A review. *J Antimicrob Chemother* 58, 266–72.

Flint, H.J., Duncan, S.H., Scott, K.P. and Louis, P. (2007) Interactions and competition within the microbial community of the human colon: Links between diet and health. *Environ Microbiol* 9, 1101–11.

Gordon, J., Ley, R.E., Wilson, R., Mardis, E., Xu, J., Fraser, C.M. and Relman, D.A. (2005) *Extending Our View of Self: The Human Gut Microbiome Initiative.* Available: http://www.genome.gov/Pages/Research/Sequencing/SeqProposals/HGMISeq.pdf. Accessed 9 October 2007.

Hull, M.W. and Chow, A.W. (2007) Indigenous microflora and innate immunity of the head and neck. *Infect Dis Clin North Am* 21, 265–82.

Jones, J.L. and Foxx-Orenstein, A.E. (2007) The role of probiotics in inflammatory bowel disease. *Dig Dis Sci* 52, 607–11.

Jones, B.V. and Marchesi, J.R. (2007) Accessing the mobile metagenome of the human gut microbiota. *Mol Biosyst* 3, 749–58.

Kelly, D., King, T. and Aminov, R. (2007) Importance of microbial colonization of the gut in early life to the development of immunity. *Mutat Res* 622, 58–69.

Kurokawa, K., Itoh, T., Kuwahara, T., Oshima, K., Toh, H., Toyoda, A., Takami, H., Morita, H., Sharma, V.K., Srivastava, T.P., Taylor, T.D., Noguchi, H., Mori, H., Ogura, Y., Ehrlich, D.S., Itoh, K., Takagi, T., Sakaki, Y., Hayashi, T. and Hattori, M. (2007) Comparative metagenomics revealed commonly enriched gene sets in human gut microbiomes. *DNA Res* Oct 3; (Epub ahead of print).

Lesbros-Pantoflickova, D., Corthesy-Theulaz, I. and Blum, A.L. (2007) *Helicobacter pylori* and probiotics. *J Nutr* 137 (3 Suppl 2), 812S–18S.

Ley, R.E., Turnbaugh, P.J., Klein, S. and Gordon, J.I. (2006) Microbial ecology: Human gut microbes associated with obesity. *Nature* 444, 1022–3.

MacDonald, T.T. and Gordon, J.N. (2005) Bacterial regulation of intestinal immune responses. *Gastroenterol Clin North Am* 34, 401–12.

Manley, K.J., Fraenkel, M.B., Mayall, B.C. and Power, DA. (2007) Probiotic treatment of vancomycin-resistant enterococci: A randomised controlled trial. *Med J Aust* 186, 454–7.

Martin, F.P., Dumas, M.E., Wang, Y., Legido-Quigley, C., Yap, I.K., Tang, H., Zirah, S., Murphy, G.M., Cloarec, O., Lindon, J.C., Sprenger, N., Fay, L.B., Kochhar, S., van Bladeren, P., Holmes, E. and Nicholson, J.K. (2007) A top-down systems biology view of microbiome-mammalian metabolic interactions in a mouse model. *Mol Syst Biol* 3, 112.

Neu, J. (2007) Perinatal and neonatal manipulation of the intestinal microbiome: a note of caution. *Nutr Rev* 65, 282–5.

O'Hara, A.M. and Shanahan, F. (2007) Gut microbiota: Mining for therapeutic potential. *Clin Gastroenterol Hepatol* 5, 274–84.

Oien, T., Storro, O. and Johnsen, R. (2006) Intestinal microbiota and its effect on the immune system – A nested case-cohort study on prevention of atopy among small children in Trondheim: The IMPACT study. *Contemp Clin Trials* 27, 389–95.

Palmer, C., Bik, E.M., Digiulio, D.B., Relman, D.A. and Brown, P.O. (2007) Development of the human infant intestinal microbiota. *PLoS Biol* 5, e177.

Palmer, C., Bik, E.M., Eisen, M.B., Eckburg, P.B., Sana, T.R., Wolber, P.K., Relman, D.A. and Brown, P.O. (2006) Rapid quantitative profiling of complex microbial populations. *Nucleic Acids Res* 34, e5.

Prescott, S.L. and Bjorksten, B. (2007) Probiotics for the prevention or treatment of allergic diseases. *J Allergy Clin Immunol* 120, 255–62.

Rajilic-Stojanovic, M., Smidt, H. and de Vos, W.M. (2007) Diversity of the human gastrointestinal tract microbiota revisited. *Environ Microbiol* 9, 2125–36.

Saavedra, J.M. (2007) Use of probiotics in pediatrics: Rationale, mechanisms of action, and practical aspects. *Nutr Clin Pract* 22, 351–65.

Turnbaugh, P.J., Ley, R.E., Hamady, M., Fraser-Liggett, C.M., Knight, R. and Gordon, J.I. (2007) The human microbiome project. *Nature* 449, 804–10.

Versalovic, J. and Relman, D. (2006) How bacterial communities expand functional repertoires. *PLoS Biol* 4, e430.

Wilks, M. (2007) Bacteria and early human development. *Early Hum Dev* 83, 165–70.

Wilson, J.D., Lee, R.A., Balen, A.H. and Rutherford, A.J. (2007) Bacterial vaginal flora in relation to changing oestrogen levels. *Int J STD AIDS* 18, 308–11.

Xu, J., Mahowald, M.A., Ley, R.E., Lozupone, C.A., Hamady, M., Martens, E.C., Henrissat, B., Coutinho, P.M., Minx, P., Latreille, P., Cordum, H., Van Brunt, A., Kim, K., Fulton, R.S., Fulton, L.A., Clifton, S.W., Wilson, R.K., Knight, R.D. and Gordon, J.I. (2007) Evolution of symbiotic bacteria in the distal human intestine. *PLoS Biol* 5, e156.

Zoetendal, E.G., Vaughan, E.E. and de Vos, W.M. (2006) A microbial world within us. *Mol Microbiol* 59, 1639–50.

INDEX

Note: Page numbers in **bold** refer to Tables; those in *italics* to Figures.

QR
171
A1
W55
2008

Wilson, Michael, 1947
  Apr. 12-

Bacteriology of
  humans.

15787
$114.95   07/01/2011

DATE			

QR
171
A1
Bacteriology of humans.

15787